Springer Series in Statistics

For further volumes:
http://www.springer.com/series/692

Satoshi Aoki • Hisayuki Hara • Akimichi Takemura

Markov Bases in Algebraic Statistics

Springer

Satoshi Aoki
Department of Mathematics
and Computer Science
Graduate School of Science
and Engineering
Kagoshima University
Kagoshima, Japan

Hisayuki Hara
Faculty of Economics
Niigata University
Niigata, Japan

Akimichi Takemura
Department of Mathematical Informatics
Graduate School of Information Science
and Technology
University of Tokyo
Tokyo, Japan

ISSN 0172-7397
ISBN 978-1-4899-9909-2 ISBN 978-1-4614-3719-2 (eBook)
DOI 10.1007/978-1-4614-3719-2
Springer New York Heidelberg Dordrecht London

Printed on acid-free paper

Springer is part of Springer Science+Business Media (www.springer.com)

Preface

Algebraic statistics is a rapidly developing field, where ideas from statistics and algebra meet and stimulate new research directions. Statistics has been relying on classical asymptotic theory as a basis for statistical inferences. This classical basis is still very useful. However, when the validity of asymptotic theory is in doubt, for example, when the sample size is small, statisticians rely more and more on various computational methods. Similarly, algebra has long been considered as the purest field of mathematics, far apart from practical computations. However, due mainly to the development of Gröbner basis technology, algebra is now becoming a field where computations for practical applications are feasible. It is an interesting trend, because historically algebra was invented to speed up various calculations.

These two trends meet in the field of algebraic statistics. Algebraic algorithms are now very useful and essential for some practical statistical computations such as Markov chain Monte Carlo tests for discrete exponential families, which is the main topic of this book. On the other hand algebraic structures and computational needs of statistical models provide new challenging problems to algebraists. Some algebraic structures are naturally motivated from statistical modeling, but not necessarily from pure mathematical considerations.

Algebraic statistics has two origins. One origin is the work by Pistone and Wynn in 1996 on the use of Gröbner bases for studying confounding relations in factorial designs of experiments. Another origin is the work by Diaconis and Sturmfels in 1998 on the use of Gröbner bases for constructing a connected Markov chain for performing conditional tests of a discrete exponential family. These two works opened up the whole new field of algebraic statistics. In this book we take up the second topic. We give a detailed treatment of results following the seminal work of Diaconis and Sturmfels. We also briefly consider the first topic in Chap. 15 of this book.

As a general reference to the first origin of algebraic statistics we mention Pistone et al. [118]. For the second origin we mention Drton et al. [55], Pachter and Sturmfels [116], and our review paper [15]. For Japanese people the following two books are very useful: Hibi [86], and JST CREST Hibi team [93]. The Markov bases database (`http://markov-bases.de/`) provides very useful online material for studying Markov bases.

Algebraic statistics gave us some exciting opportunities for research and collaboration. In particular we enjoyed working with Takayuki Hibi and Hidefumi Ohsugi, who are the leading researchers on Gröbner bases in Japan. Since 2008 Takayuki Hibi has a project, "Harmony of Gröbner Bases and the Modern Industrial Society," in the mathematics program of the Japan Science and Technology Agency. Algebraic statistics offers a rare ground where algebraists and statisticians can talk about the same problems, albeit often with different terminologies. This book is intended for statisticians with minimal backgrounds in algebra. As we ourselves learned algebraic notions through working on statistical problems, we hope that this book with many practical statistical problems is useful for statisticians to start working on algebraic statistics.

In preparing this book we very much benefited from comments of Takayuki Hibi, Hidehiko Kamiya, Kei Kobayashi, Satoshi Kuriki, Mitsunori Ogawa, Hidefumi Ohsugi, Toshio Sakata, Tomonari Sei, Kentaro Tanaka, and Ruriko Yoshida.

Finally we acknowledge great editorial help from John Kimmel.

Kagoshima, Japan — Satoshi Aoki
Niigata, Japan — Hisayuki Hara
Tokyo, Japan — Akimichi Takemura

Contents

Part II Properties of Markov Bases

Part III Markov Bases for Specific Models

Part I
Introduction and Some Relevant Preliminary Material

In Part I of this book we give introductory material on performing exact tests using Markov basis and a short survey on Gröbner basis.

In Chap. 1, using the example of Fisher's exact test for the independence model in two-way contingency tables, we give an introduction to exact tests. We also discuss conditional independence model for three-way contingency tables.

In Chap. 2 we discuss basic notions of Markov chain and Markov bases. In particular we explain the Metropolis-Hastings procedure for adjusting transition probabilities to achieve a desired stationary distribution.

Chapter 3 is a brief summary of results in the theory of Gröbner basis. In this chapter we collect relevant facts on ideals in polynomial rings and their Gröbner bases, which are often needed for discussion of Markov bases.

In this book, $\mathbb{R}, \mathbb{Q}, \mathbb{Z}, \mathbb{N} = \{0, 1, \dots\}$ stand for the set of reals, rationals, integers and nonnegative integers, respectively. For a positive integer n, we denote the set of n-dimensional vectors of elements from $\mathbb{R}, \mathbb{Q}, \mathbb{Z}, \mathbb{N}$, by $\mathbb{R}^n, \mathbb{Q}^n, \mathbb{Z}^n, \mathbb{N}^n$, respectively.

Chapter 1
Exact Tests for Contingency Tables and Discrete Exponential Families

1.1 Independence Model of 2 × 2 Two-Way Contingency Tables

The theory of exact tests for discrete exponential families is best explained by Fisher's exact test of homogeneity of two binomial populations and the independence model of 2×2 contingency tables. We begin with the test of homogeneity of two binomial populations. An excellent introduction to contingency tables is given in [59]. We also refer to Agresti [3] as a survey paper of the exact methods.

Fisher's exact test can be applied to three different sampling schemes: (i) test of homogeneity of two binomial populations, (ii) test of independence in multinomial sampling for 2×2 tables, (iii) the main effect model for logarithms of mean parameters of independent Poisson random variables in 2×2 tables. We discuss these three sampling schemes in this order. With this example we confirm that the same Markov basis can be used for different sampling schemes.

Let X be distributed according to a binomial distribution $\mathrm{Bin}(n_1, p_1)$, where n_1 is the number of trials and p_1 is the success probability. Let Y be distributed according to the binomial distribution $\mathrm{Bin}(n_2, p_2)$. Suppose that X and Y are independent. We can display X and Y in the following 2×2 contingency table:

X	$n_1 - X$	n_1
Y	$n_2 - Y$	n_2
t	$n - t$	n

where $t = X + Y$ and $n = n_1 + n_2$. The hypothesis of homogeneity of two binomial populations is specified as

$$H : p_1 = p_2.$$

S. Aoki et al., *Markov Bases in Algebraic Statistics*, Springer Series in Statistics 199, DOI 10.1007/978-1-4614-3719-2_1,

The joint probability function of X and Y is written as

$$p(x,y) = \binom{n_1}{x} p_1^x (1-p_1)^{n_1-x} \binom{n_2}{y} p_2^y (1-p_2)^{n_2-y}.$$

Note that here we are using the conventional notational distinction between random variables X, Y in capital letters and their values x, y in lower-case letters. However, for the rest of this book for notational simplicity we do not necessarily stick to this convention.

Under the null hypothesis H, the joint probability is written as

$$p(x,y) = \binom{n_1}{x}\binom{n_2}{y} p_1^{x+y}(1-p_1)^{n-(x+y)}. \tag{1.1}$$

This joint probability depends on (x,y) through $t = x+y$. Therefore from the factorization theorem for sufficient statistics (see Sect. 2.6 of Lehmann and Romano [98]), $T = X+Y$ is a sufficient statistic under the null hypothesis H. Given $T = t$, the conditional distribution of X does not depend on the value of $p_1 = p_2$. Hence by using X as the test statistic, we obtain a testing procedure, whose level does not depend on the value of $p_1 = p_2$; that is, we obtain a *similar test* (Sect. 4.3 of [98]).

Under H the distribution of $T = X+Y$ is the binomial distribution $\mathrm{Bin}(n, p_1)$. Therefore the conditional distribution of X given $T = t$ is calculated as

$$\begin{aligned} P(X = x \mid T = t) &= \frac{\binom{n_1}{x}\binom{n_2}{t-x} p_1^t (1-p_1)^{n-t}}{\binom{n_1+n_2}{t} p_1^t (1-p_1)^{n-t}} = \frac{\binom{n_1}{x}\binom{n_2}{t-x}}{\binom{n}{t}} \\ &= \frac{n_1!n_2!t!(n-t)!}{n!x!(n_1-x)!(t-x)!(n_2-t+x)!}. \end{aligned} \tag{1.2}$$

This is a *hypergeometric distribution*. Indeed the conditional distribution does not depend on the value of $p_1 = p_2$.

The null hypothesis H is rejected if the value of X is too large or too small. Because the distribution of X is not symmetric when $n_1 \neq n_2$, the rejection region is usually determined by unbiasedness consideration. For optimality of similar unbiased test see Sect. 4.4 of [98]. This testing procedure is called Fisher's exact test. It is an exact test because the significance level is computed from the hypergeometric distribution. It is also called a *conditional test* because we use the conditional null distribution given $T = t$. In contrast, the usual large-sample test is based on the large-sample normal approximation to the following "z-statistic":

$$z = \frac{\hat{p}_1 - \hat{p}_2}{\sqrt{\frac{\hat{p}_1(1-\hat{p}_1)}{n_1} + \frac{\hat{p}_2(1-\hat{p}_2)}{n_2}}}, \qquad \hat{p}_1 = \frac{X}{n_1},\ \hat{p}_2 = \frac{Y}{n_2}.$$

Table 1.1
Cross-classification of belief in afterlife by gender

	Belief in Afterlife	
Gender	Yes	No or Undecided
Females	509	116
Males	398	104

The test based on z is an unconditional test. However, when the sample size is small, it is desirable to use the exact test (Haberman [68]).

In the case of homogeneity of two binomial populations, we saw that $X+Y$ (total number of successes) is a sufficient statistic. We could also take $n-X-Y$ (total number of failures) or even the pair $(X+Y, n-X-Y)$ as a sufficient statistic. Note that the pair contains redundancy, but it is still a sufficient statistic, because fixing $(x+y, n-x-y)$ is equivalent to fixing $x+y$. Furthermore we could also include n_1 and n_2 into the sufficient statistic, although these values are fixed in the case of homogeneity of two binomial populations. Indeed $T = (X+Y, n-X-Y, n_1, n_2)$ is a sufficient statistic, because given the value of the vector T the conditional distribution of X is the hypergeometric distribution in (1.2) and it does not depend on $p_1 = p_2$.

Next we discuss the multinomial sampling scheme. Let x_{ij}, $i = 1,2$, $j = 1,2$, be frequencies of four *cells* of a 2×2 contingency table. The row sums and the column sums (i.e., the marginal frequencies) are denoted as x_{i+}, x_{+j}, $i, j = 1,2$. The total sample size is $n = x_{11} + x_{12} + x_{21} + x_{22}$. The data are displayed as follows.

$$\begin{array}{|c|c|c} \hline x_{11} & x_{12} & x_{1+} \\ \hline x_{21} & x_{22} & x_{2+} \\ \hline x_{+1} & x_{+2} & n \end{array} \tag{1.3}$$

At this point we mention some customary terminology of contingency tables. We look at the frequencies in (1.3) as the frequencies of a two-dimensional random variable $Y = (Y_1, Y_2)$, such that both Y_1 and Y_2 take the values 1 or 2. For example, in Table 1.1 taken from Chap. 2 of [5], Y_1 is the gender and Y_2 is the belief in afterlife. The values taken by a variable are often called *levels* of the variable. For example, in Table 1.1 two levels of the variable "gender" are "female" and "male". In this terminology x_{ij} is the joint frequency such that Y_1 takes the level i and Y_2 takes the level j. The row and the column of the contingency table are sometimes called *axes* of the table. Then Y_1 is the random variable for the first axis and Y_2 is the random variable for the second axis.

Let

$$p_{ij} \geq 0, \quad i = 1,2, \quad j = 1,2, \quad \sum_{i,j=1}^{2} p_{ij} = 1$$

be the probabilities of the cells. In a single multinomial trial, we observe one of the four cells according to the probabilities. With n independent and identical multinomial trials, the joint probability function of $\boldsymbol{X} = (X_{11}, X_{12}, X_{21}, X_{22})$ is given as

$$p(\boldsymbol{x}) = \binom{n}{x_{11}, x_{12}, x_{21}, x_{22}} p_{11}^{x_{11}} p_{12}^{x_{12}} p_{21}^{x_{21}} p_{22}^{x_{22}}. \tag{1.4}$$

As in this example, we use the boldface letter $\boldsymbol{x}$ for the vector of frequencies and call $\boldsymbol{x}$ the *frequency vector*. When necessary, we make the notational distinction between column vector and row vector. For example, $\boldsymbol{x}$ is meant as a column vector when we write $\boldsymbol{x} = (x_{11}, x_{12}, x_{21}, x_{22})'$. We use $'$ for denoting the transpose of a vector or a matrix in this book.

Let $p_{i+} = p_{i1} + p_{i2}$, $i = 1, 2$, denote the marginal probability of the first variable of the contingency table and similarly let $p_{+j} = p_{1j} + p_{2j}$, $j = 1, 2$, denote the marginal probability of the second variable. The hypothesis of independence H in the multinomial sampling scheme is specified as follows:

$$H: \ p_{ij} = p_{i+}p_{+j}, \quad i = 1, 2, \quad j = 1, 2. \tag{1.5}$$

On the other hand, if there is no restriction on the probability vector $\boldsymbol{p} = (p_{11}, p_{12}, p_{21}, p_{22})$, except that the elements of $\boldsymbol{p}$ are nonnegative and sum to one, we call the model *saturated*.

Write $r_i = p_{i+}$ and $c_j = p_{+j}$. Then $p_{ij} = r_i c_j$ under H. Note that in (1.5),

$$1 = \sum_{i=1}^{2} p_{i+} = \sum_{j=1}^{2} p_{+j}.$$

However, when we write $r_i = p_{i+}$ and $c_j = p_{+j}$, we can remove the restriction $1 = r_1 + r_2 = c_1 + c_2$ and only assume that r_i and c_j are nonnegative such that the total probability is 1:

$$1 = \sum_{i,j=1}^{2} r_i c_j = (r_1 + r_2)(c_1 + c_2).$$

Furthermore we can incorporate the total probability into the normalizing constant and write the probability as

$$p_{ij} = \frac{1}{(r_1 + r_2)(c_1 + c_2)} r_i c_j, \qquad i, j = 1, 2, \tag{1.6}$$

where we only assume that r_i and c_j are nonnegative without any further restrictions. In this example of 2×2 tables, the normalizing constant is obvious and the above discussion may be pedantic. However, for more general models of contingency tables, it is best to consider the joint probability in the form of (1.6).

Under H, with the normalization $1 = (r_1 + r_2)(c_1 + c_2)$, the joint probability function $p(\boldsymbol{x})$ is written as

$$\begin{aligned}
p(\boldsymbol{x}) &= \binom{n}{x_{11},x_{12},x_{21},x_{22}}(r_1c_1)^{x_{11}}(r_1c_2)^{x_{12}}(r_2c_1)^{x_{21}}(r_2c_2)^{x_{22}} \\
&= \binom{n}{x_{11},x_{12},x_{21},x_{22}} r_1^{x_{1+}} r_2^{x_{2+}} c_1^{x_{+1}} c_2^{x_{+2}} \\
&= \binom{n}{x_{11},x_{12},x_{21},x_{22}} p_{1+}^{x_{1+}} p_{2+}^{x_{2+}} p_{+1}^{x_{+1}} p_{+2}^{x_{+2}}.
\end{aligned} \tag{1.7}$$

Hence the sufficient statistic under H is given as

$$T = (x_{1+}, x_{2+}, x_{+1}, x_{+2}).$$

Given T, in the case of the 2×2 table, there is only one degree of freedom in $\boldsymbol{x}$. Namely, if x_{11} is given, then the other values x_{12}, x_{21}, x_{22} are automatically determined as

$$x_{12} = x_{1+} - x_{11}, \qquad x_{21} = x_{+1} - x_{11}, \qquad x_{22} = n - x_{1+} - x_{+1} + x_{11}.$$

As mentioned above, let us consider (i, j) as the pair of levels of two random variables Y_1 and Y_2. Under the null hypothesis H of independence in (1.5), Y_1 and Y_2 are independent. Suppose that we observe n independent realizations $(y_1^1, y_2^1), \ldots, (y_1^n, y_2^n)$ of (Y_1, Y_2). Then x_{i+} is the number of times that Y_1 takes the value i. Hence x_{1+} is distributed according to the binomial distribution $\mathrm{Bin}(n, p_{1+})$. Similarly x_{+1} is distributed according to the binomial distribution $\mathrm{Bin}(n, p_{+1})$. Furthermore they are independent. Therefore the joint distribution of x_{1+} and x_{+1} is written as

$$p(x_{1+}, x_{+1}) = \binom{n}{x_{1+}} p_{1+}^{x_{1+}} p_{2+}^{x_{2+}} \binom{n}{x_{+1}} p_{+1}^{x_{+1}} p_{+2}^{x_{+2}}. \tag{1.8}$$

From (1.7) and (1.8) it follows that the conditional distribution of X_{11} given the sufficient statistic is computed as follows.

$$\begin{aligned}
p(x_{11} \mid x_{1+}, x_{2+}, x_{+1}, x_{+2}) &= \frac{\binom{n}{x_{11},x_{12},x_{21},x_{22}} p_{1+}^{x_{1+}} p_{2+}^{x_{2+}} p_{+1}^{x_{+1}} p_{+2}^{x_{+2}}}{\binom{n}{x_{1+}} p_{1+}^{x_{1+}} p_{2+}^{x_{2+}} \binom{n}{x_{+1}} p_{+1}^{x_{+1}} p_{+2}^{x_{+2}}} \\
&= \frac{\binom{n}{x_{11},x_{12},x_{21},x_{22}}}{\binom{n}{x_{1+}}\binom{n}{x_{+1}}} = \frac{x_{1+}!\,x_{2+}!\,x_{+1}!\,x_{+2}!}{n!\,x_{11}!\,x_{12}!\,x_{21}!\,x_{22}!}.
\end{aligned} \tag{1.9}$$

This is again a hypergeometric distribution. Equation (1.9) is clearly the same as (1.2) if we write the row sums and the column sums as $n_1 = x_{1+}$, $n_2 = x_{2+}$, $t = x_{+1}$, $n - t = x_{+2}$. Therefore Fisher's exact test is the same in this multinomial sampling scheme as in the case of testing the homogeneity of two binomial populations.

Note that in this scheme n is fixed and $x_{2+} = n - x_{1+}$ and $x_{+2} = n - x_{+1}$ can be omitted from the sufficient statistic $T = (x_{1+}, x_{2+}, x_{+1}, x_{+2})$. However, as in the first scheme we can allow the redundancy in the sufficient statistic.

Finally we consider the sampling scheme of Poisson random variables. Let X_{ij}, $i, j = 1,2$, be independently distributed according to the Poisson distribution with mean λ_{ij}. The joint probability of $\boldsymbol{X}$ is written as

$$p(\boldsymbol{x}) = \prod_{i,j=1}^{2} \frac{\lambda_{ij}^{x_{ij}}}{x_{ij}!} e^{-\lambda_{ij}}.$$

Consider the null hypothesis H that λ_{ij} can be factored as

$$H : \lambda_{ij} = r_i c_j, \qquad i, j = 1,2,$$

where r_i, c_j are nonnegative. Again by writing down the joint probability under the null hypothesis H, we can easily check that a sufficient statistic under H is given by $T = (x_{1+}, x_{2+}, x_{+1}, x_{+2})$, where now the redundancy is only in $x_{+2} = x_{1+} + x_{2+} - x_{+1}$. Instead of writing out the joint probability, we use the following property of independent Poisson random variables for verifying that T is a sufficient statistic under H. Let $n = X_{11} + X_{12} + X_{21} + X_{22}$. Then n is distributed as the Poisson random variable with mean $\mu = \sum_{i,j=1}^{2} \lambda_{ij}$. Under H, $\mu = (r_1 + r_2)(c_1 + c_2)$. Given n, the conditional distribution of $(X_{11}, X_{12}, X_{21}, X_{22})$ is the multinomial distribution with cell probabilities $p_{ij} = \lambda_{ij}/\mu$. Under H, the cell probability is written as

$$p_{ij} = \frac{1}{(r_1 + r_2)(c_1 + c_2)} r_i c_j, \qquad i, j = 1,2,$$

which is the same as (1.6). From this fact we see that $T = (x_{1+}, x_{2+}, x_{+1}, x_{+2})$ is a sufficient statistic under H. Given T, the conditional distribution of $\boldsymbol{x}$ is the same as the multinomial case; that is, X_{11} follows the hypergeometric distribution in (1.9).

We now note the relation between the cell frequencies and the sufficient statistic. The column vector of cell frequencies $\boldsymbol{x} = (x_{11}, x_{12}, x_{21}, x_{22})'$ and the column vector of the sufficient statistic $(x_{1+}, x_{2+}, x_{+1}, x_{+2})'$ are related as follows:

$$\begin{pmatrix} x_{1+} \\ x_{2+} \\ x_{+1} \\ x_{+2} \end{pmatrix} = \begin{pmatrix} 1 & 1 & 0 & 0 \\ 0 & 0 & 1 & 1 \\ 1 & 0 & 1 & 0 \\ 0 & 1 & 0 & 1 \end{pmatrix} \begin{pmatrix} x_{11} \\ x_{12} \\ x_{21} \\ x_{22} \end{pmatrix}. \tag{1.10}$$

We write this as $\boldsymbol{t} = A\boldsymbol{x}$ and call the matrix A the *configuration* for the above three models.

1.2 2×2 Contingency Table Models as Discrete Exponential Family

In the previous section we explained three sampling schemes for 2×2 contingency tables and pointed that they share the same sufficient statistic when redundancies are allowed. In this section we present the standard formulation of the sampling schemes as discrete exponential family models. We confirm that the sufficient statistics under the null hypothesis correspond to nuisance parameters. Hence fixing the sufficient statistic has the effect of eliminating the nuisance parameters and the resulting conditional test is a similar test. Here we only consider the multinomial scheme of the previous section, because the other cases can be treated in a similar manner.

A family of joint probability functions $p(\boldsymbol{x}) = p(\boldsymbol{x};\boldsymbol{\theta})$, $\boldsymbol{\theta} \in \Theta$, is said to form an *exponential family* (see Sect. 2.7 of [98]) if $p(\boldsymbol{x},\boldsymbol{\theta})$ is written in the following form.

$$p(\boldsymbol{x};\boldsymbol{\theta}) = h(\boldsymbol{x})\exp\left(\sum_{j=1}^{k} T_j(\boldsymbol{x})\phi_j(\boldsymbol{\theta}) - \psi(\boldsymbol{\theta})\right). \tag{1.11}$$

By the factorization theorem (Sect. 2.6 of [98]), $T = (T_1(\boldsymbol{x}),\dots,T_k(\boldsymbol{x}))$ is a sufficient statistic of this family. Note that $p(\boldsymbol{x};\boldsymbol{\theta})$ and $\psi(\boldsymbol{\theta})$ depend on $\boldsymbol{\theta}$ only through $\boldsymbol{\phi} = (\phi_1,\dots,\phi_k)$ and we can write $\psi(\boldsymbol{\phi})$ instead of $\psi(\boldsymbol{\theta})$. In Chap. 4 we simply denote $\phi_j(\boldsymbol{\theta})$ itself as θ_j.

Let p_{ij}, $i,j = 1,2$, denote the cell probabilities in the multinomial sampling of a 2×2 contingency table. Now consider the following transformation:

$$\phi_1 = \log\frac{p_{12}}{p_{22}}, \qquad \phi_2 = \log\frac{p_{21}}{p_{22}}, \qquad \lambda = \log\frac{p_{11}p_{22}}{p_{12}p_{21}}. \tag{1.12}$$

In the region where the elements of the probability vector $\boldsymbol{p} = (p_{11},p_{12},p_{21},p_{22})$ are positive, the transformation is one-to-one and the inverse transformation is written as

$$\begin{aligned}
p_{11} &= \frac{e^{\phi_1+\phi_2+\lambda}}{1+e^{\phi_1}+e^{\phi_2}+e^{\phi_1+\phi_2+\lambda}},\\
p_{12} &= \frac{e^{\phi_1}}{1+e^{\phi_1}+e^{\phi_2}+e^{\phi_1+\phi_2+\lambda}},\\
p_{21} &= \frac{e^{\phi_2}}{1+e^{\phi_1}+e^{\phi_2}+e^{\phi_1+\phi_2+\lambda}},\\
p_{22} &= \frac{1}{1+e^{\phi_1}+e^{\phi_2}+e^{\phi_1+\phi_2+\lambda}}.
\end{aligned} \tag{1.13}$$

Substituting this into (1.4) we can write the joint probability function of $\boldsymbol{x}$ as

$$\begin{aligned}
p(\boldsymbol{x}) = \binom{n}{x_{11},x_{12},x_{21},x_{22}} &\exp((x_{11}+x_{12})\phi_1 + (x_{11}+x_{21})\phi_2 + x_{11}\lambda\\
&- n\log(1+e^{\phi_1}+e^{\phi_2}+e^{\phi_1+\phi_2+\lambda})).
\end{aligned} \tag{1.14}$$

This is written in the form (1.11) and hence the family of $p(\boldsymbol{x})$ forms an exponential family. By putting $r_1 = e^{\phi_1}, r_2 = 1, c_1 = e^{\phi_2}, c_2 = 1$ we see that the null hypothesis of the independence (1.5) is equivalently written as

$$H : \lambda = 0.$$

Note that λ is the parameter of interest for the null hypothesis and ϕ_1, ϕ_2 are the nuisance parameters under the null hypothesis. Under the null hypothesis, $\lambda = 0$ is no longer a parameter of the family of distributions and the distributions under the null hypothesis are parametrized by the nuisance parameters ϕ_1, ϕ_2. In (1.14) the sufficient statistic corresponding to (ϕ_1, ϕ_2) is

$$x_{1+} = x_{11} + x_{12}, \quad x_{+1} = x_{11} + x_{21}.$$

In (1.11) and (1.14) we considered the joint probability of the frequency vector. In fact, when we consider a single observation $n = 1$, then the cell probabilities are already in the exponential family form. Write

$$\log \boldsymbol{p} = (\log p_{11}, \log p_{12}, \log p_{21}, \log p_{22}),$$
$$\psi(\phi_1, \phi_2) = \log(1 + e^{\phi_1} + e^{\phi_2} + e^{\phi_1 + \phi_2}).$$

Taking the logarithms of p_{ij} in (1.13) with $\lambda = 0$, in a matrix form we can write

$$\log \boldsymbol{p} = (\phi_1, 0, \phi_2, 0) \begin{pmatrix} 1 & 1 & 0 & 0 \\ 0 & 0 & 1 & 1 \\ 1 & 0 & 1 & 0 \\ 0 & 1 & 0 & 1 \end{pmatrix} - \psi(\phi_1, \phi_2) \times (1, 1, 1, 1). \tag{1.15}$$

Note that the matrix on the right-hand side is the configuration A appearing in the right-hand side of (1.10).

1.3 Independence Model of General Two-Way Contingency Tables

Generalizing the discussion of the previous section we now consider the independence model of general $I \times J$ two-way contingency tables. The discussion on three sampling schemes is entirely the same as in the case of 2×2 tables. Therefore we only discuss the multinomial sampling.

Let p_{ij}, $i = 1, \ldots, I$, $j = 1, \ldots, J$, denote the cell probabilities of an $I \times J$ contingency table. Let p_{i+} and p_{+j} denote the marginal probabilities. The null hypothesis of independence is written as

$$H : p_{ij} = p_{i+} p_{+j}, \quad i = 1, \ldots, I, \qquad j = 1, \ldots, J.$$

We can also write $p_{ij} = r_i c_j$ without requiring that r_is and c_js correspond to probabilities. Let x_{ij} denote the frequency of the cell (i, j). A sufficient statistic

T under the null hypothesis H is the set of the row sums x_{i+}, $i = 1,\dots,I$ and the column sums x_{+j}, $j = 1,\dots,J$. Let n denote the total sample size.

Under the null hypothesis the joint probability of $\boldsymbol{x} = \{x_{ij}\}$ is written as

$$p(\boldsymbol{x}) = \binom{n}{x_{11},\dots,x_{IJ}} \prod_{i=1}^{I}\prod_{j=1}^{J}(p_{i+}p_{+j})^{x_{ij}}$$

$$= \binom{n}{x_{11},\dots,x_{IJ}} \prod_{i=1}^{I} p_{i+}^{x_{i+}} \prod_{j=1}^{J} p_{+j}^{x_{+j}}.$$

Also, under the null hypothesis, as in the case of 2×2 tables, the vector of row sums $\{x_{i+}\}$ and the vector of column sums $\{x_{+j}\}$ are independently distributed according to multinomial distributions:

$$p(\{x_{i+}\}) = \binom{n}{x_{1+},\dots,x_{I+}} p_{1+}^{x_{1+}} \cdots p_{I+}^{x_{I+}},$$

$$p(\{x_{+j}\}) = \binom{n}{x_{+1},\dots,x_{+J}} p_{+1}^{x_{+1}} \cdots p_{+J}^{x_{+J}}.$$

From this fact, the conditional distribution of $\boldsymbol{x} = \{x_{ij}\}$ given the sufficient statistic $\boldsymbol{t}$ is written as

$$p(\boldsymbol{x} \mid T = \boldsymbol{t}) = \frac{p(\{x_{ij}\})}{p(\{x_{i+}\})p(\{x_{+j}\})} = \frac{\binom{n}{x_{11},\dots,x_{IJ}}}{\binom{n}{x_{1+},\dots,x_{I+}}\binom{n}{x_{+1},\dots,x_{+J}}}$$

$$= \frac{\prod_{i=1}^{I} x_{i+}!\,\prod_{j=1}^{J} x_{+j}!}{n!\prod_{i,j} x_{ij}!}. \tag{1.16}$$

This distribution is often called the *multivariate hypergeometric distribution*. However in this book we show many variations of distributions of this type and we often refer to them simply as hypergeometric distributions.

Given the row sums and the column sums, the degrees of freedom in the frequency vector $\boldsymbol{x}$ is $(I-1)\times(J-1)$ because the elements of the last row and the last column are determined uniquely from the other elements. This degrees of freedom is also the dimension of the parameter of interest when the joint probability distribution is written in the exponential family form. More precisely let

$$\phi_{1i} = \log\frac{p_{iJ}}{p_{IJ}}, \qquad i = 1,\dots,I-1,$$

$$\phi_{2j} = \log\frac{p_{Ij}}{p_{IJ}}, \qquad j = 1,\dots,J-1,$$

$$\lambda_{ij} = \log\frac{p_{ij}p_{IJ}}{p_{iJ}p_{Ij}}, \qquad i = 1,\dots,I-1,\ j = 1,\dots,J-1. \tag{1.17}$$

Then the null hypothesis is written as

$$H : \lambda_{ij} = 0, \qquad i = 1,\dots,I-1, \qquad j = 1,\dots,J-1.$$

One consequence of the multidimensionality of the parameter of interest is that there is no unique best choice for a test statistic, even under the requirement of similarity and unbiasedness.

Let

$$\hat{m}_{ij} = n\hat{p}_{ij} = \frac{x_{i+}x_{+j}}{n}$$

denote the "expected frequency" of the cell (i, j), where $\hat{p}_{ij}$ is the maximum likelihood estimate (MLE) of p_{ij}. For testing the null hypothesis of independence, popular test statistics are *Pearson's chi-square test*

$$\chi^2(\boldsymbol{x}) = \sum_i \sum_j \frac{(x_{ij} - \hat{m}_{ij})^2}{\hat{m}_{ij}} \geq c_\alpha \Rightarrow \text{reject } H$$

and the (twice log) *likelihood ratio test*

$$G^2(\boldsymbol{x}) = 2\sum_i \sum_j x_{ij} \log \frac{x_{ij}}{\hat{m}_{ij}} \geq c_\alpha \Rightarrow \text{reject } H,$$

where c_α is the critical value for the respective test statistic. $G^2(\boldsymbol{x})$ is actually twice the logarithm of the likelihood ratio. In the usual asymptotic theory, c_α is approximated by the upper α-quantile of the chi-square distribution with $(I-1)(J-1)$ degrees of freedom. In this book we denote the chi-square distribution with m degrees of freedom by χ^2_m.

These two statistics are "omnibus test statistics" in the sense that all possible alternative hypotheses are roughly equally treated. When some specific deviations from the null hypothesis are expected, then a more suitable test statistic, which is sensitive against the deviation, can be used. For performing a test of H, once a test statistic is chosen, it only remains to evaluate its null distribution. As in the previous section, in this book we consider exact tests; that is, we are interested in the distribution of a test statistic under the hypergeometric distribution (1.16).

At this point we investigate the conditional sample space; that is, the set of contingency tables given the sufficient statistic for $I \times J$ case. As in the 2×2 case, the relation between the sufficient statistic and the frequency vector is written in a matrix form. Let $\boldsymbol{t} = (x_{1+},\dots,x_{I+},x_{+1},\dots,x_{+J})'$ denote the (column) vector of the sufficient statistic and let $\boldsymbol{x} = (x_{11},x_{12},\dots,x_{1J},x_{21},\dots,x_{IJ})'$ denote the frequency vector. Then

$$\boldsymbol{t} = A\boldsymbol{x}, \tag{1.18}$$

where the configuration A is an $(I+J) \times IJ$ matrix consisting of 0s and 1s as in (1.10).

An explicit form of A can be given using the Kronecker product notation. For two matrices, $C = \{c_{ij}\} : m_1 \times n_1$ and $D : m_2 \times n_2$, their Kronecker product $C \otimes D$ is an $m_1 m_2 \times n_1 n_2$ matrix of the following block form

$$C \otimes D = \begin{pmatrix} c_{11}D & \dots & c_{1n_1}D \\ \vdots & & \vdots \\ c_{m_1 1}D & \dots & c_{m_1 n_1}D \end{pmatrix}. \tag{1.19}$$

Let $\mathbf{1}_n = (1,\dots,1)'$ denote the n-dimensional vector consisting of 1s and let E_m denote an $m \times m$ identity matrix. Then A in (1.18) is written as

$$A = \begin{pmatrix} E_I \otimes \mathbf{1}_J' \\ \mathbf{1}_I' \otimes E_J \end{pmatrix}.$$

Alternatively let $\boldsymbol{e}_{j,n} = (0,\dots,0,1,0,\dots,0)' \in \mathbb{R}^n$ denote the jth standard basis vector of $\mathbb{R}^n$. When the dimension n is clear from the context, we simply write the standard basis vector as $\boldsymbol{e}_j$ instead of $\boldsymbol{e}_{j,n}$. Then the columns of A are of the form

$$\begin{pmatrix} \boldsymbol{e}_{i,I} \\ \boldsymbol{e}_{j,J} \end{pmatrix}, \quad i = 1,\dots,I, \quad j = 1,\dots,J. \tag{1.20}$$

We sometimes denote the stacked vector in (1.20) as

$$\boldsymbol{e}_{i,I} \oplus \boldsymbol{e}_{j,J} = \begin{pmatrix} \boldsymbol{e}_{i,I} \\ \boldsymbol{e}_{j,J} \end{pmatrix}. \tag{1.21}$$

It is easily checked that the rank of A is

$$\operatorname{rank} A = I + J - 1.$$

Hence the dimension of the kernel of A is given as

$$\dim \ker A = IJ - (I + J - 1) = (I-1)(J-1).$$

As mentioned above, this dimension corresponds to the fact that, if we ignore the requirement of nonnegativity, we can choose the elements of the first $I-1$ rows and the first $J-1$ columns freely. With the additional requirement of nonnegativity, the *conditional sample space* given the sufficient statistic is defined as

$$\mathscr{F}_{\boldsymbol{t}} = \{\boldsymbol{x} \in \mathbb{Z}^{IJ} \mid \boldsymbol{x} \geq \boldsymbol{0}, \boldsymbol{t} = A\boldsymbol{x}\}, \tag{1.22}$$

where $\boldsymbol{x} \geq \boldsymbol{0}$ means that the elements of $\boldsymbol{x}$ are nonnegative. We call $\mathscr{F}_{\boldsymbol{t}}$ the *fiber* of $\boldsymbol{t}$ (or also call it the $\boldsymbol{t}$-fiber). The hypergeometric distribution in (1.16) is a probability distribution over the fiber $\mathscr{F}_{\boldsymbol{t}}$. When a test statistic $\phi(\boldsymbol{x})$ is given, we want to evaluate the distribution of $\phi(\boldsymbol{x})$, where $\boldsymbol{x}$ is distributed according to the hypergeometric distribution over $\mathscr{F}_{\boldsymbol{t}}$.

Suppose that ϕ is chosen such that a larger value of ϕ indicates more deviation from the null hypothesis, as in Pearson's chi-square statistic or the likelihood ratio statistic. Then testing can be conveniently performed via *p-value*. Let $\boldsymbol{x}^o$ denote the observed contingency table. The p-value of $\boldsymbol{x}^o$ is defined as

$$p = P(\phi(\boldsymbol{x}) \geq \phi(\boldsymbol{x}^o) \mid H) = \sum_{\boldsymbol{x}\in\mathscr{F}_{\boldsymbol{t}},\phi(\boldsymbol{x})\geq\phi(\boldsymbol{x}^o)} p(\boldsymbol{x} \mid \boldsymbol{t} = A\boldsymbol{x}^o, H), \tag{1.23}$$

which is the probability under the hypergeometric distribution of observing the value $\phi(\boldsymbol{x})$ which is larger than or equal to $\phi(\boldsymbol{x}^o)$. Given the level of significance α, we reject H if $p \leq \alpha$.

There are three methods to evaluate the p-value in (1.23).

1. By enumerating $\mathscr{F}_{\boldsymbol{t}}$, $\boldsymbol{t} = A\boldsymbol{x}^o$, and performing the sum in (1.23) for all $\boldsymbol{x} \in \mathscr{F}_{\boldsymbol{t}}$ such that $\phi(\boldsymbol{x}) \geq \phi(\boldsymbol{x}^o)$.
2. Directly sampling $\boldsymbol{x}$ from the hypergeometric distribution and approximating (1.23) by Monte Carlo simulation.
3. By sampling $\boldsymbol{x}$ by a Markov chain whose stationary distribution is the hypergeometric distribution, that is, by a Markov chain Monte Carlo method.

Clearly the enumeration is the best if it is feasible. However, when the row sums and the column sums become large, the size of the fiber $\mathscr{F}_{\boldsymbol{t}}$ becomes large and the enumeration becomes infeasible. In the case of the independence model of this section, direct sampling of a frequency vector from the hypergeometric distribution is easy to carry out. In more complicated models treated later in the book, though, direct sampling is not easy. On the other hand, there exists a general theory of constructing a Markov chain having the hypergeometric distribution as the stationary distribution. Hence the subject of this book is the Markov chain sampling from the fiber $\mathscr{F}_{\boldsymbol{t}}$.

In the next chapter, again employing the independence model of $I \times J$ contingency tables, we discuss how to perform Markov chain sampling from the fiber $\mathscr{F}_{\boldsymbol{t}}$.

1.4 Conditional Independence Model of Three-Way Contingency Tables

In this section we discuss the conditional independence model for three-way contingency tables. It is a relatively simple model in the sense that for each level of the conditioning variable, the problem reduces to the case of an independence model of two-way contingency tables for the other variables. However, it is a convenient model for introducing a notation for general m-way contingency tables in the next section.

Consider an $I_1 \times I_2 \times I_3$ three-way contingency table $\boldsymbol{x}$. We denote each cell of the table by a multi-index $\boldsymbol{i} = (i_1, i_2, i_3)$. For a positive integer J write

$$[J] = \{1, \ldots, J\}.$$

The set of the cells is the following direct product

$$\mathscr{I} = \{\boldsymbol{i} = (i_1, i_2, i_3) \mid i_1 \in [I_1], i_2 \in [I_2], i_3 \in [I_3]\} = [I_1] \times [I_2] \times [I_3].$$

With this notation the three-way contingency table, or the frequency vector, is denoted as

$$\boldsymbol{x} = \{x(\boldsymbol{i}) \mid \boldsymbol{i} \in \mathscr{I}\}.$$

Note that this notation is somewhat heavy and in fact for three-way tables we prefer to use subscripts i, j, k. The merit of this notation is that it can be used for general m-way tables.

For a subset $D \subset \{1,2,3\}$ of the variables, let $\boldsymbol{i}_D$ denote the set of indices in D. For example,

$$\boldsymbol{i}_{\{1,2\}} = (i_1, i_2).$$

Note that $\boldsymbol{i}_D$ corresponds to the D-marginal cell of the contingency table. The set of D-marginal cells is denoted by

$$\mathscr{I}_D = \prod_{k \in D} [I_k]. \tag{1.24}$$

For example $\mathscr{I}_{\{1,2\}} = \{(i_1, i_2) \mid i_1 \in [I_1], i_2 \in [I_2]\}$. The D-marginal frequencies of $\boldsymbol{x}$ are written as

$$x_D(\boldsymbol{i}_D) = \sum_{\boldsymbol{i}_{D^C} \in \mathscr{I}_{D^C}} x(\boldsymbol{i}_D, \boldsymbol{i}_{D^C}), \tag{1.25}$$

where D^C denotes the complement of D. Note that in $x(\boldsymbol{i}_D, \boldsymbol{i}_{D^C})$, for notational simplicity, the indices in $\mathscr{I}_D$ are collected to the left. Also we are writing $x(\boldsymbol{i}_D, \boldsymbol{i}_{D^C})$ instead of $x((\boldsymbol{i}_D, \boldsymbol{i}_{D^C}))$. In the two-way case

$$x_{i+} = x_{\{1\}}(i) = \sum_j x_{ij}.$$

For a probability distribution $\{p(\boldsymbol{i}), \boldsymbol{i} \in \mathscr{I}\}$, we denote the D-marginal probability as $p_D(\boldsymbol{i}_D)$. Note that in $x_D(\boldsymbol{i}_D)$ and $p_D(\boldsymbol{i}_D)$, the subset D is indicated twice. If there is no notational confusion we alternatively write

$$x(\boldsymbol{i}_D), x_D(\boldsymbol{i}), p(\boldsymbol{i}_D) \quad \text{or} \quad p_D(\boldsymbol{i}) \tag{1.26}$$

for simplicity.

We call a D-marginal probability distribution *saturated* if there is no restriction on the probability vector $\{p_D(\boldsymbol{i}_D), \boldsymbol{i}_D \in \mathscr{I}_D\}$.

Let Y_1, Y_2, Y_3 be random variables corresponding to the three axes of the contingency table. We consider the model that Y_1 and Y_3 are conditionally independent given the level i_2 of Y_2. The relevant conditional probabilities are written as

$$p(i_1, i_3 \mid i_2) = \frac{p(\boldsymbol{i})}{p_{\{2\}}(i_2)}, \quad p(i_1 \mid i_2) = \frac{p_{\{1,2\}}(i_1, i_2)}{p_{\{2\}}(i_2)}, \quad p(i_3 \mid i_2) = \frac{p_{\{2,3\}}(i_2, i_3)}{p_{\{2\}}(i_2)}.$$

In the following we omit subscripts to p and write, for example, $p(i_1, i_2)$ instead of $p_{\{1,2\}}(i_1, i_2)$. Similarly we write $x(i_1, i_2)$ instead of $x_{\{1,2\}}(i_1, i_2)$. The null hypothesis of conditional independence is written as

$$H : \frac{p(\boldsymbol{i})}{p(i_2)} = \frac{p(i_1, i_2)}{p(i_2)} \times \frac{p(i_2, i_3)}{p(i_2)}, \qquad \forall \boldsymbol{i} \in \mathscr{I}, \tag{1.27}$$

or equivalently as

$$H : p(\boldsymbol{i}) = \frac{1}{p(i_2)} p(i_1, i_2) p(i_2, i_3), \qquad \forall \boldsymbol{i} \in \mathscr{I}. \tag{1.28}$$

Here we are assuming $p(i_2) > 0$. In the case $p(i_2) = 0$ for a particular level i_2, we have $p(\boldsymbol{i}) = p(i_1, i_2) = p(i_2, i_3) = 0$ for indices containing this level i_2 of Y_2. Hence in this case we understand (1.28) as $0 = 0 \times 0/0$. Let

$$\alpha(i_1, i_2) = \frac{p(i_1, i_2)}{p(i_2)}, \quad \beta(i_2, i_3) = p(i_2, i_3).$$

Then the conditional independence model is written as

$$H : p(\boldsymbol{i}) = \alpha(i_1, i_2) \beta(i_2, i_3). \tag{1.29}$$

Note that there is some indeterminacy in specifying α and β. For example we can include the factor $1/p(i_2)$ into $\beta(i_2, i_3)$ instead of into $\alpha(i_1, i_2)$.

We can show that (1.27), (1.28), and (1.29) are in fact equivalent. Suppose that $p(\boldsymbol{i}) = p(i_1, i_2, i_3)$ can be written as $p(\boldsymbol{i}) = \alpha(i_1, i_2)\beta(i_2, i_3)$. Then

$$p(i_2) = \sum_{i_1, i_3} p(i_1, i_2, i_3) = \sum_{i_1, i_3} \alpha(i_1, i_2) \beta(i_2, i_3) = \left(\sum_{i_1} \alpha(i_1, i_2)\right) \left(\sum_{i_3} \beta(i_2, i_3)\right),$$

$$p(i_1, i_2) = \sum_{i_3} p(i_1, i_2, i_3) = \alpha(i_1, i_2) \sum_{i_3} \beta(i_2, i_3),$$

$$p(i_2, i_3) = \sum_{i_1} p(i_1, i_2, i_3) = \left(\sum_{i_1} \alpha(i_1, i_2)\right) \beta(i_2, i_3).$$

Therefore

$$\begin{aligned}\frac{p(i_1,i_2)p(i_2,i_3)}{p(i_2)} &= \frac{\alpha(i_1,i_2)\beta(i_2,i_3)(\sum_{i'_1}\alpha(i'_1,i_2))(\sum_{i'_3}\beta(i_2,i'_3))}{(\sum_{i'_1}\alpha(i'_1,i_2))(\sum_{i'_3}\beta(i_2,i'_3))} \\ &= \alpha(i_1,i_2)\beta(i_2,i_3) \\ &= p(\boldsymbol{i})\end{aligned}$$

and hence (1.28) holds. This shows that the null hypothesis of conditional independence can be written in any one of (1.27), (1.28), and (1.29).

Now suppose that we observe a contingency table $\boldsymbol{x}$ of sample size n from the conditional independence model. The joint probability function is written as

$$\begin{aligned}p(\boldsymbol{x}) &= \frac{n!}{\prod_{\boldsymbol{i}\in\mathscr{I}}x(\boldsymbol{i})!}\prod_{\boldsymbol{i}\in\mathscr{I}}(\alpha(i_1,i_2)\beta(i_2,i_3))^{x(\boldsymbol{i})} \\ &= \frac{n!}{\prod_{\boldsymbol{i}\in\mathscr{I}}x(\boldsymbol{i})!}\prod_{\boldsymbol{i}_{\{1,2\}}\in\mathscr{I}_{\{1,2\}}}\alpha(i_1,i_2)^{x(i_1,i_2)}\prod_{\boldsymbol{i}_{\{2,3\}}\in\mathscr{I}_{\{2,3\}}}\beta(i_2,i_3)^{x(i_2,i_3)}. \qquad (1.30)\end{aligned}$$

Hence a sufficient statistic T is the set of $\{1,2\}$-marginals and $\{2,3\}$-marginals of $\boldsymbol{x}$:

$$T = (\{x(\boldsymbol{i}_{\{1,2\}}) \mid \boldsymbol{i}_{\{1,2\}} \in \mathscr{I}_{\{1,2\}}\},\ \{x(\boldsymbol{i}_{\{2,3\}}) \mid \boldsymbol{i}_{\{2,3\}} \in \mathscr{I}_{\{2,3\}}\}).$$

In this case the marginal distribution of T is not immediately clear and hence the conditional probability of $\boldsymbol{x}$ given $T=\boldsymbol{t}$ is also not immediately clear. However, without worrying about the marginal distribution of T at this point, we can proceed as follows. Let A be the configuration relating the frequency vector to the sufficient statistic: $\boldsymbol{t}=A\boldsymbol{x}$. Define $\mathscr{F}_{\boldsymbol{t}} = \{\boldsymbol{x}\geq 0 \mid \boldsymbol{t}=A\boldsymbol{x}\}$ as in (1.22). The terms containing the parameters α,β on the right-hand side of (1.30) are fixed by the sufficient statistic, therefore these terms do not appear in the conditional distribution of $\boldsymbol{x}$ given $\boldsymbol{t}$. It follows that the conditional distribution of $\boldsymbol{x}$ given $\boldsymbol{t}$ is written as

$$p(\boldsymbol{x}\mid\boldsymbol{t}) = c\times\frac{1}{\prod_{\boldsymbol{i}\in\mathscr{I}}x(\boldsymbol{i})!}, \qquad c=\left[\sum_{\boldsymbol{x}\in\mathscr{F}_{\boldsymbol{t}}}\frac{1}{\prod_{\boldsymbol{i}\in\mathscr{I}}x(\boldsymbol{i})!}\right]^{-1}. \qquad (1.31)$$

As in the previous examples, an exact test of the null hypothesis H of conditional independence can be performed if either we can enumerate the elements of $\mathscr{F}_{\boldsymbol{t}}$ or if we can sample from this distribution. Note that we often call (1.31) the hypergeometric distribution over $\mathscr{F}_{\boldsymbol{t}}$.

In general, the normalizing constant c cannot be written explicitly. The Markov chain sampling discussed in the next chapter can be performed without knowing the explicit form of the normalizing constant. This is one of the major advantages of Markov chain Monte Carlo methods.

It turns out that for the conditional independence model the marginal distribution of the sufficient statistic T and the normalizing constant c can be written down explicitly. This is a special case of the result of Sundberg [140] for decomposable models, which is studied in Chap. 8. In the following section, we explain the marginal distribution of T. The following section can be skipped, because the normalizing constant c is not needed for performing Markov chain Monte Carlo methods.

1.4.1 Normalizing Constant of Hypergeometric Distribution for the Conditional Independence Model

For illustration let us explicitly write out the configuration for relating the frequency vector to the sufficient statistic for the case of $2 \times 2 \times 2$ tables. We order the elements of T according to the level of Y_2. Then $\boldsymbol{t} = A\boldsymbol{x}$ is written as

$$\begin{pmatrix} x_{\{1,2\}}(1,1) \\ x_{\{1,2\}}(2,1) \\ x_{\{2,3\}}(1,1) \\ x_{\{2,3\}}(1,2) \\ x_{\{1,2\}}(1,2) \\ x_{\{1,2\}}(2,2) \\ x_{\{2,3\}}(2,1) \\ x_{\{2,3\}}(2,2) \end{pmatrix} = \begin{pmatrix} \begin{matrix} 1\,1\,0\,0 \\ 0\,0\,1\,1 \\ 1\,0\,1\,0 \\ 0\,1\,0\,1 \end{matrix} & \mathbf{0} \\ \mathbf{0} & \begin{matrix} 1\,1\,0\,0 \\ 0\,0\,1\,1 \\ 1\,0\,1\,0 \\ 0\,1\,0\,1 \end{matrix} \end{pmatrix} \begin{pmatrix} x(1,1,1) \\ x(1,1,2) \\ x(2,1,1) \\ x(2,1,2) \\ x(1,2,1) \\ x(1,2,2) \\ x(2,2,1) \\ x(2,2,2) \end{pmatrix}, \tag{1.32}$$

where the big 0 is the 4×4 zero matrix. Note that the 8×8 matrix on the right-hand side is a block diagonal with identical blocks. Furthermore, the diagonal block is the same as on the right-hand side of (1.10). Partition $\boldsymbol{x}$ on the right-hand side of (1.32) into two 4-dimensional subvectors $\boldsymbol{x}_1, \boldsymbol{x}_2$. We call each $\boldsymbol{x}_{i_2}$, $i_2 = 1,2$, the *slice* of the contingency table $\boldsymbol{x}$ by fixing the level i_2 of the second variable. Similarly we partition $\boldsymbol{t}$ on the left-hand side of (1.32) into two 4-dimensional subvectors $\boldsymbol{t}_1, \boldsymbol{t}_2$. Then clearly

$$\boldsymbol{x} \in \mathscr{F}_{\boldsymbol{t}} \quad \Leftrightarrow \quad \boldsymbol{x}_1 \in \mathscr{F}_{\boldsymbol{t}_1} \quad \text{and} \quad \boldsymbol{x}_2 \in \mathscr{F}_{\boldsymbol{t}_2}, \tag{1.33}$$

where $\mathscr{F}_{\boldsymbol{t}_1}$ and $\mathscr{F}_{\boldsymbol{t}_2}$ are fibers in (1.22) for the independence model of 2×2 contingency tables.

We have thus far looked at the $2 \times 2 \times 2$ case. However, it is clear that a similar result holds for the general $I_1 \times I_2 \times I_3$ case. Namely, when we sort the cells according to the levels of Y_2, then the configuration is in a block diagonal form with identical blocks, which correspond to the configuration of the independence model for $I_1 \times I_3$ contingency tables.

Also from (1.16) it follows that for $I \times J$ contingency tables we have

$$\sum_{x \in \mathscr{F}_t} \frac{1}{\prod_{i,j} x_{ij}!} = \frac{n!}{\prod_{i=1}^{I} x_{i+}! \prod_{j=1}^{J} x_{+j}!}.$$

Combining this with (1.33) and by summing for each slice separately, we have the following expression of c^{-1} in (1.31) for the conditional independence model.

$$\begin{aligned} \frac{1}{c} &= \prod_{i_2=1}^{I_2} \frac{x(i_2)!}{\prod_{i_1=1}^{I_1} x(i_1,i_2)! \prod_{i_3=1}^{I_3} x(i_2,i_3)!} \\ &= \frac{\prod_{i_2 \in \mathscr{I}_{\{2\}}} x(i_2)!}{\prod_{(i_1,i_2) \in \mathscr{I}_{\{1,2\}}} x(i_1,i_2)! \prod_{(i_2,i_3) \in \mathscr{I}_{\{2,3\}}} x(i_2,i_3)!}. \end{aligned}$$

If we apply this sum to (1.30), we see that the joint probability distribution of T is given as

$$\begin{aligned} p(T) = &\frac{n! \prod_{i_2 \in \mathscr{I}_{\{2\}}} x(i_2)!}{\prod_{(i_1,i_2) \in \mathscr{I}_{\{1,2\}}} x(i_1,i_2)! \prod_{(i_2,i_3) \in \mathscr{I}_{\{2,3\}}} x(i_2,i_3)!} \\ &\times \prod_{i_{\{1,2\}} \in \mathscr{I}_{\{1,2\}}} \alpha(i_1,i_2)^{x(i_1,i_2)} \prod_{i_{\{2,3\}} \in \mathscr{I}_{\{2,3\}}} \beta(i_2,i_3)^{x(i_2,i_3)}. \end{aligned}$$

Then the conditional probability in (1.31) is explicitly written as

$$p(\boldsymbol{x} \mid \boldsymbol{t}) = \frac{\prod_{(i_1,i_2) \in \mathscr{I}_{\{1,2\}}} x(i_1,i_2)! \prod_{(i_2,i_3) \in \mathscr{I}_{\{2,3\}}} x(i_2,i_3)!}{\prod_{i_2 \in \mathscr{I}_{\{2\}}} x(i_2)! \prod_{\boldsymbol{i} \in \mathscr{I}} x(\boldsymbol{i})!}. \tag{1.34}$$

1.5 Notation of Hierarchical Models for *m*-Way Contingency Tables

In this section we introduce notation for general m-way contingency tables and hierarchical models for these tables. The notation introduced here is used extensively later in this book, such as Chaps. 8 and 9. Readers may skip this section and check the notation when it is needed later in the book.

In the previous section we considered three-way contingency tables. We generalize the notation to m-way tables. The set of the cells for an m-way table is the direct product

$$\mathscr{I} = \{\boldsymbol{i} = (i_1,\ldots,i_m) \mid i_1 \in [I_1],\ldots,i_m \in [I_m]\} = [I_1] \times \cdots \times [I_m].$$

An m-way contingency table, or the frequency vector, is denoted by $\boldsymbol{x} = \{x(\boldsymbol{i}) \mid \boldsymbol{i} \in \mathscr{I}\}$. We denote the set of m variables as $\Delta = [m] = \{1,\ldots,m\}$. The notation for marginal cells, marginal frequencies, and marginal probabilities was already given in (1.24), (1.25), and (1.26).

Consider the logarithm of the probability function of the conditional independence model in (1.29):

$$\log p(\boldsymbol{i}) = \log \alpha(i_1,i_2) + \log \beta(i_2,i_3). \tag{1.35}$$

Here $\log p(\boldsymbol{i})$ is written as a sum of two functions, one of which depends only on (i_1,i_2) and the other depends only on (i_2,i_3).

Generalizing this formulation we now define a hierarchical model for m-way contingency tables. Let $D_1,\ldots,D_r$ be subsets of Δ, such that there is no inclusion relation between D_i and D_j, $1 \le i \ne j \le m$. Denote $\mathscr{D} = \{D_1,\ldots,D_r\}$. A hierarchical model with the generating class $\mathscr{D}$ is defined as follows.

$$\log p(\boldsymbol{i}) = \sum_{D \in \mathscr{D}} \mu_D(\boldsymbol{i}_D), \tag{1.36}$$

where μ_D is a function depending only on the marginal cell $\boldsymbol{i}_D$. For a general hierarchical model let

$$\mathscr{K} = \mathscr{K}(\mathscr{D}) = \{D \mid D \subset D_i \text{ for some } D_i \in \mathscr{D}\}$$

denote the set of subsets of $D_1,\ldots,D_r$. Note that $\mathscr{K}$ has the following property,

$$A \in \mathscr{K},\ B \subset A \ \Rightarrow\ B \in \mathscr{K}.$$

A family of subsets of Δ satisfying this property is called a *simplicial complex* ([96]). Note that $D_1,\ldots,D_r$ are maximal elements of $\mathscr{K}$ with respect to set inclusion. Maximal elements of a simplicial complex $\mathscr{K}$ are called *facets* of $\mathscr{K}$. From a statistical viewpoint, the facets correspond to maximal interaction terms in the hierarchical model. In a hierarchical model, when an interaction term is present in the model, then all smaller interaction terms and the main effects included in the interaction term are also present in the model. This is a natural assumption, because, for example, a two-variable interaction is usually interpreted only in the presence of main effects of the variables.

A sufficient statistic for a hierarchical model is given by the set of marginal frequencies for $D_1,\ldots,D_r$:

$$T = \{x_D(\boldsymbol{i}_D) \mid \boldsymbol{i}_D \in \mathscr{I}_D, D \in \mathscr{D}\}.$$

Finally we discuss indeterminacy of μ_Ds in (1.36). As an example consider again the conditional independence model in (1.35). There is some indeterminacy on the right-hand side of (1.35). One way of resolving this indeterminacy is to use the standard ANOVA (analysis of variance) decomposition (e.g., Scheffé [132]). In (1.35) we can write

$$\log p(\boldsymbol{i}) = \mu_0 + \mu_1(i_1) + \mu_2(i_2) + \mu_3(i_3) + \mu_{12}(i_1,i_2) + \mu_{23}(i_2,i_3), \qquad (1.37)$$

where we require

$$0 = \sum_{i_j=1}^{I_j} \mu_j(i_j),\ j = 1,2,3, \quad 0 = \sum_{i_1=1}^{I_1} \mu_{12}(i_1,i_2) = \sum_{i_2=1}^{I_2} \mu_{12}(i_1,i_2),\ \forall i_1,i_2,$$

$$0 = \sum_{i_2=1}^{I_2} \mu_{23}(i_2,i_3) = \sum_{i_3=1}^{I_3} \mu_{23}(i_2,i_3),\ \forall i_2,i_3.$$

Under these requirements the right-hand side of (1.37) is unique. Similarly, for a general hierarchical model we can uniquely express $\log p(\boldsymbol{i})$ as

$$\log p(\boldsymbol{i}) = \sum_{D \in \mathscr{K}} \mu_D(\boldsymbol{i}_D), \qquad (1.38)$$

where for every $D = \{j_1,\ldots,j_l\} \in \mathscr{K}$ we require

$$\sum_{i_{j_h}=1}^{I_{j_h}} \mu_D(i_{j_1},\ldots,i_{j_l}) = 0, \quad h = 1,\ldots,l.$$

Another popular method for avoiding the indeterminacy is to treat a particular level, for example, the last level I_j, $j = 1,\ldots,m$, as the "base level" and require for every $D \in \mathscr{K}$, $D \neq \emptyset$, that

$$\mu_D(i_{j_1},\ldots,i_{j_l}) = 0, \qquad \text{if } i_{j_h} = I_{j_h} \text{ for some } h = 1,\ldots,l. \qquad (1.39)$$

In the case of a complete independence model of two-way tables, this corresponds to the lattice basis in (2.5).

Chapter 2
Markov Chain Monte Carlo Methods over Discrete Sample Space

2.1 Constructing a Connected Markov Chain over a Conditional Sample Space: Markov Basis

In the previous chapter we discussed exact tests for some simple models of contingency tables. As we discussed at the end of Sect. 1.3, the Markov chain Monte Carlo method is general and useful when the cardinality of conditional sample space (fiber) is large. We first consider connectivity of a Markov chain, without fully specifying the transition probabilities.

Consider the independence model of general two-way contingency tables in Sect. 1.3. The fiber is the set of $I \times J$ contingency tables with fixed row sums and column sums:

$$\mathscr{F}_{\boldsymbol{t}} = \{\boldsymbol{x} \geq 0 \mid x_{i+}, i \in [I],\ x_{+j}, j \in [J] \text{ are fixed according to } \boldsymbol{t}\}. \tag{2.1}$$

Let A be the configuration in (1.18). The kernel of A is denoted by $\ker A$. The set of integer vectors in $\ker A$ is called the *integer kernel* of A and is denoted by

$$\ker_{\mathbb{Z}} A = \{\boldsymbol{z} \mid A\boldsymbol{z} = 0, \boldsymbol{z} \in \mathbb{Z}^{\eta}\}, \quad \eta = IJ.$$

An element of $\ker_{\mathbb{Z}} A$ is called a *move* for the configuration A. If $\boldsymbol{x}$ and $\boldsymbol{y}$ belong to the same fiber $\mathscr{F}_{\boldsymbol{t}}$, then $\boldsymbol{y} - \boldsymbol{x}$ is a move, because

$$A(\boldsymbol{y} - \boldsymbol{x}) = A\boldsymbol{y} - A\boldsymbol{x} = \boldsymbol{t} - \boldsymbol{t} = 0. \tag{2.2}$$

Now consider the following integer matrix $\boldsymbol{z} = \boldsymbol{z}(i_1, i_2; j_1, j_2) = \{z_{ij}\}$,

$$z_{ij} = \begin{cases} +1, & (i,j) = (i_1, j_1), (i_2, j_2), \\ -1, & (i,j) = (i_1, j_2), (i_2, j_1), \\ 0, & \text{otherwise}. \end{cases} \tag{2.3}$$

S. Aoki et al., *Markov Bases in Algebraic Statistics*, Springer Series in Statistics 199, DOI 10.1007/978-1-4614-3719-2_2,

The nonzero elements of $\boldsymbol{z}(i_1,i_2;j_1,j_2)$ are depicted as

$$\begin{array}{c|cc|} & j_1 & j_2 \\ \cline{2-3} i_1 & +1 & -1 \\ i_2 & -1 & +1 \\ \cline{2-3} \end{array}\ . \tag{2.4}$$

Adding $\boldsymbol{z}(i_1,i_2;j_1,j_2)$ to a contingency table $\boldsymbol{x}$ does not alter the row sums and the column sums. Hence $\boldsymbol{z}(i_1,i_2;j_1,j_2)$ is a move for A in (1.18); that is, $\boldsymbol{z}(i_1,i_2;j_1,j_2) \in \ker_{\mathbb{Z}} A$. We call a move of the form (2.4) a *basic move* for the independence model of two-way contingency tables. Because of the elements -1 in $\boldsymbol{z}(i_1,i_2;j_1,j_2)$, $\boldsymbol{x}+\boldsymbol{z}$ contains a negative element if $x_{i_2 j_1} = 0$ or $x_{i_1 j_2} = 0$. If both of these elements are positive, then $\boldsymbol{x}+\boldsymbol{z}$ is in $\mathscr{F}_{\boldsymbol{t}}$ if $\boldsymbol{x} \in \mathscr{F}_{\boldsymbol{t}}$. We have "moved" from $\boldsymbol{x}$ to $\boldsymbol{x}+\boldsymbol{z}$ in $\mathscr{F}_{\boldsymbol{t}}$. This is why we call $\boldsymbol{z}(i_1,i_2;j_1,j_2)$ a move. The following is an example of adding a move for the case of $I=J=3$, $i_1=j_1=1$, $i_2=j_2=2$.

$$\begin{array}{|c|c|c|c} \hline 2&1&1&4\\ \hline 2&0&2&4\\ \hline 1&2&0&3\\ \hline 5&3&3& \end{array} + \begin{array}{|c|c|c|} \hline 1&-1&0\\ \hline -1&1&0\\ \hline 0&0&0\\ \hline \end{array} = \begin{array}{|c|c|c|c} \hline 3&0&1&4\\ \hline 1&1&2&4\\ \hline 1&2&0&3\\ \hline 5&3&3& \end{array}\ .$$

Suppose that we always use the last row I and the last column J in the move and let $i_2 = I$ and $j_2 = J$. Then

$$\{\boldsymbol{z}(i_1,I;j_1,J) \mid 1 \le i_1 \le I-1, 1 \le j_1 \le J-1\}$$

forms a basis of $\ker_{\mathbb{Z}} A$. More precisely the set forms a *lattice basis* of $\ker_{\mathbb{Z}} A$ in the sense that every $\boldsymbol{z} \in \ker_{\mathbb{Z}} A$ is uniquely written as an integer combination of $\boldsymbol{z}(i_1,I;j_1,J)$s. In fact the elements of the last row and the last column of $\boldsymbol{z} = \{z_{ij}\} \in \ker_{\mathbb{Z}} A$ are uniquely determined from the other elements. Hence $\boldsymbol{z} \in \ker_{\mathbb{Z}} A$ can be uniquely written as

$$\boldsymbol{z} = \sum_{i_1=1}^{I-1} \sum_{j_1=1}^{J-1} z_{i_1 j_1} \times \boldsymbol{z}(i_1,I;j_1,J), \tag{2.5}$$

because both sides have the same elements in the first $I-1$ rows and the first $J-1$ columns. This is related to use of the last level as the base level discussed at the end of Chap. 1.

Note that the lattice basis is very simple for the independence model of $I \times J$ tables. However, for the fiber in (2.1) we are requiring nonnegativeness of the frequency vectors. As an example consider the following two elements of the fiber for $I=J=3$ with $1 = x_{1+} = x_{2+} = x_{+1} = x_{+2}$, $0 = x_{3+} = x_{+3}$.

$$\begin{array}{|c|c|c|c} \hline 1&0&0&1\\ \hline 0&1&0&1\\ \hline 0&0&0&0\\ \hline 1&1&0&2 \end{array}\ , \qquad \begin{array}{|c|c|c|c} \hline 0&1&0&1\\ \hline 1&0&0&1\\ \hline 0&0&0&0\\ \hline 1&1&0&2 \end{array}\ .$$

We see that we cannot add or subtract any of $z(i_1,3;j_1,3)$ to/from these tables without making some cell frequency negative. However, obviously these two tables are connected by the following move:

$$\begin{array}{|c|c|c|}\hline 1 & -1 & 0 \\ \hline -1 & 1 & 0 \\ \hline 0 & 0 & 0 \\ \hline \end{array}\,.$$

This example suggests that we can move around a fiber if we can use all moves of the form (2.3).

Let $\mathscr{B} \subset \ker_{\mathbb{Z}} A$ be a finite set of moves for a configuration A. $\mathscr{B}$ is called a *Markov basis* if for all fibers $\mathscr{F}_t$ and for all elements $\boldsymbol{x},\boldsymbol{y} \in \mathscr{F}_t$, $\boldsymbol{x} \neq \boldsymbol{y}$, there exist $K > 0$, $\boldsymbol{z}_1,\ldots,\boldsymbol{z}_K \in \mathscr{B}$ and $\varepsilon_1,\ldots,\varepsilon_K \in \{-1,1\}$, such that

$$\boldsymbol{y} = \boldsymbol{x} + \sum_{k=1}^{K} \varepsilon_k \boldsymbol{z}_k, \qquad \boldsymbol{x} + \sum_{k=1}^{L} \varepsilon_k \boldsymbol{z}_k \in \mathscr{F}_t, \qquad L = 1,\ldots,K-1. \tag{2.6}$$

The first condition says that by adding or subtracting elements of $\mathscr{B}$, we can move from $\boldsymbol{x}$ to $\boldsymbol{y}$. The second condition says that on the way from $\boldsymbol{x}$ to $\boldsymbol{y}$ we never encounter a negative frequency. Therefore if a Markov basis $\mathscr{B}$ is given, then we can move all over any fiber by adding or subtracting moves from $\mathscr{B}$. Thus connectivity of every fiber is guaranteed by a Markov basis. We define Markov basis again in Chap. 4 for a general configuration A. In this introductory explanation, we give a proof that a Markov basis for the $I \times J$ independence model of two-way contingency tables is given by the set of moves $\boldsymbol{z}(i_1,i_2;j_1,j_2)$. We state this as a theorem.

Theorem 2.1. *Let*

$$\mathscr{B} = \{\boldsymbol{z}(i_1,i_2;j_1,j_2) \mid 1 \leq i_1 < i_2 \leq I, 1 \leq j_1 < j_2 \leq J\}.$$

$\mathscr{B}$ forms a Markov basis for the $I \times J$ independence model of two-way contingency tables.

The following proof is a typical "distance reducing argument," that is frequently used in later chapters of this book.

Proof. We argue by contradiction. Suppose that $\mathscr{B}$ is not a Markov basis. Then there exists a fiber $\mathscr{F}_t$ and two elements $\boldsymbol{x},\boldsymbol{y} \in \mathscr{F}_t$ of the fiber, such that we cannot move from $\boldsymbol{x}$ to $\boldsymbol{y}$ by the moves of $\mathscr{B}$ as in (2.6). Let

$$\mathscr{N}_{\boldsymbol{x}} = \{\boldsymbol{y} \in \mathscr{F}_t \mid \text{we cannot move from } \boldsymbol{x} \text{ to } \boldsymbol{y} \text{ by moves of } \mathscr{B}\}.$$

Then $\mathscr{N}_{\boldsymbol{x}}$ is not empty by assumption. For $\boldsymbol{z} = \{z_{ij}\} \in \ker_{\mathbb{Z}} A$, let $|\boldsymbol{z}| = \sum_{i=1}^{I}\sum_{j=1}^{J} |z_{ij}|$ denote its 1-norm. In Sect. 4.3 we define $\deg \boldsymbol{z}$ as $|\boldsymbol{z}|/2$.

Define

$$\boldsymbol{y}^* = \underset{\boldsymbol{y}\in\mathscr{N}_{\boldsymbol{x}}}{\arg\min} |\boldsymbol{x}-\boldsymbol{y}|. \tag{2.7}$$

$\boldsymbol{y}^*$ is one of the closest elements of $\mathscr{F}_{\boldsymbol{t}}$ that cannot be reached from $\boldsymbol{x}$ by $\mathscr{B}$:

$$|\boldsymbol{x}-\boldsymbol{y}^*| = \min_{\boldsymbol{y}\in\mathscr{N}_{\boldsymbol{x}}} |\boldsymbol{x}-\boldsymbol{y}|.$$

Now let $\boldsymbol{w} = \boldsymbol{x}-\boldsymbol{y}^*$ and consider the signs of elements of $\boldsymbol{w}$. Because $\boldsymbol{w}$ contains a positive element, let $w_{i_1j_1} > 0$. Then because $\boldsymbol{w}$ is a move, there exist $j_2 \neq j_1$ with $w_{i_1j_2} < 0$ and $i_2 \neq i_1$ with $w_{i_2j_1} < 0$. Hence for $\boldsymbol{y}^* = \{y^*_{ij}\}$ we have $y^*_{i_1j_2} > 0$, $y^*_{i_2j_1} > 0$. Then

$$\boldsymbol{y}^* + \boldsymbol{z}(i_1,i_2;j_1,j_2) \in \mathscr{F}_{\boldsymbol{t}}.$$

$\boldsymbol{y}^*$ cannot be reached from $\boldsymbol{x}$ by $\mathscr{B}$, therefore $\boldsymbol{y}^* + \boldsymbol{z}(i_1,i_2;j_1,j_2)$ cannot be reached from $\boldsymbol{x}$ by $\mathscr{B}$ either and $\boldsymbol{y}^* + \boldsymbol{z}(i_1,i_2;j_1,j_2) \in \mathscr{N}_{\boldsymbol{x}}$. Now we check the value of $|\boldsymbol{x} - (\boldsymbol{y}^* + \boldsymbol{z}(i_1,i_2;j_1,j_2))|$.

- If $w_{i_2j_2} > 0$, then $|\boldsymbol{x} - (\boldsymbol{y}^* + \boldsymbol{z}(i_1,i_2;j_1,j_2))| = |\boldsymbol{x}-\boldsymbol{y}^*| - 4$,
- If $w_{i_2j_2} \leq 0$, then $|\boldsymbol{x} - (\boldsymbol{y}^* + \boldsymbol{z}(i_1,i_2;j_1,j_2))| = |\boldsymbol{x}-\boldsymbol{y}^*| - 2$.

Therefore for both cases, $|\boldsymbol{x} - (\boldsymbol{y}^* + \boldsymbol{z}(i_1,i_2;j_1,j_2))| < |\boldsymbol{x}-\boldsymbol{y}^*|$. However, this contradicts the minimality in (2.7) of $\boldsymbol{y}^*$. □

By this theorem, we can construct a connected Markov chain over any fiber. We choose $i_1,i_2 \in [I]$ and $j_1,j_2 \in [J]$ randomly. We add or subtract $\boldsymbol{z}(i_1,i_2;j_1,j_2)$ to/from the current state $\boldsymbol{x}$ and move to $\boldsymbol{y} = \boldsymbol{x} + \boldsymbol{z}(i_1,i_2;j_1,j_2)$ as long as there is no negative frequency in $\boldsymbol{y}$. In the case where $\boldsymbol{y}$ contains a negative element, we choose another set of indices $i_1,i_2 \in [I]$ and $j_1,j_2 \in [J]$ and continue. Then connectivity of every fiber is guaranteed by Theorem 2.1.

Note that in the above explanation we are not precisely specifying the probability distribution of choosing an element $\boldsymbol{z}(i_1,i_2;j_1,j_2)$. Also, when we say "add or subtract," we are not exactly saying which to choose. In fact, we should choose the sign of a move $\boldsymbol{z}(i_1,i_2;j_1,j_2)$ (i.e., whether we add it or subtract it) with probability $1/2$. This is related to the Markov chain symmetry for the Metropolis–Hastings algorithm in the next section. Other than the choice of the sign of a move, the distribution for choosing a move can be arbitrary.

In this section we considered the independence model of two-way contingency tables. We now briefly mention the conditional independence model of three-way contingency tables. As we saw in the previous section, the conditional independence model of three-way contingency tables can be treated as the two-way independence model given each level of the conditioning variable. Therefore a Markov basis for the conditional independence model of three-way contingency tables is given as a union of Markov bases for two-way cases in each slice of the contingency table given the level of the conditioning variable. The two-way independence model and the conditional independence model of three-way contingency tables are actually simple examples. Markov bases for more complicated models of contingency tables are in fact difficult and each model needs separate consideration. One notable exception is the decomposable model studied in Chap. 8.

On the other hand, there exists a general algorithm to compute a Markov basis in the form of the Gröbner basis for any configuration. So is the problem of obtaining a Markov basis already solved by a general algorithm? The answer is yes and no, depending on the viewpoint. The existence of a general algorithm means that the answer is yes from a certain theoretical viewpoint. On the other hand, for practical purposes, the computation of the Gröbner basis for a complicated model is often infeasible in a practical amount of time and in this sense the answer is no. Therefore, both theoretical investigations of Markov bases for specific models and the further general improvements in the algorithms for Gröbner basis computation are very much needed at present.

2.2 Adjusting Transition Probabilities by Metropolis–Hastings Algorithm

In this section we explain how to construct a Markov chain that has a specified distribution as the stationary distribution. A good reference on important facts on Markov chains is Häggström [69].

Consider a Markov chain over a finite sample space $\mathscr{F}$. Suppose that the elements of $\mathscr{F}$ are given as

$$\mathscr{F} = \{\boldsymbol{x}_1, \dots, \boldsymbol{x}_s\}. \tag{2.8}$$

Let $\{Z_t,\ t = 0,1,2,\dots\}$, $Z_t \in \mathscr{F}$, be a Markov chain over $\mathscr{F}$ with the transition probability $Q = (q_{ij})$:

$$q_{ij} = P(Z_{t+1} = \boldsymbol{x}_j \mid Z_t = \boldsymbol{x}_i), \qquad 1 \le i, j \le s.$$

A Markov chain is called *symmetric* if Q is a symmetric matrix ($q_{ij} = q_{ji}$).

Let

$$\boldsymbol{\pi} = (\pi_1, \dots, \pi_s)$$

denote the initial probability distribution of Z_0 (by standard notation, we consider $\boldsymbol{\pi}$ as a row vector). $\boldsymbol{\pi}$ is called a *stationary distribution* if

$$\boldsymbol{\pi} = \boldsymbol{\pi} Q.$$

$\boldsymbol{\pi}$ is the eigenvector from the left of Q with the eigenvalue 1.

It is known that the stationary distribution exists uniquely under the assumption that the Markov chain is irreducible and aperiodic. We only consider Markov chains satisfying these conditions. Under these conditions, starting from an arbitrary state $Z_0 = \boldsymbol{x}_i$, the distribution of Z_t for large t is close to the stationary distribution $\boldsymbol{\pi}$. Therefore if we can construct a Markov chain with the "target" stationary distribution $\boldsymbol{\pi}$, then by running a Markov chain and discarding a large number t of initial steps (called *burn-in steps*), we can consider $Z_{t+1}, Z_{t+2}, \dots$ as observations from the stationary distribution $\boldsymbol{\pi}$.

For our problem, the target distribution $\boldsymbol{\pi}$ is already given as the hypergeometric distribution over the fiber in (1.31). We want to construct a Markov chain over $\mathscr{F}_t$ just for the purpose of sampling from the hypergeometric distribution. For this purpose the Metropolis–Hastings algorithm is very useful. By the algorithm, once we can construct an arbitrary irreducible (i.e., connected) chain over $\mathscr{F}_t$, we can easily modify the stationary distribution to the given target distribution $\boldsymbol{\pi}$.

Theorem 2.2 (Metropolis–Hastings algorithm). *Let $\boldsymbol{\pi}$ be a probability distribution on $\mathscr{F}$. Let $R = (r_{ij})$ be the transition probability matrix of an irreducible, aperiodic, and symmetric Markov chain over $\mathscr{F}$. Define $Q = (q_{ij})$ by*

$$
\begin{aligned}
q_{ij} &= r_{ij} \min\left(1, \frac{\pi_j}{\pi_i}\right), \qquad i \neq j, \\
q_{ii} &= 1 - \sum_{j \neq i} q_{ij}.
\end{aligned}
\tag{2.9}
$$

Then Q satisfies $\boldsymbol{\pi} = \boldsymbol{\pi} Q$.

This result is a special case of Hastings [82] and the symmetry assumption on R can be removed relatively easily. In this book we only consider symmetric R and the simple statement of the above theorem is sufficient for our purposes.

Proof (Theorem 2.2). It suffices to show that the above Q is "reversible" in the following sense.

$$
\pi_i q_{ij} = \pi_j q_{ji}. \tag{2.10}
$$

In fact, under the reversibility

$$
\pi_i = \pi_i \sum_{j=1}^{S} q_{ij} = \sum_{j=1}^{S} \pi_j q_{ji}
$$

and we have $\boldsymbol{\pi} = \boldsymbol{\pi} Q$. Now (2.10) clearly holds for $i = j$. Also for $i \neq j$

$$
\pi_i q_{ij} = \pi_i r_{ij} \min\left(1, \frac{\pi_j}{\pi_i}\right) = r_{ij} \min(\pi_i, \pi_j);
$$

hence (2.10) holds if $r_{ij} = r_{ji}$. □

Equation (2.10) is often called the *detailed balance* or detailed balance equation.

An important advantage of the Markov chain Monte Carlo method is that it does not need the explicit evaluation of the normalizing constant of the stationary distribution $\boldsymbol{\pi}$. We only need to know $\boldsymbol{\pi}$ up to a multiplicative constant. In fact in (2.9) the stationary distribution only appears in the form of ratios of its elements π_i/π_j and the normalizing constant is canceled.

Another important point in (2.9) is how the transition probability r_{ij} is modified. It is modified by $\min(1, \pi_j/\pi_i)$, which does not depend on how r_{ij} is specified.

In fact (2.9) can be understood as follows. r_{ij} is the proposal transition probability. Suppose that we are at state i and we propose to move to j with the conditional probability r_{ij} by some random mechanism. Then after the proposal, we actually move to j with probability $\min(1, \pi_j/\pi_i)$ (or stay at i with probability $1 - \min(1, \pi_j/\pi_i)$). We can do this even without knowing the value of r_{ij}, as long as it is symmetric. This fact is relevant in the application of the Markov basis, because when a Markov basis element is chosen "randomly," the probability distribution of choosing an element can be arbitrary, as long as there is a positive probability of choosing every element. Irrespective of the distribution, the Metropolis–Hastings algorithm yields a Markov chain whose stationary distribution is $\boldsymbol{\pi}$.

By Theorem 2.2 we only need to construct one Markov chain, which is irreducible, aperiodic, and symmetric. By the Metropolis–Hastings algorithm, we can then modify the transition probability to achieve the desired stationary distribution $\boldsymbol{\pi}$.

In the previous section we obtained a Markov basis for two-way tables. Once a Markov basis is obtained for some model, it is easy to construct an irreducible and symmetric Markov chain over $\mathscr{F}_{A\boldsymbol{x}^o}$, where $\boldsymbol{x}^o$ is the observed frequency vector and $\mathscr{F}_{A\boldsymbol{x}^o}$ is the fiber containing $\boldsymbol{x}^o$. For example, at each step of the Markov chain, randomly choose an element $\boldsymbol{z} \in \mathscr{B}$ of the Markov basis and the sign $\varepsilon \in \{-1, +1\}$. If $\boldsymbol{x} + \varepsilon\boldsymbol{z} \in \mathscr{F}_{\boldsymbol{t}}$ then we move to $\boldsymbol{x} + \varepsilon\boldsymbol{z}$. If $\boldsymbol{x} + \varepsilon\boldsymbol{z} \notin \mathscr{F}_{\boldsymbol{t}}$ we stay at $\boldsymbol{x}$. Then the resulting Markov chain is irreducible and symmetric. It is important to note that this holds irrespective of the distribution of choosing an element from $\mathscr{B}$, as long as each element of $\mathscr{B}$ is chosen with positive probability. On the other hand, the sign of ε should be chosen with probability $1/2$.

We can then apply the Metropolis–Hastings algorithm of Theorem 2.2 to this Markov chain. The resulting algorithm is given as follows.

Algorithm 2.1

Input: Observed frequency vector $\boldsymbol{x}^o$, Markov basis $\mathscr{B}$, number of steps N, configuration A, the null distribution $f(\cdot)$, test statistic $T(\cdot)$, $\boldsymbol{t} = A\boldsymbol{x}^o$.
Output: Estimate of the p-value.
Variables: `obs`, `count`, `sig`, $\boldsymbol{x}$, $\boldsymbol{x}_{next}$.
Step 1: `obs` $= T(\boldsymbol{x}^o)$, $\boldsymbol{x} = \boldsymbol{x}^o$, `count` $= 0$, `sig` $= 0$.
Step 2: Choose $\boldsymbol{z} \in \mathscr{B}$ randomly. Choose $\varepsilon \in \{-1, +1\}$ with probability $\frac{1}{2}$.
Step 3: If $\boldsymbol{x} + \varepsilon\boldsymbol{z} \notin \mathscr{F}_{\boldsymbol{t}}$ then $\boldsymbol{x}_{next} = \boldsymbol{x}$ and go to Step 5. If $\boldsymbol{x} + \varepsilon\boldsymbol{z} \in \mathscr{F}_{\boldsymbol{t}}$ then let u be a uniform random number between 0 and 1.
Step 4: If $u \le \frac{f(\boldsymbol{x}+\varepsilon\boldsymbol{z})}{f(\boldsymbol{x})}$ then let $\boldsymbol{x}_{next} = \boldsymbol{x} + \varepsilon\boldsymbol{z}$ and go to Step 5. If $u > \frac{f(\boldsymbol{x}+\varepsilon\boldsymbol{z})}{f(\boldsymbol{x})}$ then let $\boldsymbol{x}_{next} = \boldsymbol{x}$ and go to Step 5.
Step 5: If $T(\boldsymbol{x}_{next}) \ge$ `obs` then let `sig` $=$ `sig` $+1$.
Step 6: $\boldsymbol{x} = \boldsymbol{x}_{next}$, `count` $=$ `count` $+1$.
Step 7: If `count` $< N$ then go to Step 2.
Step 8: The estimate of p-value is `sig`$/N$.

We should mention one important point concerning the counting of steps. There are two cases where we stay at the same state $\boldsymbol{x}_{next} = \boldsymbol{x}$. One case is that we reject a move $\boldsymbol{z}$ because $\boldsymbol{x} + \varepsilon\boldsymbol{z} \notin \mathscr{F}_{\boldsymbol{t}}$ in Step 3. Another case is that the proposed state is

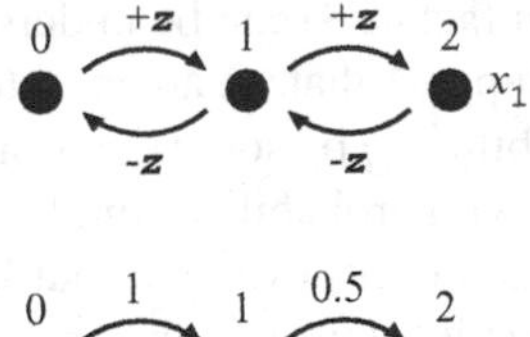

Fig. 2.1 The fiber $\mathscr{F}_2$

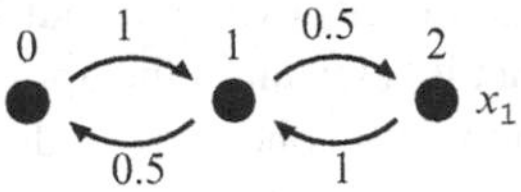

Fig. 2.2 Transition probabilities ignoring rejections in Step 3

rejected because of $u > f(\boldsymbol{x}+\varepsilon \boldsymbol{z})/f(\boldsymbol{x})$ in Step 4. In both cases, we evaluate the value of the test statistic $T(\boldsymbol{x}_{next}) = T(\boldsymbol{x})$ and the counter `count` is increased. For unbiased estimation of the p-value, we need to include both cases in evaluation of T and the counting of the steps.

In Step 3, if $\boldsymbol{x}$ is close to the boundary of $\mathscr{F}_t$, then it may be the case that $\boldsymbol{x}+\varepsilon\boldsymbol{z} \notin \mathscr{F}_t$ with high probability. In this case we might be tempted to choose $\boldsymbol{z}$ depending on $\boldsymbol{x}$ such that the probability of $\boldsymbol{x}+\varepsilon\boldsymbol{z} \in \mathscr{F}_t$ is higher. This is an interesting topic for investigation, although it is not trivial to guarantee the symmetry $r_{ij} = r_{ji}$ if we choose a move depending on the state.

The above point can be illustrated by the following very simple example. Consider a configuration $A = (1,1)$, which is a 1×2 matrix. Let $t = A\boldsymbol{x}$, $\boldsymbol{x} = (x_1, x_2)'$ and consider the fiber with $t = 2$:

$$\mathscr{F}_2 = \{(x_1,x_2) \mid x_1 + x_2 = 2, x_1, x_2 \in \mathbb{N}\}, \qquad \mathbb{N} = \{0,1,2,\dots\}.$$

Then $\boldsymbol{z} = (1,-1)'$ is a move, which obviously connects $\mathscr{F}_2$. The fiber is depicted as in Fig. 2.1, where the states are labeled by the values of x_1.

Note that $\boldsymbol{z}$ cannot be subtracted from $(0,2)$ and $\boldsymbol{z}$ cannot be added to $(2,0)$, because these operations produce -1. Therefore if we are at $(0,2)$ we can only add $\boldsymbol{z}$. Similarly if we are at $(2,0)$ we can only subtract $\boldsymbol{z}$. Now suppose that we want to sample from the uniform distribution over $\mathscr{F}_2$. Then in the Metropolis–Hastings algorithm, $\min(1, \pi_j/\pi_i) \equiv 1$. Therefore we stay at the same state only because of Step 3 of Algorithm 1. If we ignore the rejections in Step 3 for this example, the transition probabilities of the chain are depicted in Fig. 2.2. The stationary distribution of this chain is given by

$$(\pi(0,2), \pi(1,1), \pi(2,0)) = \left(\frac{1}{4}, \frac{1}{2}, \frac{1}{4}\right),$$

which is not uniform.

On the other hand if we count the rejections in Step 3, then the Markov chain has self-loops and the transition probabilities of the chain are depicted in Fig. 2.3. For this chain the stationary distribution is the uniform distribution, which was our target.

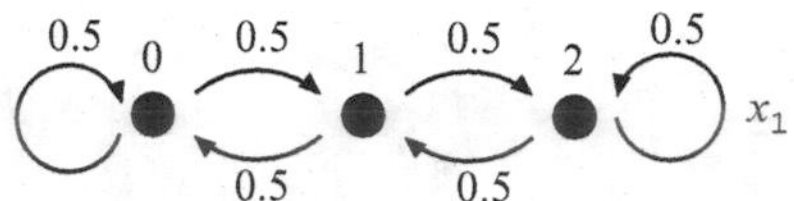

Fig. 2.3 Transition probabilities taking rejections in Step 3 into account

Algorithm 2.1 is a very simple algorithm and various improvements are possible. For example, grouping several steps of Algorithm 2.1 in one step makes the convergence to the stationary distribution faster. This can be achieved as follows.

Algorithm 2.2 Modify Steps 2, 3, 4 in Algorithm 2.1 as follows.

Step 2 : Choose $\boldsymbol{z} \in \mathscr{B}$ randomly.
Step 3 : Let $I = \{n \mid \boldsymbol{x} + n\boldsymbol{z} \in \mathscr{F}_{\boldsymbol{t}}\}$.
Step 4 : Choose $\boldsymbol{x}_{\text{next}}$ from $\{\boldsymbol{x} + n\boldsymbol{z} \mid n \in I\}$ according to the probability

$$p_n = \frac{f(\boldsymbol{x}+n\boldsymbol{z})}{\sum_{n \in I} f(\boldsymbol{x}+n\boldsymbol{z})}.$$

Note that both in Algorithms 2.1 and 2.2, the target distribution $f(\cdot)$ appears in the form of the ratio. Hence we do not need to compute the normalizing constant for $f(\cdot)$. Often the computation of the normalizing constant is difficult, therefore this is an important advantage of the Markov chain Monte Carlo method.

Chapter 3
Toric Ideals and Their Gröbner Bases

Readers can skip this chapter and come back to individual results when they are referenced in later chapters. There are many good textbooks on computational algebra and Gröbner basis. This chapter is based on a great deal Chapter 1 of [93] by Takayuki Hibi, although for individual results we cite Cox et al. [42] as a reference. Sturmfels [139] gives more specific results relevant for algebraic statistics and toric ideals.

In presenting results on polynomial rings, the difference of standard notation for statistics and algebra is annoying. For example, in statistics n usually denotes the sample size, whereas in the notation for polynomial rings n usually stands for the number of indeterminates, which corresponds to the total number of cells $|\mathscr{I}|$. In this book $x(\boldsymbol{i})$ stands for the frequency of the cell $\boldsymbol{i}$, whereas in the polynomial ring, $x_1,\dots,x_n$ usually denote indeterminates.

In this chapter we use a mixture of these different notations to make the correspondences easier to understand.

3.1 Polynomial Ring

Let $\mathbb{Q},\mathbb{R},\mathbb{C}$ denote the fields of rational numbers, real numbers and complex numbers, respectively. Let k stand for any of these fields. We denote the indeterminates by $u_1,\dots,u_\eta$, where $\eta = |\mathscr{I}|$ is the total number of the cells. A *monomial* in $u_1,\dots,u_\eta$ is a product of powers of us (with the coefficient $1 \in k$). For $k = \mathbb{Q}$ and $\eta = 3$,

$$u_1^2 u_2 u_3^3 \tag{3.1}$$

is an example of a monomial in u_1,u_2,u_3. We write this as $\boldsymbol{u}^{\boldsymbol{x}}$ where $\boldsymbol{x} = (2,1,3)$ and $\boldsymbol{u} = (u_1,u_2,u_3)$. Note that the elements of $\boldsymbol{x}$ are used as powers, rather than as indeterminates. This notation is used because frequencies in a contingency table correspond to the powers of a monomial. Each indeterminate $u_{\boldsymbol{i}}$ stands for a cell

S. Aoki et al., *Markov Bases in Algebraic Statistics*, Springer Series in Statistics 199, DOI 10.1007/978-1-4614-3719-2_3,

$\boldsymbol{i}$ in a contingency table. Hence the monomial $u_1^2u_2u_3^3$ in (3.1) corresponds to the following 1×3 contingency table,

2	1	3

.

A monomial is called *square-free* if the power for each $u_{\boldsymbol{i}}$ is at most one.

A polynomial is a finite sum of monomials multiplied by coefficients in k. Again for $k = \mathbb{Q}$ and $\eta = 3$, an example of a polynomial is

$$\frac{3}{2}u_1^2u_2u_3^3 + \frac{1}{3}u_1u_2^3u_3^2.$$

A polynomial with more than two terms does not correspond to a contingency table.

In Chap. 1 we denoted the vector of frequencies as $\boldsymbol{x} = (x(1),\ldots,x(\eta)) \in \mathbb{N}^\eta$. Accordingly we also often denote $u_{\boldsymbol{i}}$ as $u(\boldsymbol{i})$. Below we use these two indexing notations interchangeably. Then a monomial is written as

$$\boldsymbol{u}^{\boldsymbol{x}} = \prod_{\boldsymbol{i}\in\mathscr{I}} u(\boldsymbol{i})^{x(\boldsymbol{i})}.$$

A polynomial f is written as

$$f = \sum_{\boldsymbol{x}\in\mathbb{N}^\eta} c_{\boldsymbol{x}}\boldsymbol{u}^{\boldsymbol{x}},$$

where the sum is finite; that is $c_{\boldsymbol{x}} \in k$ is zero except for a finite number of $\boldsymbol{x}$.

The set of polynomials in $\boldsymbol{u}$ with coefficients from k is written as

$$k[\boldsymbol{u}] = k[u_1,\ldots,u_\eta].$$

$k[\boldsymbol{u}]$ is called the *polynomial ring* in $u_1,\ldots,u_\eta$ over k. It is called a ring because the operations of addition $f+g$ and multiplication fg of polynomials are defined for $k[\boldsymbol{u}]$.

Let $\mathfrak{M}$ denote the set of monomials in $k[\boldsymbol{u}]$. Let $v = \boldsymbol{u}^{\boldsymbol{x}}$ and $w = \boldsymbol{u}^{\boldsymbol{y}}$ be two monomials in $\mathfrak{M}$. Then w divides v if $\boldsymbol{y} \le \boldsymbol{x}$:

$$x(\boldsymbol{i}) \le y(\boldsymbol{i}) \quad \text{for all } \boldsymbol{i} \in \mathscr{I}.$$

We write $w|v$ if w divides v. Let $M \subset \mathfrak{M}$ be a subset of monomials. $v \in M$ is called a minimal element of M if $w \in M, w|v$ implies $v = w$. We present Dickson's lemma (Sect. 2.4 of [42]) in the following form.

Lemma 3.1 (Dickson's Lemma). *Let $M \subset \mathfrak{M}$ be a nonempty set of monomials. The set of minimal elements of M is finite.*

Another important notion on the polynomial ring is the notion of an ideal. A subset $I \subset k[\boldsymbol{u}]$ is called an *ideal* of $k[\boldsymbol{u}]$ if

- $f, g \in I \Rightarrow f+g \in I$.
- $f \in I,\, g \in k[\boldsymbol{u}] \Rightarrow fg \in I$.

Let $\{f_\lambda \mid \lambda \in \Lambda\} \subset k[\boldsymbol{u}]$ be a nonempty set of polynomials. Let I denote the set of polynomials of the form

$$\sum_{\lambda \in \Lambda} g_\lambda f_\lambda, \quad g_\lambda \in k[\boldsymbol{u}],\ \forall \lambda \in \Lambda,$$

where the sum is finite, that is, $g_\lambda = 0$ except for a finite number of λ. Clearly this I is an ideal. This I is called the ideal generated by $\{f_\lambda \mid \lambda \in \Lambda\}$ and is denoted by

$$I = \langle \{f_\lambda \mid \lambda \in \Lambda\} \rangle.$$

$\{f_\lambda \mid \lambda \in \Lambda\}$ is called a set or a system of *generators* of I. In particular if $\{f_\lambda \mid \lambda \in \Lambda\} = \{f_1, \dots, f_s\}$ is a finite set, then $I = \langle \{f_\lambda \mid \lambda \in \Lambda\} \rangle$ is simply denoted as

$$I = \langle f_1, \dots, f_s \rangle.$$

An ideal I is called a *monomial ideal* if it is generated by a subset $M \subset \mathfrak{M}$ of monomials. By Dickson's lemma, a monomial ideal $I = \langle M \rangle$ is generated by the (finite) set of minimal monomials $v_1, \dots, v_s \in M$:

$$I = \langle v_1, \dots, v_s \rangle. \tag{3.2}$$

Note that the set of minimal monomials of M is unique.

3.2 Term Order and Gröbner Basis

Term order (term ordering, monomial ordering) is a total order $\prec$ on the set $\mathfrak{M}$ of monomials, such that

- $1 \prec v$ for $1 \neq v$.
- $v \prec w$ implies $vt \prec wt$ for every $t \in \mathfrak{M}$.

An example of a term order is *pure lexicographic* term order $\prec_{\text{lex}}$, where $v = \boldsymbol{u}^{\boldsymbol{x}}$ is ordered by the lexicographic order of the exponents $\boldsymbol{x} = (x_1, \dots, x_\eta)$. In the lexicographic order we order the indeterminates as

$$u_1 \succ u_2 \succ \cdots \succ u_\eta.$$

Another example is the *graded lexicographic* term order $\prec_{\text{grlex}}$, where monomials $v = \boldsymbol{u}^{\boldsymbol{x}}$ are first compared by the total degree $|\boldsymbol{x}|_1 = \sum_{i \in \mathscr{I}} x(\boldsymbol{i})$ and then (in the case of the same degree) by the lexicographic order of $\boldsymbol{x}$. The third example is the *graded reverse lexicographic* term order $\prec_{\text{grevlex}}$, where monomials are first compared by the total degree and then (in the case of the same degree) $\boldsymbol{u}^{\boldsymbol{x}} \succ_{\text{grevlex}} \boldsymbol{u}^{\boldsymbol{y}}$ if the last nonzero element of $\boldsymbol{x} - \boldsymbol{y}$ is negative.

Let $f = c_1 v_1 + \cdots + c_s v_s \in k[\boldsymbol{u}]$ be a nonzero polynomial, where $0 \neq c_j \in k$, $v_j \in \mathfrak{M}$, $j = 1, \ldots, s$. Given a term order $\prec$, we can take v_1 such that $v_1 \succ v_j$, $j = 2, \ldots, s$. v_1 is called the *initial monomial* (leading monomial, leading term) of f and written as $\text{in}_\prec(f)$. For example, with the pure lexicographic term order, the initial monomial of $f = (3/2)u_1^2 u_2 u_3^3 + (1/3)u_1 u_2^3 u_3^2$ is given by

$$\text{in}_\prec(f) = \text{in}_\prec\left(\frac{3}{2}u_1^2 u_2 u_3^3 + \frac{1}{3}u_1 u_2^3 u_3^2\right) = u_1^2 u_2 u_3^3.$$

For an ideal $I \neq \{0\}$, its *initial ideal* $\text{in}_\prec(I)$ is defined as

$$\text{in}_\prec(I) = \langle \{\text{in}_\prec(f) \mid 0 \neq f \in I\} \rangle,$$

which is the monomial ideal generated by initial monomials of $f \in I$, $f \neq 0$. By (3.2) there exist $f_1, \ldots, f_s \in I$ such that $\text{in}_\prec(I) = \langle \text{in}_\prec(f_1), \ldots, \text{in}_\prec(f_s) \rangle$. Based on this fact a Gröbner basis is defined as follows.

Definition 3.1. Fix a term order $\prec$. A finite subset $G = \{f_1, \ldots, f_s\}$ of nonzero elements of an ideal I is a Gröbner basis of I with respect to the term order if

$$\text{in}_\prec(I) = \langle \text{in}_\prec(f_1), \ldots, \text{in}_\prec(f_s) \rangle.$$

More informally, $\{f_1, \ldots, f_s\}$ is a Gröbner basis of I if the initial monomial of any $f \in I$ is divisible by the initial monomial of some f_j, $j = 1, \ldots, s$.

From this definition it is not immediately clear that G is indeed a set of generators of I. However, again based on Dickson's lemma, it can be shown that G is indeed a set of generators of I (Sect. 2.5 of [42]).

Proposition 3.1. *I is generated by any Gröbner basis G of I.*

The following "Hilbert basis theorem" (Theorem 4 in Sect. 2.5 of [42]) is an immediate consequence of this proposition.

Corollary 3.1 (Hilbert basis theorem). *Every ideal I of the polynomial ring $k[\boldsymbol{u}]$ has a finite set of generators.*

If $\text{in}_\prec(f_1), \ldots, \text{in}_\prec(f_s)$ are minimal monomials of $\{\text{in}_\prec(f) \mid 0 \neq f \in I\}$, then $G = \{f_1, \ldots, f_s\}$ is a *minimal* Gröbner basis, in the sense that any proper subset of G is no longer a Gröbner basis. Given a minimal Gröbner basis $G = \{\text{in}_\prec(f_1), \ldots, \text{in}_\prec(f_s)\}$, no initial monomial $\text{in}_\prec(f_i)$ is divisible by any one of the other initial monomials $\text{in}_\prec(f_j)$, $j \neq i$. However, other (noninitial) monomials appearing in f_i may be divisible by some $\mathrm{in}_{\prec}(f_j)$, $j \neq i$. By replacing noninitial monomials appearing in f_i with the remainders by f_j, $j \neq i$, we arrive at the *reduced* Gröbner basis. More formally, the reduced Gröbner basis is defined as follows.

Definition 3.2. A Gröbner basis G is reduced if (i) for each $f \in G$, the coefficient of $\mathrm{in}_{\prec}(f)$ is one, and (ii) for each $f \in G$, no monomial appearing in f lies in $\langle\{\mathrm{in}_{\prec}(g) \mid g \in G, g \neq f\}\rangle$.

It is known that the reduced Gröbner basis is *unique* for any given term order $\prec$.

Given a term order $\prec$ and an ideal I, a monomial v is called a *standard monomial* if $v \notin \mathrm{in}_{\prec}(I)$. Every polynomial $f \in k[\boldsymbol{u}]$ is a unique finite linear combination of monomials with coefficients from k, thus $k[\boldsymbol{u}]$ is an infinite-dimensional vector space over k, where the monomials are the basis vectors. Similarly I can be regarded as a vector space over k. Regarding them as vector spaces means that we only consider addition of polynomials and ignore multiplication of polynomials. Then we can regard the "quotient ring" $k[\boldsymbol{u}]/I$ as a complementary linear subspace of I in $k[\boldsymbol{u}]$. Concerning this quotient vector space $k[\boldsymbol{u}]/I$, the following theorem holds (cf. Sect. 5.3 of [42], Chapter 1 of [139]).

Theorem 3.1. *The set of standard monomials forms a basis of the vector space $k[\boldsymbol{u}]/I$ over k.*

An ideal I is *zero-dimensional* if the vector space $k[\boldsymbol{u}]/I$ over k is finite-dimensional (Appendix D of [42]). By Theorem 3.1, I is zero-dimensional if and only if the number of standard monomials is finite. Theorem 3.1 for the case of a zero-dimensional ideal is fundamental for a Gröbner basis approach to design of experiments in Chap. 15.

We now consider dividing a polynomial by a set of polynomials. For univariate polynomials $f(u)$ and $g(u)$, the division of f by g can be performed by repeatedly eliminating the leading term (i.e., the initial monomial) of $f(u)$ by the leading term of g. The resulting expression is

$$f(u) = q(u)g(u) + r(u),$$

where $q(u)$ is the quotient and $r(u)$ is the remainder with $\deg r < \deg g$. A generalization of this division to more than one indeterminate is given as follows. Fix a term order $\prec$. Let $f \neq 0$ be the dividend and let $g_1, \ldots, g_s \neq 0$ be the divisors. Then f can be written as follows.

$$f = q_1 g_1 + \cdots + q_s g_s + r, \qquad q_1, \ldots, q_s \in k[\boldsymbol{u}], \tag{3.3}$$

where (i) if $r \neq 0$, then every monomial appearing in r is divisible by none of $\mathrm{in}_{\prec}(g_1), \ldots, \mathrm{in}_{\prec}(g_s)$, and (ii) $\mathrm{in}_{\prec}(f) \succeq \mathrm{in}_{\prec}(q_i g_i)$ for $i = 1, \ldots, s$. r is the *remainder* of this division. Actually there is an algorithm called the *division algorithm* (Chapter 1 of [42]) that yields (3.3). In general, a remainder r is not unique and the output of the division algorithm depends on the order of $g_1, \ldots, g_s$. However, if $G = \{g_1, \dots, g_s\}$ is a Gröbner basis for an ideal I, then the remainder r is uniquely determined. Also in this case we have

$$r = 0 \Leftrightarrow f \in I. \tag{3.4}$$

3.3 Buchberger's Algorithm

The importance of the theory of Gröbner basis lies in the fact that there is an algorithm to compute the Gröbner basis. For two monomials $v = \boldsymbol{u}^{\boldsymbol{x}}$, $w = \boldsymbol{u}^{\boldsymbol{y}}$, the least common multiple of them is defined as $\mathrm{lcm}(v,w) = \boldsymbol{u}^{\max(\boldsymbol{x},\boldsymbol{y})}$, where max denotes the elementwise maximum. For nonzero polynomials $f, g \in k[\boldsymbol{u}]$, let c_f and c_g denote coefficients of their initial monomials. The *S-polynomial* of f and g is defined as

$$S(f,g) = \frac{\mathrm{lcm}(\mathrm{in}_\prec(f), \mathrm{in}_\prec(g))}{c_f \cdot \mathrm{in}_\prec(f)} f - \frac{\mathrm{lcm}(\mathrm{in}_\prec(f), \mathrm{in}_\prec(g))}{c_g \cdot \mathrm{in}_\prec(g)} g. \tag{3.5}$$

The right-hand side looks somewhat complicated, but the purpose of the operation on the right-hand side is to cancel the initial monomials of f and g. For example, with the pure lexicographic term order,

$$\begin{aligned} S\left(\frac{1}{2}u_1u_2 - 2u_3u_4, u_1u_3 - u_2u_5\right) &= \frac{u_1u_2u_3}{\frac{1}{2}u_1u_2}\left(\frac{1}{2}u_1u_2 - 2u_3u_4\right) \\ &\quad - \frac{u_1u_2u_3}{u_1u_3}(u_1u_3 - u_2u_5) \\ &= -4u_3^2u_4 + u_2^2u_5. \end{aligned} \tag{3.6}$$

Fix a term order $\prec$. The following theorem is called *Buchberger's criterion* (Theorem 6, Sect. 2.5 of [42]).

Theorem 3.2 (Buchberger's Criterion). *Let $G = \{g_1, \dots, g_s\}$ be a set of generators of an ideal $I \neq \{0\}$. G is a Gröbner basis of I if and only if for all pairs $i \neq j$, the remainder on division of $S(g_i, g_j)$ by G (listed in some order) is zero.*

Based on this criterion, the following simple idea can be implemented as an algorithm, called *Buchberger's algorithm* to compute a Gröbner basis of $I = \langle g_1, \dots, g_s \rangle$:

> As long as $G = \{g_1, \dots, g_s\}$ is not a Gröbner basis of I, keep adding to G a remainder of some $S(g_i, g_j)$ by G.

Note that Buchberger's algorithm can be used when a finite set of generators of I is known. In contrast, toric ideals, which are important for algebraic statistics, are defined implicitly and the problem is to obtain a set of generators for the ideals. In this case Buchberger's algorithm cannot be used directly. However, it is also fundamental to toric ideals via the elimination theory described in the next section.

3.4 Elimination Theory

In the polynomial ring $k[\boldsymbol{u}]$, $\boldsymbol{u} = (u_1,\ldots,u_\eta)$, consider the set of polynomials involving only $u_{\zeta+1},\ldots,u_\eta$, where $1 \leq \zeta < \eta$. Let $k[u_{\zeta+1},\ldots,u_\eta]$ denote the set of these polynomials. For an ideal I of $k[\boldsymbol{u}]$, it is easy to check that $I \cap k[u_{\zeta+1},\ldots,u_\eta]$ is an ideal of $k[u_{\zeta+1},\ldots,u_\eta]$. $I \cap k[u_{\zeta+1},\ldots,u_\eta]$ is called an *elimination ideal* because the indeterminates $u_1,\ldots,u_\zeta$ are eliminated.

The main use of the elimination ideal is for solving a set of polynomial equations. For a Markov basis, its use is to give a general algorithm for computing a Gröbner basis of a toric ideal (see Proposition 3.2 below).

When a Gröbner basis G of I with respect to the lexicographic order $u_1 \succ \cdots \succ u_\eta$ is given, it is straightforward to obtain a Gröbner basis of the elimination ideal (Sect. 3.1 of [42]).

Theorem 3.3 (The Elimination Theorem). *Let G be a Gröbner basis of I with respect to the lexicographic order $u_1 \succ \cdots \succ u_\eta$. Then*

$$G \cap k[u_{\zeta+1},\ldots,u_\eta]$$

is a Gröbner basis of the elimination ideal $I \cap k[u_{\zeta+1},\ldots,u_\eta]$.

3.5 Toric Ideals

So far we summarized relevant facts on a Gröbner basis of a general ideal I of $k[\boldsymbol{u}]$. For the theory of Markov basis we only need to consider a special kind of ideal, called a toric ideal. In this section we give more detailed explanations than in previous sections, because toric ideals are not covered in [42].

For defining a toric ideal we start with a $\nu \times \eta$ integer matrix A. We call A a *configuration*. In this book we assume that the row vector $(1,1,\ldots,1)$ is in the real vector space spanned by the rows of A; that is, there exists a ν-dimensional real vector θ such that

$$\theta' A = (1,1,\ldots,1). \tag{3.7}$$

This assumption is called *homogeneity* of the toric ideal (Lemma 4.14 of [139]). Under the assumption of homogeneity, for $\boldsymbol{z} \in \ker_{\mathbb{Z}} A$

$$0 = \theta' A\boldsymbol{z} = (1,1,\ldots,1)\boldsymbol{z} = \sum_{\boldsymbol{i}\in\mathscr{I}} z(\boldsymbol{i}). \tag{3.8}$$

In algebraic statistics, the rows of A are indexed by sufficient statistics and the columns of A are indexed by the cells $\boldsymbol{i}$ of the sample space $\mathscr{I}$. Hence let us denote the elements of A as $a_j(\boldsymbol{i})$, $j = 1,\ldots,\nu$, $\boldsymbol{i} = 1,\ldots,\eta$. Let $\boldsymbol{a}(\boldsymbol{i}) = (a_1(\boldsymbol{i}),\ldots,a_\nu(\boldsymbol{i}))'$ denote the $\boldsymbol{i}$th column of A. We often consider A also as a set of its column vectors $\{\boldsymbol{a}(\boldsymbol{i}),\ \boldsymbol{i} = 1,\ldots,\eta\}$. Note that under the homogeneity, each column $\boldsymbol{a}(\boldsymbol{i})$ is a nonzero vector, because otherwise $\theta' A$ has 0 in the $\boldsymbol{i}$th position.

A difference of two monomials (with coefficient 1 and -1)

$$f = w - v, \quad v = \boldsymbol{u}^{\boldsymbol{x}}, \quad w = \boldsymbol{u}^{\boldsymbol{y}}$$

is called a *binomial*. For a monomial $v = \boldsymbol{u}^{\boldsymbol{x}}$ or the frequency vector $\boldsymbol{x}$, the *support* is defined as the set of (indices of) indeterminates with positive powers:

$$\operatorname{supp}(\boldsymbol{u}^{\boldsymbol{x}}) = \operatorname{supp}(\boldsymbol{x}) = \{\boldsymbol{i} \mid x(\boldsymbol{i}) > 0\}. \tag{3.9}$$

A binomial $w - v$ is called square-free if both w and v are square-free monomials. Consider two nonnegative integer vectors $\boldsymbol{x}, \boldsymbol{y} \in \mathbb{N}^{\eta}$ such that

$$A\boldsymbol{x} = A\boldsymbol{y}.$$

Then $\boldsymbol{z} = \boldsymbol{y} - \boldsymbol{x}$ belongs to $\ker_{\mathbb{Z}} A$. Now we give the first definition of toric ideal $I_A \subset k[\boldsymbol{u}]$.

Definition 3.3. The toric ideal $I_A = \langle \{\boldsymbol{u}^{\boldsymbol{y}} - \boldsymbol{u}^{\boldsymbol{x}} \mid A\boldsymbol{x} = A\boldsymbol{y},\ \boldsymbol{x}, \boldsymbol{y} \in \mathbb{N}^{\eta}\} \rangle$ is the ideal generated by binomials $\boldsymbol{u}^{\boldsymbol{y}} - \boldsymbol{u}^{\boldsymbol{x}}$ such that $\boldsymbol{y} - \boldsymbol{x} \in \ker_{\mathbb{Z}} A$.

For example, for the configuration A in (1.18) for the independence model of $I \times J$ two-way contingency tables, I_A is the ideal generated by $\boldsymbol{u}^{\boldsymbol{x}} - \boldsymbol{u}^{\boldsymbol{y}}$, where $\boldsymbol{x}, \boldsymbol{y}$ share the common row sums and column sums.

So far we allowed negative elements in A. We now argue that without loss of generality we can assume that the elements of A are nonnegative under the assumption of homogeneity. Let $\boldsymbol{a} \in \mathbb{N}^{\nu}$ be a column vector whose elements are large enough. We add $\boldsymbol{a}\,(1,1,\ldots,1)$ to A. Let $\tilde{A} = A + \boldsymbol{a}(1,1,\ldots,1)$ be the resulting matrix whose elements are nonnegative. Note that $\theta'\tilde{A}$ is written as

$$\theta'\tilde{A} = (1,1,\ldots,1) + \theta'\boldsymbol{a}(1,1,\ldots,1) = (1 + \theta'\boldsymbol{a})(1,1,\ldots,1).$$

By appropriately choosing $\boldsymbol{a}$ we can make $0 \neq (1 + \theta'\boldsymbol{a})$. Hence $\tilde{A}$ also satisfies the assumption of homogeneity. Now by (3.8) we have $\ker_{\mathbb{Z}} A = \ker_{\mathbb{Z}} \tilde{A}$ and $I_A = I_{\tilde{A}}$. Therefore we can assume that the elements of A are nonnegative.

We are now going to present another definition of a toric ideal. Introduce indeterminates $q_1, \ldots, q_\nu$ corresponding to the rows of A. Let $\boldsymbol{q} = \{q_1, \ldots, q_\nu\}$ and let $k[\boldsymbol{q}]$ denote the polynomial ring in $\boldsymbol{q}$ over k. Consider a map π_A from $k[\boldsymbol{u}]$ to $k[\boldsymbol{q}]$ such that each indeterminate $u(\boldsymbol{i})$ is mapped to a monomial in $k[\boldsymbol{q}]$ as

$$\pi_A : u(\boldsymbol{i}) \mapsto \boldsymbol{q}^{\boldsymbol{a}(\boldsymbol{i})} = q_1^{a_1(\boldsymbol{i})} q_2^{a_2(\boldsymbol{i})} \cdots q_\nu^{a_\nu(\boldsymbol{i})},$$

where $\boldsymbol{a}(\boldsymbol{i})$ is the $\boldsymbol{i}$th column of A. For a polynomial $f \in k[\boldsymbol{u}]$, $\pi_A(f)$ is obtained by substituting $\boldsymbol{q}^{\boldsymbol{a}(\boldsymbol{i})}$ into the indeterminate $u(\boldsymbol{i})$, $\boldsymbol{i} \in \mathscr{I}$. Then, for a monomial $\boldsymbol{u}^{\boldsymbol{x}} = \prod_{\boldsymbol{i} \in \mathscr{I}} u(\boldsymbol{i})^{x(\boldsymbol{i})}$,

$$\pi_A(\boldsymbol{u}^{\boldsymbol{x}}) = \pi_A(\prod_{\boldsymbol{i}\in\mathscr{I}} u(\boldsymbol{i})^{x(\boldsymbol{i})}) = \prod_{\boldsymbol{i}\in\mathscr{I}} \pi_A(u(\boldsymbol{i}))^{x(\boldsymbol{i})} = \prod_{\boldsymbol{i}\in\mathscr{I}} \boldsymbol{q}^{\boldsymbol{a}(\boldsymbol{i})x(\boldsymbol{i})}$$
$$= \prod_{j=1}^{\nu} q_j^{\sum_{\boldsymbol{i}\in\mathscr{I}} x(\boldsymbol{i})a_j(\boldsymbol{i})} = \boldsymbol{q}^{A\boldsymbol{x}} \tag{3.10}$$

and for a polynomial $f = \sum_{\boldsymbol{x}} c_{\boldsymbol{x}} \boldsymbol{u}^{\boldsymbol{x}} \in k[\boldsymbol{u}]$

$$\pi_A(f) = \sum_{\boldsymbol{x}} c_{\boldsymbol{x}} \pi_A(\boldsymbol{u}^{\boldsymbol{x}}) = \sum_{\boldsymbol{x}} c_{\boldsymbol{x}} \boldsymbol{q}^{A\boldsymbol{x}} \quad \in k[\boldsymbol{q}].$$

We see that π_A is a homomorphism from $k[\boldsymbol{u}]$ to $k[\boldsymbol{q}]$.

Let us illustrate π_A for the case of the independence model of two-way tables under the multinomial sampling scheme. Let (i,j) denote the cell of a two-way table and consider the probability p_{ij} of the cell as an indeterminate (instead of $u(\boldsymbol{i})$). Under the independence model $p_{ij} = r_i c_j$. We can understand this as "substituting $r_i c_j$ into p_{ij}" and consider

$$\pi_A : p_{ij} \mapsto r_i c_j.$$

For $I = J = 2$, consider the following contingency table:

$$\boldsymbol{x} = \begin{array}{|c|c|}\hline 1 & 2 \\ \hline 1 & 0 \\ \hline\end{array}\,. \tag{3.11}$$

The probability, without the hypothesis of independence, of this contingency table $\boldsymbol{x}$ is written as

$$p(\boldsymbol{x}) = \binom{4}{1,2,1,0} \boldsymbol{p}^{\boldsymbol{x}} = 12 p_{11} p_{12}^2 p_{21}.$$

Furthermore under the hypothesis H of independence, by substituting $c_i r_j$ into p_{ij}, the probability of $\boldsymbol{x}$ is given by

$$\pi_A(12 p_{11} p_{12}^2 p_{21}) = 12 \pi_A(p_{11}) \pi_A(p_{12}^2) \pi_A(p_{21}) = 12 (r_1 c_1)(r_1 c_2)^2 (r_2 c_1)$$
$$= 12 r_1^3 r_2^1 c_1^2 c_2^2.$$

By (3.10) the exponents of r_i and c_j on the right-hand side correspond to the marginal frequencies of $\boldsymbol{x}$. Indeed, for example, r_1^3 on the right-hand side shows that the first row sum of $\boldsymbol{x}$ is $x_{1+} = 3$.

Now the second definition of I_A is given as the kernel of this π_A.

Definition 3.4.
$$I_A = \{ f \in k[\boldsymbol{u}] \mid \pi_A(f) = 0 \}. \tag{3.12}$$

It can be shown that Definitions 3.3 and 3.4 are equivalent. See Lemma 4.1 of [139]. By (3.10) it is easily seen that

$$A\boldsymbol{y} = A\boldsymbol{x} \;\Leftrightarrow\; \pi_A(\boldsymbol{u}^{\boldsymbol{y}} - \boldsymbol{u}^{\boldsymbol{x}}) = 0. \tag{3.13}$$

Hence the ideal in Definition 3.3 is clearly a subset of the ideal in Definition 3.4. We need some extra argument to show that they are the same.

Again consider an example of 2×2 table. Let $\boldsymbol{x}$ be as in (3.11) and let

$$\boldsymbol{y} = \begin{array}{|c|c|}\hline 2 & 1 \\ \hline 0 & 1 \\ \hline \end{array},$$

which has the same marginal frequencies as $\boldsymbol{x}$. Then

$$\begin{aligned}\pi_A(\boldsymbol{u}^{\boldsymbol{y}} - \boldsymbol{u}^{\boldsymbol{x}}) &= \pi_A(\boldsymbol{u}^{\boldsymbol{y}}) - \pi_A(\boldsymbol{u}^{\boldsymbol{x}}) \\ &= (r_1c_1)^2(r_1c_2)(r_2c_2) - (r_1c_1)(r_1c_2)^2(r_2c_1) \\ &= r_1^3r_2c_1^2c_2^2 - r_1^3r_2c_1^2c_2^2 = 0.\end{aligned}$$

Hence $\boldsymbol{u}^{\boldsymbol{y}} - \boldsymbol{u}^{\boldsymbol{x}} \in I_A$.

A Gröbner basis of I_A can be obtained by the elimination theory of the previous section and Definition 3.4. Let $k[\boldsymbol{q},\boldsymbol{u}] = k[q_1,\dots,q_\nu,u_1,\dots u_\eta]$ be the polynomial ring in $q_1,\dots,q_\nu,u_1,\dots u_\eta$ over k. Consider an ideal

$$J_A = \langle \{u(\boldsymbol{i}) - \boldsymbol{q}^{\boldsymbol{a}(\boldsymbol{i})}, \boldsymbol{i} \in \mathscr{I}\}\rangle$$

of $k[\boldsymbol{q},\boldsymbol{u}]$. Then I_A is characterized as an elimination ideal of J_A. Combining this fact with Buchberger's algorithm and the elimination theory, a Gröbner basis of I_A can be computed as follows (Algorithm 4.5 of [139]).

Proposition 3.2. *The toric ideal I_A is written as*

$$I_A = J_A \cap k[\boldsymbol{u}].$$

Let $\prec$ be the pure lexicographic term order such that $q_1 \succ \cdots \succ q_\nu \succ u_1 \succ \cdots \succ u_\eta$. By Buchberger's algorithm compute a Gröbner basis G of J_A. Then $G \cap k[\boldsymbol{u}]$ is a Gröbner basis of I_A.

In this algorithm, as long as $q_j \succ u_i$ for all $1 \le j \le \nu$, $1 \le i \le \eta$, the orders within $\boldsymbol{q}$ and $\boldsymbol{u}$ can be different from the lexicographic term order.

As mentioned above, in Definitions 3.3 and 3.4 no finite set of generators of I_A is given and Proposition 3.2 gives a general algorithm for obtaining a set of generators of I_A in the form of a Gröbner basis.

However, it is also of interest to consider Buchberger's algorithm for a toric ideal, when a finite set of generators is known. Consider the operation of forming an S-polynomial in (3.5) and (3.6). In (3.6), for illustrative purpose, we had coefficients $\frac{1}{2}$ and 2. Suppose we compute the S-polynomial of two binomials. For example,

$$\begin{aligned}S(u_1u_2 - u_3u_4, u_1u_3 - u_2u_5) &= \frac{u_1u_2u_3}{u_1u_2}(u_1u_2 - u_3u_4) - \frac{u_1u_2u_3}{u_1u_3}(u_1u_3 - u_2u_5) \\ &= -u_3^2u_4 + u_2^2u_5.\end{aligned} \tag{3.14}$$

We note that the right-hand side is a binomial. It is clear that this holds for two arbitrary binomials and the following lemma holds.

Lemma 3.2. *The S-polynomial of two binomials is a binomial.*

This lemma is important for proving the fundamental theorem of Markov basis in Sect. 4.4. Furthermore, from this lemma the following result can be shown (Corollary 4.4 of [139]).

Proposition 3.3. *For any term order, the reduced Gröbner basis of a toric ideal I_A consists of binomials.*

This is consistent with Proposition 3.2 and Definition 3.3, because the reduced Gröbner basis G of J_A consists of binomials and hence $G \cap k[\boldsymbol{u}]$ consists of binomials as well. Also by Definition 3.2 it is obvious that $G \cap k[\boldsymbol{u}]$ is reduced if G is reduced.

Part II
Properties of Markov Bases

In Part II of this book, we define Markov bases more precisely and develop a general theory of Markov bases.

In Chap. 4 we define Markov bases for discrete exponential family models and discuss other relevant bases, such as lattice bases and the Graver basis.

In Chap. 5 we consider minimality of Markov bases and establish basic structures of minimal Markov bases. We define the notion of indispensable moves and establish a condition for the existence of the unique minimal Markov basis.

In Chap. 6 we give a formal presentation of the distance reduction argument, which is often very useful for proving that a given set of moves forms a Markov basis.

Finally in Chap. 7 we define the notion of invariance of Markov bases using the notion of action of groups on the set of cells.

Chapter 4
Definition of Markov Bases and Other Bases

4.1 Discrete Exponential Family

As in Chap. 1 let $\mathscr{I}$ denote a finite sample space. Because of many applications to contingency table models, we call an element $\boldsymbol{i} \in \mathscr{I}$ a *cell*. We consider a family $\{p(\boldsymbol{i};\boldsymbol{\theta})\}$, $\boldsymbol{\theta} = (\theta_1,\dots,\theta_\nu)$ of distributions over $\mathscr{I}$ of the form

$$\log p(\boldsymbol{i};\boldsymbol{\theta}) = \sum_{j=1}^{\nu} \theta_j a_j(\boldsymbol{i}) - \psi(\boldsymbol{\theta}), \tag{4.1}$$

where $\exp(-\psi(\boldsymbol{\theta}))$ is the *normalizing constant* of the exponential family. Note that θ_j was denoted as $\phi_j(\boldsymbol{\theta})$ in (1.11). Equation (4.1) corresponds to the multinomial sampling scheme of Chap. 1. This model is often called a log affine model of probability distributions over $\mathscr{I}$. Let $\eta = |\mathscr{I}|$ and let

$$A = \{a_j(\boldsymbol{i})\}, \quad j = 1,\dots,\nu,\ \boldsymbol{i} \in \mathscr{I}$$

denote a $\nu \times \eta$ matrix with the $(j,\boldsymbol{i})$ element $a_j(\boldsymbol{i})$. A is the configuration introduced in the previous chapter. In many regression settings, A corresponds to the design matrix of regressors. Then it is natural to call A a *design matrix*. Actually the transpose A' of A is called the design matrix in regression settings. In our setting a configuration A has more columns than rows. In a regression setting the design matrix usually has more rows than columns.

It is also useful to look at the rows $\boldsymbol{a}_j = (a_j(\boldsymbol{i}), \boldsymbol{i} \in \mathscr{I})$ of the configuration A. Except for the (negative logarithm of the) normalizing constant ψ, (4.1) implies that the logarithm of the probability vector $\{p(\boldsymbol{i};\boldsymbol{\theta}), \boldsymbol{i} \in \mathscr{I}\}$ lies in the linear space spanned by the rows of A. In this sense we call the linear space spanned by the rows of A the *model space* and denote it by rowspan(A). In this book, we assume homogeneity (3.7); that is, the constant row vector $(1,1,\dots,1)$ is in rowspan(A).

In many contingency table models we often allow linear dependence among the rows of A for symmetry of describing the model. For example, in the two-way

S. Aoki et al., *Markov Bases in Algebraic Statistics*, Springer Series in Statistics 199, DOI 10.1007/978-1-4614-3719-2_4,

independence model we usually take $\nu = I+J$ for the independence model of $I\times J$ two-way tables, although the degrees of the model (the dimension of the model) is $I+J-1$. Therefore ν is not necessarily the dimension of rowspan(A).

Let $q_j = e^{\theta_j}$, $j=1,\ldots,\nu$. Taking the exponential, (4.1) is written in the monomial form

$$p(\boldsymbol{i};\boldsymbol{\theta}) = e^{-\psi(\boldsymbol{\theta})} \times \prod_{j=1}^{\nu} q_j^{a_j(\boldsymbol{i})} = e^{-\psi(\boldsymbol{\theta})}\boldsymbol{q}^{\boldsymbol{a}(\boldsymbol{i})}. \tag{4.2}$$

The exponential family notation in (4.1) is more traditional in statistics. The monomial form (4.2) is often called a "toric model" in algebraic statistics. In the exponential form it is assumed that $p(\boldsymbol{i};\boldsymbol{\theta})>0$ for all $\boldsymbol{i}$. However in (4.2) $p(\boldsymbol{i};\boldsymbol{\theta})=0$ is allowed.

Suppose that we obtain n observations from the distribution (4.1) under the multinomial sampling and let $x(\boldsymbol{i})$, $\boldsymbol{i}\in\mathscr{I}$ denote the frequencies of the cells. Then the joint probability function of the frequency vector $\boldsymbol{x}=\{x(\boldsymbol{i})\}_{\boldsymbol{i}\in\mathscr{I}}$ is written as

$$\begin{aligned} p(\boldsymbol{x}) &= \frac{n!}{\prod_{\boldsymbol{i}\in\mathscr{I}} x(\boldsymbol{i})!}\prod_{\boldsymbol{i}\in\mathscr{I}} p(\boldsymbol{i};\boldsymbol{\theta})^{x(\boldsymbol{i})} \\ &= \frac{n!}{\prod_{\boldsymbol{i}\in\mathscr{I}} x(\boldsymbol{i})!}\exp\left\{\sum_{j=1}^{\nu}\theta_j\sum_{\boldsymbol{i}\in\mathscr{I}} a_j(\boldsymbol{i})x(\boldsymbol{i}) - n\psi(\boldsymbol{\theta})\right\}. \end{aligned} \tag{4.3}$$

Then a sufficient statistic of the model is given by $t_j = \sum_{\boldsymbol{i}\in\mathscr{I}} a_j(\boldsymbol{i})x(\boldsymbol{i})$, $j=1,\ldots,\nu$. We write this relation as

$$\boldsymbol{t} = A\boldsymbol{x}, \tag{4.4}$$

where $\boldsymbol{t}=(t_1,\ldots,t_\nu)'$ is the ν-dimensional column vector of t_js and $\boldsymbol{x}$ is the η-dimensional column vector of frequencies. Then (4.3) is written as

$$p(\boldsymbol{x}) = \frac{n!}{\prod_{\boldsymbol{i}\in\mathscr{I}} x(\boldsymbol{i})!}\exp(\boldsymbol{\theta}'A\boldsymbol{x} - n\psi(\boldsymbol{\theta})). \tag{4.5}$$

We denote the set of frequency vectors as $\mathscr{X}=\mathbb{N}^\eta$ and the set of frequency vectors with the common value of the sufficient statistic by

$$\mathscr{F}_{\boldsymbol{t}} = \{\boldsymbol{x}\in\mathscr{X} \mid A\boldsymbol{x}=\boldsymbol{t}\}$$

and call it a ***t*-*fiber***. We denote the set of possible values of the sufficient statistic as

$$\mathscr{T} = \mathscr{T}_A = \{\boldsymbol{t} \mid \boldsymbol{t}=A\boldsymbol{x},\ \boldsymbol{x}\in\mathbb{N}^\eta\}. \tag{4.6}$$

$\mathscr{T}_A$ is often referred to as a *semigroup* generated by A. In the notation of Sect. 3.5, let $\boldsymbol{q}=\{q_1,\ldots,q_\nu\}$ be indeterminates corresponding to the rows of A and let $k[\boldsymbol{q}]$ denote the polynomial ring in $\boldsymbol{q}$. The image $\pi_A(k[\boldsymbol{u}])$, which is a subring of $k[\boldsymbol{q}]$, is called the *semigroup ring* associated with the configuration A.

Under the assumption of homogeneity, let $\boldsymbol{\theta}'$ be a row vector such that $\boldsymbol{\theta}'A = (1,1,\ldots,1)$. Then for two frequency vectors $\boldsymbol{x},\tilde{\boldsymbol{x}}$ in the same fiber we have

$$\mathbf{0} = A\tilde{\boldsymbol{x}} - A\boldsymbol{x} \quad \Rightarrow \quad 0 = \boldsymbol{\theta}'A\tilde{\boldsymbol{x}} - \boldsymbol{\theta}'A\boldsymbol{x} = \sum_{\boldsymbol{i}} \tilde{x}(\boldsymbol{i}) - \sum_{\boldsymbol{i}} x(\boldsymbol{i}).$$

Therefore the sample size n of frequency vectors in the same fiber is common. We call this sample size n the *degree* of $\boldsymbol{x}$ as well as the degree of $\boldsymbol{t} = A\boldsymbol{x}$:

$$n = \deg \boldsymbol{x} = \sum_{\boldsymbol{i}} x(\boldsymbol{i}) \quad \text{and} \quad \deg \boldsymbol{t} = \deg \boldsymbol{x} \quad (\text{for any } \boldsymbol{x} \text{ such that } \boldsymbol{t} = A\boldsymbol{x}).$$

In this book we sometimes write $|\boldsymbol{x}| = \deg \boldsymbol{x}$ and $|\boldsymbol{t}| = \deg \boldsymbol{t}$, although $\deg \boldsymbol{t}$ is not the 1-norm of the vector $\boldsymbol{t}$. Also note that each fiber is finite, because

$$\mathscr{F}_{\boldsymbol{t}} \subset \{\boldsymbol{x} \in \mathbb{N}^{\eta} \mid \deg \boldsymbol{x} = \deg \boldsymbol{t}\}$$

and the right-hand side is a finite set.

Given the values of the sufficient statistic $\boldsymbol{t}$, the conditional distribution of $\boldsymbol{x}$ does not depend on the parameters $\boldsymbol{\theta} = (\theta_1,\ldots,\theta_\nu)$. Note that the marginal probability function of $\boldsymbol{t}$ is written as

$$p(\boldsymbol{t}) = \sum_{\boldsymbol{x}\in\mathscr{F}_{\boldsymbol{t}}} \frac{n!}{\prod_{\boldsymbol{i}\in\mathscr{I}} x(\boldsymbol{i})!} \exp\left\{ \sum_{j=1}^{\nu} \theta_j t_j - n\psi(\theta_1,\ldots,\theta_\nu) \right\}.$$

Therefore the conditional distribution of $\boldsymbol{x}$ given $\boldsymbol{t}$ is written as

$$p(\boldsymbol{x} \mid \boldsymbol{t}) = c \times \frac{1}{\prod_{\boldsymbol{i}\in\mathscr{I}} x(\boldsymbol{i})!}, \qquad \boldsymbol{x} \in \mathscr{F}_{\boldsymbol{t}}, \tag{4.7}$$

where c is the normalizing constant. We call (4.7) the *hypergeometric distribution* over the fiber $\mathscr{F}_{\boldsymbol{t}}$.

If we can sample from the hypergeometric distribution, we can perform the conditional tests of the fit of the model (4.3). The Markov basis allows us to construct a Markov chain over the fiber $\mathscr{F}_{\boldsymbol{t}}$ for this purpose. The normalizing constant

$$c = c_{\boldsymbol{t}} = \left[\sum_{\boldsymbol{x}\in\mathscr{F}_{\boldsymbol{t}}} \frac{1}{\prod_{\boldsymbol{i}\in\mathscr{I}} x(\boldsymbol{i})!} \right]^{-1}$$

cannot be expressed in a closed form except for special cases, such as the decomposable models for contingency tables. In this respect, the Markov chain Monte Carlo method is especially useful, because a Markov chain can be constructed without knowing the normalizing constant.

4.2 Definition of Markov Basis

A Markov basis is a set of "moves" for constructing a Markov chain over any fiber. Let $\boldsymbol{z} \in \mathbb{Z}^\eta$ denote an η-dimensional column vector of integers. $\boldsymbol{z}$ is called a *move* if $A\boldsymbol{z} = 0$; that is, $\boldsymbol{z}$ belongs to the integer kernel

$$\ker_{\mathbb{Z}} A = \ker A \cap \mathbb{Z}^\eta$$

of A. If $A\boldsymbol{x} = \boldsymbol{t}$ and $\boldsymbol{z}$ is a move, then

$$A(\boldsymbol{x}+\boldsymbol{z}) = A\boldsymbol{x} = \boldsymbol{t}.$$

Therefore by adding $\boldsymbol{z}$ to $\boldsymbol{x}$ we remain in the same fiber as long as $\boldsymbol{x}+\boldsymbol{z}$ does not contain a negative element. If $\boldsymbol{x}+\boldsymbol{z}$ contains a negative element, then we have to choose another move $\boldsymbol{z}$ to add to $\boldsymbol{x}$. Suppose that we have a set of moves $\mathscr{B}$, then by adding moves from $\mathscr{B}$ to the current frequency vector we can "move around" a fiber. Our purpose is to find a finite set of moves $\mathscr{B} = \{\boldsymbol{z}_1, \ldots, \boldsymbol{z}_L\}$, such that we can move *all over* the fiber. For the Markov basis we also require that $\boldsymbol{z}_1, \ldots, \boldsymbol{z}_L$ allow us to move all over *every* fiber $\boldsymbol{t}$, namely, for every possible value of the sufficient statistic $\boldsymbol{t}$.

Note that $-\boldsymbol{z}$ is a move if $\boldsymbol{z}$ is a move. When $\boldsymbol{x}+\boldsymbol{z}$ contains a negative element, we might try $\boldsymbol{x}-\boldsymbol{z}$ instead. So we can also subtract a move from $\boldsymbol{x}$. For convenience we often ignore the sign of $\boldsymbol{z}$ and think of $\pm\boldsymbol{z}$ as a move.

Suppose that we are given a finite set of moves $\mathscr{B} = \{\boldsymbol{z}_1, \ldots, \boldsymbol{z}_L\}$. We consider an undirected graph $\mathscr{G} = \mathscr{G}_{\boldsymbol{t},\mathscr{B}}$ whose vertices are the elements of a fiber $\mathscr{F}_{\boldsymbol{t}}$. We draw an (undirected) edge between $\boldsymbol{x}$ and $\boldsymbol{y}$ if there exists $\boldsymbol{z} \in \mathscr{B}$ such that $\boldsymbol{y} = \boldsymbol{x}+\boldsymbol{z}$ or $\boldsymbol{y} = \boldsymbol{x}-\boldsymbol{z}$. Being able to move all over $\mathscr{F}_{\boldsymbol{t}}$ corresponds to the connectedness of $\mathscr{G}_{\boldsymbol{t},\mathscr{B}}$. Therefore we are led to the following definition of a Markov basis.

Definition 4.1. A finite set $\mathscr{B} = \{\boldsymbol{z}_1, \ldots, \boldsymbol{z}_L\}$ of moves is called a *Markov basis* if $\mathscr{G}_{\boldsymbol{t},\mathscr{B}}$ is connected for every $\boldsymbol{t} \in \mathscr{T}$.

In this definition we require the finiteness of $\mathscr{B}$. This causes no difficulty because the existence of a Markov basis is guaranteed by Hilbert's basis theorem (see Corollary 3.1 of Chap. 3). Note that the definition of a Markov basis in Definition 4.1 is equivalent to the earlier definition given in (2.6).

How about uniqueness in the definition of a Markov basis? Except for some special cases, Markov bases are not unique. First, if $\mathscr{B}$ is a Markov basis, then $\mathscr{B} \cup \{\boldsymbol{z}\}$ is a Markov basis for every move $\boldsymbol{z}$. Therefore when we ask the question of uniqueness, we naturally should consider Markov bases that are minimal in the sense of set inclusion. Even with the requirement of minimality, Markov bases are not unique in general. This fact leads to various notions and classes of Markov bases.

Another point in the definition of a Markov basis is that it is common for every fiber $\mathscr{F}_{\boldsymbol{t}}$, $\forall \boldsymbol{t} \in \mathscr{T}$. Given a particular data set $\boldsymbol{x} \in \mathscr{F}_{A\boldsymbol{x}}$, the set of moves connecting $\mathscr{F}_{A\boldsymbol{x}}$ alone may be smaller and need not be a Markov basis. However, at present there is no general methodology for obtaining a set of moves connecting a particular fiber. Therefore we use Definition 4.1, except for Chap. 13.

At this point we discuss the basic relation between the model space rowspan(A) and the kernel of A. It is a standard fact in linear algebra that

$$\text{rowspan}(A)^\perp = \ker A, \qquad \text{rowspan}(A) = (\ker A)^\perp, \tag{4.8}$$

where $\perp$ denotes the orthogonal complement. The first equality can be seen as follows. For $\boldsymbol{w} \in \mathbb{R}^\eta$,

$$\begin{aligned} \boldsymbol{w} \in \ker A &\Leftrightarrow A\boldsymbol{w} = \boldsymbol{0} \\ &\Leftrightarrow \boldsymbol{\theta}' A\boldsymbol{w} = 0, \ \forall \boldsymbol{\theta} \in \mathbb{R}^\nu \\ &\Leftrightarrow \boldsymbol{w} \in \text{rowspan}(A)^\perp. \end{aligned} \tag{4.9}$$

The second equality holds because $(L^\perp)^\perp = L$ for any subspace L of $\mathbb{R}^\eta$. It is useful to remember that the model space of A and the kernel of A are equivalent in the sense of (4.8). In studying a statistical model, we can use either rowspan(A) or $\ker A$.

4.3 Properties of Moves and the Lattice Basis

In this section we summarize basic properties of moves. For a move $\boldsymbol{z} \in \ker_{\mathbb{Z}} A$, we distinguish its positive elements and negative elements. Collect the positive elements of $\boldsymbol{z}$ into its *positive part* $\boldsymbol{z}^+ \in \mathbb{N}^\eta$ as

$$z^+(\boldsymbol{i}) = \max(0, z(\boldsymbol{i})), \qquad \boldsymbol{i} \in \mathscr{I}.$$

Similarly define the *negative part* $\boldsymbol{z}^-$ of $\boldsymbol{z}$ by $z^-(\boldsymbol{i}) = -\min(0, z(\boldsymbol{i}))$, $\boldsymbol{i} \in \mathscr{I}$. Then $\boldsymbol{z}$ is written as the difference of its positive part and negative part

$$\boldsymbol{z} = \boldsymbol{z}^+ - \boldsymbol{z}^-.$$

Note that $A\boldsymbol{z} = 0$ means that $A\boldsymbol{z}^+ = A\boldsymbol{z}^-$; that is, the positive part and the negative part of a move belong to the same fiber. In view of this fact we sometimes say that $\boldsymbol{z}$ is a move belonging to the fiber $\mathscr{F}_{\boldsymbol{t}}$, where $\boldsymbol{t} = A\boldsymbol{z}^+$. We define the degree of $\boldsymbol{z}$ ($\deg \boldsymbol{z}$) by the degree of $\boldsymbol{z}^+$ (or $\boldsymbol{z}^-$). Note that

$$|\boldsymbol{z}| = 2\deg \boldsymbol{z},$$

where $|\boldsymbol{z}|$ is the 1-norm of $\boldsymbol{z}$.

For $\boldsymbol{z} \in \mathbb{Z}^\eta$, the *support* of $\boldsymbol{z}$ is defined to be the set of cells where $\boldsymbol{z}$ is nonzero:

$$\text{supp}(\boldsymbol{z}) = \{\boldsymbol{i} \mid z(\boldsymbol{i}) \neq 0\}. \tag{4.10}$$

Note that supports of $\boldsymbol{z}^+$ and $\boldsymbol{z}^-$ are disjoint:

$$\text{supp}(\boldsymbol{z}^+) \cap \text{supp}(\boldsymbol{z}^-) = \emptyset.$$

Now let $\boldsymbol{x}$ and $\boldsymbol{y}$ be two frequency vectors of the same fiber with disjoint supports ($\mathrm{supp}(\boldsymbol{x}) \cap \mathrm{supp}(\boldsymbol{y}) = \emptyset$). Then

$$\boldsymbol{z} = \boldsymbol{y} - \boldsymbol{x}$$

is a move with $\boldsymbol{z}^+ = \boldsymbol{y}$ and $\boldsymbol{z}^- = \boldsymbol{x}$. Therefore a move is in a one-to-one relation to an ordered pair of two frequency vectors in the same fiber with disjoint supports.

Note that $\boldsymbol{z} = \boldsymbol{y} - \boldsymbol{x}$ for $\boldsymbol{x}, \boldsymbol{y} \in \mathscr{F}_{\boldsymbol{t}}$ is always a move. However, if supports of $\boldsymbol{x}$ and $\boldsymbol{y}$ have a nonempty intersection, then $\boldsymbol{x}$ is larger than $\boldsymbol{z}^-$ in some cell $\boldsymbol{i}$. Similarly $\boldsymbol{y}$ is larger than $\boldsymbol{z}^+$ in some cell $\boldsymbol{i}$. In fact we can write

$$\boldsymbol{z}^+ = \boldsymbol{y} - \min(\boldsymbol{x}, \boldsymbol{y}), \qquad \boldsymbol{z}^- = \boldsymbol{x} - \min(\boldsymbol{x}, \boldsymbol{y}), \tag{4.11}$$

where $\min(\boldsymbol{x}, \boldsymbol{y})$ is the elementwise minimum of $\boldsymbol{x}$ and $\boldsymbol{y}$.

In the definition of Markov basis, we are concerned whether adding a move $\boldsymbol{z}$ to a frequency vector $\boldsymbol{x}$ produces a negative cell. Write

$$\boldsymbol{x} + \boldsymbol{z} = (\boldsymbol{x} - \boldsymbol{z}^-) + \boldsymbol{z}^+.$$

Here we are subtracting the frequencies $\boldsymbol{z}^-$ from $\boldsymbol{x}$ and then adding the frequencies $\boldsymbol{z}^+$ in different cells. Therefore $\boldsymbol{x} + \boldsymbol{z}$ contains a negative cell if and only if $\boldsymbol{x} - \boldsymbol{z}^-$ contains a negative cell. In other words, $\boldsymbol{x} + \boldsymbol{z}$ does not contain a negative cell if and only if

$$\boldsymbol{x} \geq \boldsymbol{z}^-, \tag{4.12}$$

where the inequality is elementwise. When (4.12) holds, we say that $\boldsymbol{z}$ can be added to $\boldsymbol{x}$. When $\boldsymbol{z}$ can be added to $\boldsymbol{x}$ or can be subtracted from $\boldsymbol{x}$, we say that $\boldsymbol{z}$ is *applicable* to $\boldsymbol{x}$.

By (4.11), in the notation of monomials and binomials of the previous chapter, $\boldsymbol{x} + \varepsilon \boldsymbol{z} = \boldsymbol{y}$, $\varepsilon = \pm 1$, if and only if

$$\boldsymbol{u}^{\boldsymbol{y}} = \boldsymbol{u}^{\boldsymbol{x}} + \varepsilon \boldsymbol{u}^{\min(\boldsymbol{x},\boldsymbol{y})}(\boldsymbol{u}^{\boldsymbol{z}^+} - \boldsymbol{u}^{\boldsymbol{z}^-}). \tag{4.13}$$

A move $\boldsymbol{z}$ is called square-free if $\boldsymbol{u}^{\boldsymbol{z}^+} - \boldsymbol{u}^{\boldsymbol{z}^-}$ is a square-free binomial; that is, if the elements of $\boldsymbol{z}$ are $-1, 0$, or 1.

Markov bases are difficult exactly because we are worried about producing negative elements. Suppose that we just ignore the nonnegativeness of elements of frequency vector. Then the notion of a basis is simple. Note that $\ker_{\mathbb{Z}} A$ as a subset of $\mathbb{Z}^\eta$ is closed under integer multiplication and addition:

$$\boldsymbol{z}_1, \boldsymbol{z}_2 \in \ker_{\mathbb{Z}} A \Rightarrow a\boldsymbol{z}_1 + b\boldsymbol{z}_2 \in \ker_{\mathbb{Z}} A, \quad \forall a, b \in \mathbb{Z}.$$

A subset of $\mathbb{Z}^\eta$ with this property is called an *integer lattice*. Let $d = \dim \ker_{\mathbb{Z}} A = \eta - \mathrm{rank} A$ denote the dimension of linear space spanned by the elements of $\ker_{\mathbb{Z}} A$ in $\mathbb{R}^\eta$. It is a standard fact [134] that an integer lattice contains a *lattice basis* $\{\boldsymbol{z}_1, \ldots, \boldsymbol{z}_d\}$, such that every $\boldsymbol{z} \in \ker_{\mathbb{Z}} A$ is a unique integer combination of $\boldsymbol{z}_1, \ldots, \boldsymbol{z}_d$.

Given A, it is fairly easy to obtain a lattice basis of $\ker_{\mathbb{Z}} A$ using the Hermite normal form of A. In statistical applications, the configuration A often has redundant rows as in the case of two-way contingency tables in Sect. 1.3. However, $\ker A$ and hence $\ker_{\mathbb{Z}} A$ are defined by linearly independent rows of A. Therefore consider $A : \nu \times \eta$ with linearly independent rows; rank $A = \nu$. Then there exists an integer matrix U with $\det U = \pm 1$ (called a *unimodular* matrix) such that

$$AU = (B, 0),$$

where B is a $\nu \times \nu$ upper triangular matrix with positive elements. Then the columns $\nu + 1, \ldots, \eta$ of U give a lattice basis of $\ker_{\mathbb{Z}} A$. See Sect. 4.1 of [134].

In the case of the independence model of $I \times J$ two-way contingency tables, as elements of the lattice basis we can take the moves

$$\begin{array}{cc} +1 & -1 \\ -1 & +1 \end{array}$$

where the lower-right $+1$ is in the (I, J) cell. However, as we saw in Sect. 2.1, these moves do not form a Markov basis.

On the other hand, it is easy to see that a Markov basis $\mathscr{B}$ always contains a lattice basis. Given any $\boldsymbol{z} \neq \boldsymbol{0} \in \ker_{\mathbb{Z}} A$, we can move from $\boldsymbol{z}^-$ to $\boldsymbol{z}^+$ by a sequence of moves from $\mathscr{B}$, namely we can write

$$\boldsymbol{z}^+ = \boldsymbol{z}^- + \varepsilon_1 \boldsymbol{z}_{i_1} + \cdots + \varepsilon_K \boldsymbol{z}_{i_K}, \qquad \varepsilon_j = \pm 1, \quad \boldsymbol{z}_{i_j} \in \mathscr{B}, \qquad j = 1, \ldots, K.$$

Then $\boldsymbol{z} = \boldsymbol{z}^+ - \boldsymbol{z}^-$ is written as an integer combination of elements of $\mathscr{B}$.

So far we have considered the integer lattice $\ker_{\mathbb{Z}} A$. We now look at the integer lattice L generated by the columns of A:

$$L = \mathbb{Z}A = \{A\boldsymbol{z} \mid \boldsymbol{z} \in \mathbb{Z}^\eta\}.$$

Also let

$$\mathrm{cone}(A) = \mathbb{R}_{\geq 0} A = \{\sum_{\boldsymbol{i} \in \mathscr{I}} c_{\boldsymbol{i}} \boldsymbol{a}(\boldsymbol{i}) \mid c_{\boldsymbol{i}} \geq 0, \ \boldsymbol{i} \in \mathscr{I}\}$$

denote the cone generated by the column of A. Then the semigroup $\mathscr{T}_A$ in (4.6) is clearly a subset of $L \cap \mathrm{cone}(A)$:

$$\mathscr{T}_A \subset L \cap \mathrm{cone}(A). \tag{4.14}$$

We introduce some terminology concerning the semigroup $\mathscr{T}_A$. If the equality holds in (4.14) then the semigroup $\mathscr{T}_A$ is called *normal.* $L \cap \mathrm{cone}(A)$ is called the *saturation* of $\mathscr{T}_A$ (Definition 7.24 of [105]). When $\mathscr{T}_A$ is not normal, the elements of $L \cap \mathrm{cone}(A) \setminus \mathscr{T}_A$ are called *holes* of the saturation of $\mathscr{T}_A$ [85, 145, 146]. If $L \cap \mathrm{cone}(A) \setminus \mathscr{T}_A$ is a finite set, then the semigroup is called *very ample* [113].

Although in this book we do not go into details on normality of semigroups, the notion of normality is important for Proposition 5.4 and in discussing non-square-free indispensable moves in Sect. 9.5.

4.4 The Fundamental Theorem of Markov Basis

In this section we explain relations between moves and binomials of a toric ideal. Then we prove the fundamental theorem of Markov bases, which states that a Markov basis is a set of generators of a toric ideal. For readability we repeat some material from Chap. 3.

Consider the monomial form of the model (4.2):

$$p(\boldsymbol{i}) \propto \prod_{j=1}^{\nu} q_j^{a_j(\boldsymbol{i})}.$$

Let us regard $p(\boldsymbol{i})$, $\boldsymbol{i} \in \mathscr{I}$, as "symbols" or indeterminates rather than probabilities. Also let us regard q_j, $j = 1,\dots,\nu$, as indeterminates. Then the above model assigns to each indeterminate $p(\boldsymbol{i})$ a monomial $\prod_{j=1}^{\nu} q_j^{a_j(\boldsymbol{i})}$ in q_js. We formalize this consideration as follows. Let k be a field, such as the field $\mathbb{R}$ of real numbers. Let $k[\boldsymbol{p}] = k[p(\boldsymbol{i}), \boldsymbol{i} \in \mathscr{I}]$ be the polynomial ring in $\boldsymbol{p} = \{p(\boldsymbol{i}), \boldsymbol{i} \in \mathscr{I}\}$. Similarly define $k[\boldsymbol{q}] = k[q_1,\dots,q_\nu]$ to be the polynomial ring in $q_1,\dots,q_\nu$. As in Sect. 3.5, define a homomorphism $\pi_A : k[\boldsymbol{p}] \to k[\boldsymbol{q}]$ by

$$\pi_A(p(\boldsymbol{i})) = \prod_{j=1}^{\nu} q_j^{a_j(\boldsymbol{i})}.$$

π_A for a general polynomial of $k[\boldsymbol{p}]$ is defined by a homomorphism, that is, by substituting $\prod_{j=1}^{\nu} q_j^{a_j(\boldsymbol{i})}$ into each $p(\boldsymbol{i})$. The toric ideal I_A is defined as the kernel of π_A:

$$I_A = \ker \pi_A = \{f \in k[\boldsymbol{p}] \mid \pi_A(f) = 0\}.$$

These notions have already been illustrated in Chap. 1 and Sect. 3.5 with the example of two-way tables.

Let $\boldsymbol{x}$ be a frequency vector. As in Chap. 3, $\boldsymbol{x}$ is identified with the monomial

$$\boldsymbol{p}^{\boldsymbol{x}} = \prod_{\boldsymbol{i} \in \mathscr{I}} p(\boldsymbol{i})^{x(\boldsymbol{i})},$$

which corresponds to the joint probability of $\boldsymbol{x}$ except for the multinomial coefficient. A move $\boldsymbol{z} = \boldsymbol{z}^+ - \boldsymbol{z}^-$ corresponds to a binomial $\nu = \boldsymbol{p}^{\boldsymbol{z}^+} - \boldsymbol{p}^{\boldsymbol{z}^-}$. $\boldsymbol{p}^{\boldsymbol{z}^+} - \boldsymbol{p}^{\boldsymbol{z}^-}$ is in I_A if and only if $\boldsymbol{z} = \boldsymbol{z}^+ - \boldsymbol{z}^-$ is a move (see (3.13)). Now the fundamental theorem of Markov bases established by [50] states that a Markov basis corresponds to a system of generators of I_A.

Theorem 4.1 ([50]). *A finite set of moves $\mathscr{B}$ is a Markov basis for A if and only if the set of binomials $\{\boldsymbol{p}^{\boldsymbol{z}^+} - \boldsymbol{p}^{\boldsymbol{z}^-} \mid \boldsymbol{z} \in \mathscr{B}\}$ generates the toric ideal I_A.*

The "only if" (necessity) part is easy to prove. However, the proof of the "if" part (sufficiency) is somewhat hard. Because this theorem is of basic importance for the whole theory of Markov bases, we give a careful proof. For a finite set of moves $\mathscr{B} = \{z_1, \ldots, z_L\}$ for a configuration A, we write the set of corresponding binomials as $F_{\mathscr{B}} = \{p^{z_i^+} - p^{z_i^-}, i = 1, \ldots, L\}$.

Proof. We first show the necessity: if $\mathscr{B}$ is a Markov basis for A then $F_{\mathscr{B}}$ generates I_A. Binomials generate (cf. Definition 3.3) the toric ideal I_A, therefore we only need to show that any binomial $f = p^y - p^x$, $Ax = Ay$, belongs to the ideal $\langle F_{\mathscr{B}} \rangle$. Inasmuch as $\mathscr{B}$ is a Markov basis, x and $y (\in \mathscr{F}_{Ax})$ are mutually accessible by $\mathscr{B}$. Hence

$$y = x + \sum_{j=1}^{S} \varepsilon_j z_{i_j}, \quad x + \sum_{j=1}^{s} \varepsilon_j z_{i_j} \in \mathscr{F}_{Ax}, \quad 1 \le s \le S,$$

for some $S > 0$, $\varepsilon_j \in \{-1, 1\}$, $z_{i_j} \in \mathscr{B}$, $j = 1, \ldots, S$. Write $x_s = x + \sum_{j=1}^{s} \varepsilon_j z_{i_j}$, $0 \le s \le S$, with $x_0 = x$ and $x_S = y$. Then by (4.13), with the notation for indeterminates p instead of u, it follows that

$$p^{x_s} = p^x + \sum_{j=1}^{s} \varepsilon_j p^{\min(x_{j-1}, x_j)} (p^{z_{i_j}^+} - p^{z_{i_j}^-}), \quad s = 1, \ldots, S.$$

Hence

$$p^y - p^x = \sum_{j=1}^{S} \varepsilon_j p^{\min(x_{j-1}, x_j)} (p^{z_{i_j}^+} - p^{z_{i_j}^-}) \in \langle F_{\mathscr{B}} \rangle. \tag{4.15}$$

This proves the necessity part.

Next we prove the "if" part (sufficiency). We want to show that if I_A is generated by $F_{\mathscr{B}}$, then every $x, y (\neq x) \in \mathscr{F}_{Ax}$ is mutually accessible by $\mathscr{B}$. By the assumption $p^y - p^x$ can be written as a finite sum

$$p^y - p^x = \sum_{i=1}^{L} f_i(p)(p^{z_i^+} - p^{z_i^-}), \tag{4.16}$$

where $f_i(p) \in k[p]$, $i = 1, \ldots, L$, are polynomials in p. Expand $f_i(p)$ into monomials. Then allowing repetitions, we can write $p^y - p^x$ as a finite sum

$$p^y - p^x = \sum_{l} a_l p^{h_l} (p^{z_{i_l}^+} - p^{z_{i_l}^-}), \quad a_l \in k. \tag{4.17}$$

In Lemma 4.1 below we show that we can choose a_l as integers. Given this fact, instead of $a_l p^{h_l} (p^{z_{i_l}^+} - p^{z_{i_l}^-})$, we can write $|a_l|$ times the binomial $p^{h_l} (p^{z_{i_l}^+} - p^{z_{i_l}^-})$ and use $\varepsilon_l = \pm 1$. Then by allowing further repetitions, we can write $p^y - p^x$ as

$$p^y - p^x = \sum_{j=1}^{S} \varepsilon_j p^{h_j} (p^{z_{i_j}^+} - p^{z_{i_j}^-}), \quad \varepsilon_j = \pm 1. \tag{4.18}$$

Note that (4.18) is already similar to (4.15). The difference between them is that in (4.15) the order of the terms from x to y on the right-hand is already given. On the other hand the sum on the right-hand side of (4.18) is not ordered for moving from $\boldsymbol{x}$ to $\boldsymbol{y}$. We need to find a suitable path from $\boldsymbol{x}$ to $\boldsymbol{y}$ on the right-hand side of (4.18). The path can be found as follows. Expand (4.18) into $2S$ terms. Then at least one term has to be equal to $\boldsymbol{p}^{\boldsymbol{x}}$. Namely for some j we have $\boldsymbol{p}^{\boldsymbol{x}} = \varepsilon_j \boldsymbol{p}^{\boldsymbol{h}_j} \boldsymbol{p}^{\boldsymbol{z}_{i_j}^+}$ or $\boldsymbol{p}^{\boldsymbol{x}} = -\varepsilon_j \boldsymbol{p}^{\boldsymbol{h}_j} \boldsymbol{p}^{\boldsymbol{z}_{i_j}^-}$. Then by (4.13) we can move from $\boldsymbol{x}$ to $\boldsymbol{x}_1$ such that $\boldsymbol{p}^{\boldsymbol{x}_1} = \boldsymbol{p}^{\boldsymbol{x}} - \varepsilon_j \boldsymbol{p}^{\boldsymbol{h}_j} \boldsymbol{p}^{\boldsymbol{z}_{i_j}^+}$ or $\boldsymbol{p}^{\boldsymbol{x}_1} = \boldsymbol{p}^{\boldsymbol{x}} + \varepsilon_j \boldsymbol{p}^{\boldsymbol{h}_j} \boldsymbol{p}^{\boldsymbol{z}_{i_j}^+}$. In either case we can now write

$$\boldsymbol{p}^{\boldsymbol{y}} - \boldsymbol{p}^{\boldsymbol{x}_1} = \sum_{l \neq j} \varepsilon_l \boldsymbol{p}^{\boldsymbol{h}_l} (\boldsymbol{p}^{\boldsymbol{z}_{i_l}^+} - \boldsymbol{p}^{\boldsymbol{z}_{i_l}^-}),$$

where the sum on the right-hand side is a sum of $S-1$ terms. Now we can employ induction on S and find the steps $\boldsymbol{x}_2, \ldots, \boldsymbol{x}_S$ to move from $\boldsymbol{x}$ to $\boldsymbol{y} = \boldsymbol{x}_S$. This proves the sufficiency. □

In the above proof of sufficiency, the proof of the integerness of coefficients a_l in (4.17) is left to the following lemma.

Lemma 4.1. *If $F_{\mathscr{B}}$ generates I_A, then each binomial $\boldsymbol{p}^{\boldsymbol{y}} - \boldsymbol{p}^{\boldsymbol{x}} \in I_A$ can be written as a finite sum*

$$\boldsymbol{p}^{\boldsymbol{y}} - \boldsymbol{p}^{\boldsymbol{x}} = \sum_l a_l \boldsymbol{p}^{\boldsymbol{h}_l} (\boldsymbol{p}^{\boldsymbol{z}_{i_l}^+} - \boldsymbol{p}^{\boldsymbol{z}_{i_l}^-}), \tag{4.19}$$

where the a_ls are integers.

We give two different proofs of this lemma. The first proof is based on a Gröbner basis. The second proof is longer, but only uses linear algebra.

Proof. Denote a Gröbner basis of I_A by $\{g_1, \ldots, g_L\}$, which is obtained by Buchberger's algorithm from the set of generators $F_{\mathscr{B}}$ of I_A. Because $\boldsymbol{p}^{\boldsymbol{y}} - \boldsymbol{p}^{\boldsymbol{x}} \in I_A = \langle g_1, \ldots, g_L \rangle$, the binomial $\boldsymbol{p}^{\boldsymbol{y}} - \boldsymbol{p}^{\boldsymbol{x}}$ is written as

$$\boldsymbol{p}^{\boldsymbol{y}} - \boldsymbol{p}^{\boldsymbol{x}} = \sum_{i=1}^{L} f_i g_i, \qquad f_i \in k[\boldsymbol{p}].$$

By expanding f_i into monomials, $\boldsymbol{p}^{\boldsymbol{y}} - \boldsymbol{p}^{\boldsymbol{x}}$ is further written as

$$\boldsymbol{p}^{\boldsymbol{y}} - \boldsymbol{p}^{\boldsymbol{x}} = \sum_l a_l \boldsymbol{p}^{\boldsymbol{v}_{i_l}} g_{i_l},$$

where $a_j \in k$. Note that the sum on the right-hand side corresponds to a division by a Gröbner basis and each step of the division is an operation of eliminating the leading term by a monomial with coefficient ± 1. Hence, if we allow repetitions, we can indeed assume $a_j = \pm 1$. On the other hand, because $g_1, \ldots, g_L \in I_A = \langle F_{\mathscr{B}} \rangle$ each element of the Gröbner basis is written as

$$g_j = \sum_\ell d_\ell \boldsymbol{p}^{\boldsymbol{w}_\ell} (\boldsymbol{p}^{\boldsymbol{z}_{i_\ell}^+} - \boldsymbol{p}^{\boldsymbol{z}_{i_\ell}^-}).$$

The right-hand side corresponds to obtaining g_j by the Buchberger algorithm from the set of generators. In Lemma 3.2 we saw that the S-polynomial of two binomials is again a binomial. This implies that we can also assume $d_\ell = \pm 1$. Therefore, the binomial $\boldsymbol{p}^{\boldsymbol{y}} - \boldsymbol{p}^{\boldsymbol{x}}$ can be written as (4.17) where a_l are integers. This completes our first proof.

We now give an alternative proof. By the assumption, (4.19) holds with $a_l \in k$. At this point a_ls are not necessarily integers and we want to show that we can always replace a_l by integers. Note that there are only a finite number of monomials appearing in (4.19). Choose a sufficiently large D such that the degrees of all monomials in (4.19) are less than or equal to D. Let $\mathfrak{M}_D$ denote the set of monomials of degrees less than or equal to D. Let $M = |\mathfrak{M}_D|$ denote the cardinality of $\mathfrak{M}_D$. Then $\mathfrak{M}_D$ is a basis of the M-dimensional vector space V_D of polynomials of degree less than or equal to D.

With respect to this basis, each binomial is represented as a column vector with two nonzero elements which are 1 and -1 and other elements are zeros. Let $N = N_D$ denote the number of binomials whose degrees are less than or equal to D. Let C be an $M \times N$ matrix whose columns correspond to binomials of degree less than or equal to D. Then the right-hand side of (4.19) is written as

$$\boldsymbol{b} = C\boldsymbol{a},$$

where $\boldsymbol{b}$ corresponds to $\boldsymbol{p}^{\boldsymbol{y}} - \boldsymbol{p}^{\boldsymbol{x}}$. Now by Lemma 4.2 below, there exists an $M \times M$ permutation matrix P and an $N \times N$ unimodular matrix U such that $\tilde{C} = PCU$ is of the form (4.20) below. Then $\boldsymbol{b} = C\boldsymbol{a}$ can be equivalently written as

$$P\boldsymbol{b} = \tilde{C}\tilde{\boldsymbol{a}}, \quad \tilde{\boldsymbol{a}} = U^{-1}\boldsymbol{a}.$$

From the form of $\tilde{C}$, it is clear that $\tilde{\boldsymbol{a}}$ can be chosen to be an integer vector. Then $\boldsymbol{a} = U\tilde{\boldsymbol{a}}$ is an integer vector. This finishes our second proof. □

Lemma 4.2. *Let C be an $M \times N$ matrix, such that each column $\boldsymbol{c}$ of C has two nonzero elements which are 1 and -1; namely, $\boldsymbol{c}$ is written as $\boldsymbol{c} = \boldsymbol{e}_i - \boldsymbol{e}_j$, $i \neq j$, where $\boldsymbol{e}_i$ is the ith standard basis vector with 1 in the ith position. Then by the following three elementary operations (i) sign change of columns, (ii) addition (or subtraction) of a column to (or from) another column, and (iii) permutation of rows, C can be transformed to the following block diagonal form.*

$$\begin{pmatrix} B_{d_1} & 0 & \dots & 0 & 0 \\ 0 & B_{d_2} & \dots & 0 & 0 \\ \vdots & \vdots & \ddots & \vdots & \vdots \\ 0 & 0 & \dots & B_{d_K} & 0 \\ 0 & 0 & \dots & 0 & 0 \end{pmatrix}, \quad \textit{where} \quad B_d = \begin{pmatrix} 1 & 0 & \dots & 0 \\ 0 & 1 & \dots & 0 \\ \vdots & \vdots & \ddots & \vdots \\ 0 & 0 & \dots & 1 \\ -1 & -1 & \dots & -1 \end{pmatrix} : d \times (d-1). \quad (4.20)$$

Proof. We give a proof based on the induction on the number of columns. Let M be fixed. If $N = 1$, the result is trivial. Suppose that the result holds up to N and consider adding a new column $\boldsymbol{c}$. We can assume that the first N columns have already been transformed to the form in (4.20). Let the new column be denoted as $\boldsymbol{c} = \boldsymbol{e}_i - \boldsymbol{e}_j$. There are four cases to consider.

Case 1. If neither of i, j belongs to (the rows of the) blocks $B_{d_1}, \ldots, B_{d_K}$, then $\boldsymbol{c} = B_2$ forms a new block.

Case 2. If both of i, j belong to a block, say B_{d_1}, then $\boldsymbol{c}$ can be transformed to 0. This is obvious if $\boldsymbol{c}$ is equal to some column of B_{d_1}. Otherwise $\boldsymbol{c}$ is the difference of the ith and jth columns of B_{d_1}.

Case 3. Suppose that i belongs to, say, B_{d_1} and j does not belong to any block. Let $j = d_1 + 1$ without loss of generality. Subtracting the ith column from $\boldsymbol{c}$, we obtain

$$(\underbrace{0, \ldots, 0}_{d_1 - 1}, 1, -1)'.$$

Adding this to other columns of B_{d_1} we obtain a new block of the form B_{d_1+1}.

Case 4. Suppose that i and j belong to different blocks, say B_{d_1} and B_{d_2}. If $i = d_1$ and $j = d_2$, then adding $\boldsymbol{c}$ to the columns of B_{d_1} we obtain a new block of the form $B_{d_1+d_2}$. Otherwise, if $i < d_1$, subtract the ith column from $\boldsymbol{c}$. Similarly if $j < d_2$, then subtract the jth column from $\boldsymbol{c}$. Then $\boldsymbol{c}$ is transformed to $\boldsymbol{e}_{d_1} - \boldsymbol{e}_{d_2}$ and this reduces to the former case. □

This lemma shows that the diagonal block of the Smith normal form of C is the identity matrix and every elementary divisor of C is 1 (cf. Sect. 4.4 of [134]).

Remark 4.1. We can consider columns of C in Lemma 4.2 as edges of a graph. Consider a graph $\mathscr{G}$ with M vertices. For $\boldsymbol{c} = \boldsymbol{e}_i - \boldsymbol{e}_j$, draw an edge between i and j. Then it can be easily seen that the blocks in (4.20) correspond to connected components of $\mathscr{G}$.

Remark 4.2. Two proofs of Lemma 4.2 look different but they are essentially the same. The second proof is based on the relations among vectors with two nonzero elements which are 1 and -1, such as

$$\begin{pmatrix} 1 \\ -1 \\ 0 \end{pmatrix} + \begin{pmatrix} 0 \\ 1 \\ -1 \end{pmatrix} = \begin{pmatrix} 1 \\ 0 \\ -1 \end{pmatrix},$$

where the second element is canceled. This in fact corresponds to forming an S-polynomial of two binomials.

4.5 Gröbner Basis from the Viewpoint of Markov Basis

In the previous section we discussed the Gröbner basis from an algebraic viewpoint. Here we discuss it from the viewpoint of the Markov basis. In Chap. 3 we have already summarized relevant facts on the Gröbner basis. Here we discuss the Gröbner basis from a viewpoint close to our definition of the Markov basis. In the case of the Markov basis we tend to ignore the sign of a move $\boldsymbol{z}$. In the Gröbner basis it is important to keep track of the sign of the move.

As in Chap. 3 let a term order $\prec$ be given. Let $G = \{g_1, \ldots, g_L\} \subset k[\boldsymbol{p}]$ be a Gröbner basis with respect to $\prec$, such that $g_1, \ldots, g_L$ are binomials. Then $g_1, \ldots, g_L$ correspond to moves. By Proposition 3.3 the reduced Gröbner basis consists of binomials. Write

$$g_l = \boldsymbol{p}^{\boldsymbol{z}_l^+} - \boldsymbol{p}^{\boldsymbol{z}_l^-}, \quad \boldsymbol{p}^{\boldsymbol{z}_l^+} = \mathrm{in}_\prec(g_l), \quad l = 1, \ldots, L.$$

As before we identify the monomial $\boldsymbol{p}^{\boldsymbol{x}}$ with the frequency vector $\boldsymbol{x}$.

Because the term order is the total order, every fiber $\mathscr{F}_{\boldsymbol{t}}$ (which is finite as remarked in Sect. 4.1) has the unique minimum element $\boldsymbol{x}_{\boldsymbol{t}}^*$. For any other $\boldsymbol{x} \neq \boldsymbol{x}_{\boldsymbol{t}}^*$ of the fiber, $\boldsymbol{p}^{\boldsymbol{x}} - \boldsymbol{p}^{\boldsymbol{x}_{\boldsymbol{t}}^*} \in I_A$. Also $\boldsymbol{p}^{\boldsymbol{x}} \succ \boldsymbol{p}^{\boldsymbol{x}_{\boldsymbol{t}}^*}$. Therefore $\boldsymbol{p}^{\boldsymbol{x}}$ is divisible by some $\mathrm{in}_\prec(g_l)$; that is, $\boldsymbol{x} \geq \boldsymbol{z}_l^+$. Dividing $\boldsymbol{p}^{\boldsymbol{x}}$ by g_l corresponds to moving from $\boldsymbol{x}$ to $\boldsymbol{x} - (\boldsymbol{z}_l^+ - \boldsymbol{z}_l^-)$, which is a smaller element than $\boldsymbol{x}$ in $\mathscr{F}_{\boldsymbol{t}}$. On the other hand, $\boldsymbol{p}^{\boldsymbol{x}_{\boldsymbol{t}}^*}$ is divisible by none of $\mathrm{in}_\prec(g_l)$, $l = 1, \ldots, L$, because otherwise $\boldsymbol{p}^{\boldsymbol{x}_{\boldsymbol{t}}^*}$ would not be the minimum element of $\mathscr{F}_{\boldsymbol{t}}$. By the definition of the standard monomial (cf. Sect. 3.2) we have now shown the following fact.

Lemma 4.3. *Given a term order $\prec$, $\{\boldsymbol{p}^{\boldsymbol{x}_{\boldsymbol{t}}^*} \mid \boldsymbol{t} \in \mathscr{T}\}$ is the set of standard monomials of I_A.*

Now let $\mathscr{B}$ be any finite set of moves. For a given term order we always choose a sign of a move $\boldsymbol{z} = \boldsymbol{z}^+ - \boldsymbol{z}^- \in \mathscr{B}$ by

$$\boldsymbol{p}^{\boldsymbol{z}^+} \succ \boldsymbol{p}^{\boldsymbol{z}^-}.$$

In a fiber $\mathscr{F}_{\boldsymbol{t}}$, we draw a directed edge from $\boldsymbol{y}$ to $\boldsymbol{x}$ if there exists $\boldsymbol{z} \in \mathscr{B}$ such that

$$\boldsymbol{y} - \boldsymbol{x} = \boldsymbol{z}.$$

Then each fiber $\mathscr{F}_{\boldsymbol{t}}$ becomes a directed graph $\bar{\mathscr{G}}_{\boldsymbol{t},\mathscr{B}}$. As discussed in the previous section, if $\boldsymbol{y} - \boldsymbol{x} = \boldsymbol{z}$ then

$$\boldsymbol{x} = \boldsymbol{z}^- + \min(\boldsymbol{x}, \boldsymbol{y}), \qquad \boldsymbol{y} = \boldsymbol{z}^+ + \min(\boldsymbol{x}, \boldsymbol{y}).$$

Hence $\boldsymbol{p}^{\boldsymbol{y}} \succ \boldsymbol{p}^{\boldsymbol{x}}$ by the second property of the term order. Therefore by subtracting $\boldsymbol{z} \in \mathscr{B}$ from $\boldsymbol{y}$, we always move to a smaller element of the fiber and there exists no directed loop in $\bar{\mathscr{G}}_{\boldsymbol{t},\mathscr{B}}$; that is, $\bar{\mathscr{G}}_{\boldsymbol{t},\mathscr{B}}$ is a directed acyclic graph (DAG). We now have the following proposition.

Proposition 4.1. *A finite set of moves $\mathscr{B}$ is a Gröbner basis if and only if from every element of every $\bar{\mathscr{G}}_{t,\mathscr{B}}$ there exists a directed path to the minimum element $\boldsymbol{x}_t^*$ of the fiber.*

Proof. If $\mathscr{B}$ is a Gröbner basis, then by Lemma 4.3 from every element of every $\bar{\mathscr{G}}_{t,\mathscr{B}}$ there exists a directed path to the minimum element $\boldsymbol{x}_t^*$.

Conversely, suppose that from every state of every $\bar{\mathscr{G}}_{t,\mathscr{B}}$ there exists a directed path to the minimum element $\boldsymbol{x}_t^*$. Then every $\boldsymbol{x} \in \mathscr{F}_t$ is divisible by some $p^{\boldsymbol{z}^+}$, $\boldsymbol{z} \in \mathscr{B}$. Hence $\mathscr{B}$ is a Gröbner basis. □

4.6 Graver Basis, Lawrence Lifting, and Logistic Regression

Finally we discuss the Graver basis and the Lawrence lifting. Consider a sum of two moves $\boldsymbol{z} = \boldsymbol{z}_1 + \boldsymbol{z}_2$. We say that there is *no cancellation of signs* in this sum if

$$\operatorname{supp}(\boldsymbol{z}^+) = \operatorname{supp}(\boldsymbol{z}_1^+) \cup \operatorname{supp}(\boldsymbol{z}_2^+), \quad \operatorname{supp}(\boldsymbol{z}^-) = \operatorname{supp}(\boldsymbol{z}_1^-) \cup \operatorname{supp}(\boldsymbol{z}_2^-).$$

In this case we also say that $\boldsymbol{z}$ is a *conformal sum* of $\boldsymbol{z}_1$ and $\boldsymbol{z}_2$. Similarly we say that there is no cancellation of signs in the sum of m moves $\boldsymbol{z} = \boldsymbol{z}_1 + \cdots + \boldsymbol{z}_m$ (or $\boldsymbol{z}$ is a conformal sum of m moves) if

$$\operatorname{supp}(\boldsymbol{z}^+) = \operatorname{supp}(\boldsymbol{z}_1^+) \cup \cdots \cup \operatorname{supp}(\boldsymbol{z}_m^+), \quad \operatorname{supp}(\boldsymbol{z}^-) = \operatorname{supp}(\boldsymbol{z}_1^-) \cup \cdots \cup \operatorname{supp}(\boldsymbol{z}_m^-).$$

We also say that $\boldsymbol{z}_1 + \cdots + \boldsymbol{z}_m$ is a *conformal decomposition* of $\boldsymbol{z}$. We call a move $\boldsymbol{z}$ *conformally primitive* if it cannot be written as a sum of two nonzero moves $\boldsymbol{z} = \boldsymbol{z}_1 + \boldsymbol{z}_2$ with no cancellation of signs. For clarity we say "conformally" primitive, but usually a conformally primitive move is simply called a primitive move. A binomial corresponding to a conformally primitive move is called a primitive binomial.

Definition 4.2. The Graver basis is the set of conformally primitive moves.

In Sect. 5.4.3 we show that the Graver basis is finite. We first see that the Graver basis is a Markov basis. For $\boldsymbol{x}, \boldsymbol{y}$ in the same fiber let $\boldsymbol{z} = \boldsymbol{y} - \boldsymbol{x}$. If a move $\boldsymbol{z}$ is not conformally primitive, then we can recursively decompose $\boldsymbol{z}$ into a conformal sum of moves. This implies that $\boldsymbol{z}$ can be written as a conformal sum

$$\boldsymbol{z} = \boldsymbol{z}_1 + \cdots + \boldsymbol{z}_m, \tag{4.21}$$

where $\boldsymbol{z}_1, \ldots, \boldsymbol{z}_m$ are (not necessarily distinct) nonzero elements of the Graver basis. Then we can move from $\boldsymbol{x}$ to $\boldsymbol{y}$ by the above sequence of conformally primitive moves. This shows that the Graver basis is a Markov basis. Also note that because of no cancellation of signs, whenever $\boldsymbol{z}$ is applicable to some $\boldsymbol{x}$, $\boldsymbol{z}$ can be replaced by $\boldsymbol{z}_1, \ldots, \boldsymbol{z}_m$ in arbitrary order without causing negative elements on the way.

As an important example we consider the Graver basis for the independence model of $I \times J$ contingency tables.

Definition 4.3. For $2 \leq r \leq \min(I,J)$, let $i_1,\dots,i_r$ be distinct row indices and let $j_1,\dots,j_r$ be distinct column indices. Denote $\boldsymbol{i}_{[r]} = (i_1,\dots,i_r)$, $\boldsymbol{j}_{[r]} = (i_1,\dots,i_r)$. *A loop of degree r*

$$\boldsymbol{z}_r(\boldsymbol{i}_{[r]};\boldsymbol{j}_{[r]}) = \{z_{ij}\}, \quad 1 \leq i_1,\dots,i_r \leq I,\ 1 \leq j_1,\dots,j_r \leq J, \tag{4.22}$$

is a move such that

$$z_{i_1 j_1} = z_{i_2 j_2} = \cdots = z_{i_{r-1} j_{r-1}} = z_{i_r j_r} = 1,$$
$$z_{i_1 j_2} = z_{i_2 j_3} = \cdots = z_{i_{r-1} j_r} = z_{i_r j_1} = -1,$$

and all the other elements are zeros.

There is at most one $+1$ and -1 in each row and each column of a degree r loop. An example of a loop of degree $r = 3$ is depicted as follows.

1	−1	0
0	1	−1
−1	0	1

.

Then we have the following proposition.

Proposition 4.2. *Loops of degree* $2 \leq r \leq \min(I,J)$ *form the Graver basis for the independence model of* $I \times J$ *contingency tables.*

Proof. Consider any nonzero move $\boldsymbol{z}$. Let $\boldsymbol{z}(i_1, j_1) > 0$. Because the i_1-row sum of $\boldsymbol{z}$ is zero, we can find j_2 such that $\boldsymbol{z}(i_1, j_2) < 0$. The j_2-column sum of $\boldsymbol{z}$ is zero, therefore we can now find i_2 such that $\boldsymbol{z}(i_2, j_2) > 0$. Visiting cells in this way, we come back to a cell that was already visited. Among such "cycles," consider a shortest one. Then the shortest one is a loop, namely, the row indices and the column indices are distinct among themselves. Taking away this loop, we have a move of smaller degree. If we apply this procedure recursively, we can express $\boldsymbol{z}$ as a conformal sum of loops.

On the other hand, it is obvious that each loop cannot be written as a conformal sum of other nonzero moves. This proves the proposition. □

Let A be a configuration. The *Lawrence lifting* $\Lambda(A)$ of A is the configuration ($(2\nu + \eta) \times 2\eta$ matrix)

$$\Lambda(A) = \begin{pmatrix} A & 0 \\ 0 & A \\ E_\eta & E_\eta \end{pmatrix}, \tag{4.23}$$

where E_η is the $\eta \times \eta$ identity matrix. Note that

$$(0\,A) = (A\,A) - (A\,0) = A(E_\eta\ E_\eta) - (A\,0).$$

Therefore the second block $(0\,A)$ in (4.23) is redundant for defining $\ker \Lambda(A)$. Hence instead of $\Lambda(A)$ we can also use

$$\tilde{\Lambda}(A) = \begin{pmatrix} A & 0 \\ E_\eta & E_\eta \end{pmatrix}. \tag{4.24}$$

However it is more convenient to retain the second block in $\Lambda(A)$ for explanation.

From (4.23), an element of $\ker \Lambda(A)$ is of the form

$$\begin{pmatrix} \boldsymbol{z} \\ -\boldsymbol{z} \end{pmatrix},$$

where $\boldsymbol{z} \in \ker A$. Clearly $\boldsymbol{z} = \boldsymbol{z}_1 + \boldsymbol{z}_2$ is a conformal sum of two moves for A if and only if

$$\begin{pmatrix} \boldsymbol{z} \\ -\boldsymbol{z} \end{pmatrix} = \begin{pmatrix} \boldsymbol{z}_1 \\ -\boldsymbol{z}_1 \end{pmatrix} + \begin{pmatrix} \boldsymbol{z}_2 \\ -\boldsymbol{z}_2 \end{pmatrix}$$

is a conformal sum of two moves for $\Lambda(A)$. By this observation we have the following proposition.

Proposition 4.3. *Let $\{\boldsymbol{z}_1, \ldots, \boldsymbol{z}_L\}$ be the Graver basis for A. The Graver basis of $\Lambda(A)$ is given by*

$$\left\{ \begin{pmatrix} \boldsymbol{z}_1 \\ -\boldsymbol{z}_1 \end{pmatrix}, \ldots, \begin{pmatrix} \boldsymbol{z}_L \\ -\boldsymbol{z}_L \end{pmatrix} \right\}. \tag{4.25}$$

From the viewpoint of statistical models, the Lawrence lifting corresponds to a logistic regression. See Christensen [37] for a detailed treatment of logistic regression. Consider the model (4.1). We make two copies $\mathscr{I}'$ and $\mathscr{I}''$ of the sample space $\mathscr{I}$ and consider a corresponding pair of cells $(\boldsymbol{i}', \boldsymbol{i}'')$. Call $\boldsymbol{i}'$ a "success" of the cell $\boldsymbol{i}$ and $\boldsymbol{i}''$ a "failure" of the cell $\boldsymbol{i}$. Consider a Bernoulli random variable $Y_{\boldsymbol{i}} \in \{0, 1\}$, such that

$$P(Y_{\boldsymbol{i}} = 1) = p_{\boldsymbol{i}} = \frac{\exp(\sum_{j=1}^{\nu} \theta_j a_j(\boldsymbol{i}))}{1 + \exp(\sum_{j=1}^{\nu} \theta_j a_j(\boldsymbol{i}))}. \tag{4.26}$$

We let $Y_{\boldsymbol{i}} = 1$ correspond to an observation in $\boldsymbol{i}'$ and $Y_{\boldsymbol{i}} = 0$ correspond to an observation in $\boldsymbol{i}''$.

For each $\boldsymbol{i}$, observe $n_{\boldsymbol{i}}$ independent Bernoulli random variables with the success probability (4.26). Let $x(\boldsymbol{i}')$ be the frequency of the cell $\boldsymbol{i}'$ (i.e., the number of successes for the cell $\boldsymbol{i}$) and let $x(\boldsymbol{i}'') = n_{\boldsymbol{i}} - x(\boldsymbol{i}')$ be the number of failures. Under the logistic regression model, $x(\boldsymbol{i}')$ has the binomial distribution $\mathrm{Bin}(n_{\boldsymbol{i}}, p_{\boldsymbol{i}})$.

We now consider the sufficient statistic for the logistic regression model. It is easily seen that a sufficient statistic for this logistic regression model, when $n_{\boldsymbol{i}}$s are fixed and regarded as parameters, is a sufficient statistic for the original A computed from the number of successes $x(\boldsymbol{i}')$, $\boldsymbol{i} \in \mathscr{I}$. When $n_{\boldsymbol{i}}$s are fixed, we can alternatively compute a sufficient statistic for the original A from the number of failures $x(\boldsymbol{i}'')$, $\boldsymbol{i} \in \mathscr{I}$. As in the discussion on three sampling schemes for 2×2 contingency tables in Sect. 1.1, we can also regard $n_{\boldsymbol{i}}$s as a part of a sufficient statistic. If we vary $n_{\boldsymbol{i}}$s and regard them as arbitrary nonnegative integers, then the configuration of the logistic regression is given by the Lawrence lifting (4.24).

As we show in Sect. 5.4.3 the unique Markov basis of $\Lambda(A)$ coincides with the Graver basis of $\Lambda(A)$ and the latter is essentially the same as the Graver basis of A by Proposition 4.3. Hence if we allow arbitrary nonnegative $n_{\boldsymbol{i}}$s, then we need the Graver basis of A. However, it seems that many elements of the Graver basis of A are needed in order to guarantee the connectivity when some $n_{\boldsymbol{i}}$s are zeros. When all $n_{\boldsymbol{i}}$s are positive and fixed, connectivity of a particular fiber may be guaranteed by a proper subset of the Graver basis. This problem is discussed mainly in Chap. 13.

In this chapter we discussed a Markov basis, a lattice basis, a Gröbner basis, and the Graver basis. We here summarize implications among them. By definition of these bases, the inclusion relations between them are given as follows.

$$\begin{aligned} &\text{a lattice basis} \subset \text{a minimal Markov basis} \subset \text{a reduced Gröbner basis} \\ &\quad \subset \text{the Graver basis.} \end{aligned}$$

Chapter 5
Structure of Minimal Markov Bases

5.1 Accessibility by a Set of Moves

Let $\mathscr{B} = \{\boldsymbol{z}_1, \ldots, \boldsymbol{z}_L\}$ be a finite set of moves, which may not be a Markov basis. Let $\boldsymbol{x}, \boldsymbol{y}(\neq \boldsymbol{x}) \in \mathscr{F}_{\boldsymbol{t}}$. We say that $\boldsymbol{y}$ is *accessible* from $\boldsymbol{x}$ by $\mathscr{B}$ and denote this by

$$\boldsymbol{x} \sim \boldsymbol{y} \qquad (\text{mod } \mathscr{B}),$$

if there exists a sequence of moves $\boldsymbol{z}_{i_1}, \ldots, \boldsymbol{z}_{i_L}$ from $\mathscr{B}$ and $\varepsilon_j = \pm 1,\ j = 1, \ldots, L$, such that $\boldsymbol{y} = \boldsymbol{x} + \sum_{j=1}^{L} \varepsilon_j \boldsymbol{z}_{i_j}$ and

$$\boldsymbol{x} + \sum_{j=1}^{h} \varepsilon_j \boldsymbol{z}_{i_j} \in \mathscr{F}_{\boldsymbol{t}}, \qquad h = 1, \ldots, L-1; \tag{5.1}$$

that is, we can move from $\boldsymbol{x}$ to $\boldsymbol{y}$ by moves from $\mathscr{B}$ without causing negative elements on the way. Obviously the notion of accessibility is symmetric and transitive:

$$\boldsymbol{x} \sim \boldsymbol{y} \quad \Rightarrow \quad \boldsymbol{y} \sim \boldsymbol{x} \qquad (\text{mod } \mathscr{B}),$$
$$\boldsymbol{x}_1 \sim \boldsymbol{x}_2, \quad \boldsymbol{x}_2 \sim \boldsymbol{x}_3 \quad \Rightarrow \quad \boldsymbol{x}_1 \sim \boldsymbol{x}_3 \qquad (\text{mod } \mathscr{B}).$$

Allowing moves to be $\boldsymbol{0}$ also yields reflexivity. Therefore accessibility by $\mathscr{B}$ is an equivalence relation and each fiber $\mathscr{F}_{\boldsymbol{t}}$ is partitioned into disjoint equivalence classes by moves of $\mathscr{B}$. We call these equivalence classes *$\mathscr{B}$-equivalence classes of $\mathscr{F}_{\boldsymbol{t}}$*. Because the notion of accessibility is symmetric, we also say that $\boldsymbol{x}$ and $\boldsymbol{y}$ are mutually accessible by $\mathscr{B}$ if $\boldsymbol{x} \sim \boldsymbol{y}$ (mod $\mathscr{B}$).

Let $\boldsymbol{x}$ and $\boldsymbol{y}$ be elements from two different $\mathscr{B}$-equivalence classes of $\mathscr{F}_{\boldsymbol{t}}$. We say that a move $\boldsymbol{z} = \boldsymbol{x} - \boldsymbol{y}$ *connects* these two equivalence classes.

S. Aoki et al., *Markov Bases in Algebraic Statistics*, Springer Series in Statistics 199, DOI 10.1007/978-1-4614-3719-2_5,

In this chapter the set of moves with degree less than or equal to n,

$$\mathscr{B}_n = \{z \mid \deg z \le n\}, \tag{5.2}$$

is of particular importance. Consider $\mathscr{B}_{n-1}$-equivalence classes of a fiber $\mathscr{F}_t$ with $n = \deg t$. Let K_t denote the number of equivalence classes and partition $\mathscr{F}_t$ as

$$\mathscr{F}_t = \mathscr{F}_{t,1} \cup \cdots \cup \mathscr{F}_{t,K_t}. \tag{5.3}$$

We also define

$$\mathscr{B}_t = \{z \in \ker_{\mathbb{Z}} A \mid t = Az^+ = Az^-\} = \{z \in \ker_{\mathbb{Z}} A \mid z^+, z^- \in \mathscr{F}_t\}. \tag{5.4}$$

We call $z \in \mathscr{B}_t$ a move belonging to $\mathscr{F}_t$.

The equivalence classes $\mathscr{F}_{t,1}, \ldots, \mathscr{F}_{t,K_t}$ in (5.3) can be understood as follows. Let $x, y \in \mathscr{F}_t$. Suppose that there exists z with $\deg z \le n-1$, such that

$$y = x + z = (x - z^-) + z^+.$$

Then $x - z^-$ is not a zero vector and $\operatorname{supp}(x - z^-) \neq \emptyset$. Because $\operatorname{supp}(x - z^-)$ is contained in both $\operatorname{supp}(x)$ and $\operatorname{supp}(y)$, we have

$$\operatorname{supp}(x) \cap \operatorname{supp}(y) \neq \emptyset.$$

Conversely if $\operatorname{supp}(x) \cap \operatorname{supp}(y) \neq \emptyset$ then $\deg(y - x) < n$ and $y - x \in \mathscr{B}_{n-1}$. We have shown that

$$\deg(y - x) < n \Leftrightarrow \operatorname{supp}(x) \cap \operatorname{supp}(y) \neq \emptyset.$$

Now if x and y are in the same $\mathscr{B}_{n-1}$-equivalence class $\mathscr{F}_{t,k}$, then there exists a sequence of states $x = x_0, x_1, \ldots, x_L = y$, such that $\deg(x_j - x_{j-1}) < n$, $j = 1, \ldots, L$, or equivalently $\operatorname{supp}(x_j) \cap \operatorname{supp}(x_{j-1}) \neq \emptyset$. Therefore the equivalence classes in (5.3) can be understood as the connected components of the following graph G. The set of vertices of G is the fiber $\mathscr{F}_t$ and G has an undirected edge between x and $y \in \mathscr{F}_t$ if and only if $\deg(y - x) < n$.

5.2 Structure of Minimal Markov Basis and Indispensable Moves

A Markov basis $\mathscr{B}$ is *minimal* if no proper subset of $\mathscr{B}$ is a Markov basis. A minimal Markov basis always exists, because from any Markov basis, we can remove redundant elements one by one, until none of the remaining elements can be removed any further. From this definition, a minimal Markov basis is not *sign-invariant* in the sense that for each $z \in \mathscr{B}$, $-z$ is not a member of $\mathscr{B}$ when $\mathscr{B}$ is a minimal Markov basis, because $-z$ can be omitted from $\mathscr{B}$ without affecting the connectivity.

At this point, we discuss the signs of moves in a Markov basis. In discussing minimality of Markov bases, it is sometimes more convenient if both (or neither of) z and $-z$ belong to a Markov basis. We call a set of moves $\mathscr{B}$ *sign-invariant* if $z \in \mathscr{B}$ implies $-z \in \mathscr{B}$. We call a sign-invariant Markov basis minimal if no proper sign-invariant subset of $\mathscr{B}$ is a Markov basis. If a sign-invariant Markov basis $\mathscr{B}$ is minimal, then for each move $z \in \mathscr{B}$, we can leave exactly one of z and $-z$ in the Markov basis and have a minimal Markov basis without the requirement of sign invariance. Conversely if $\mathscr{B}$ is a minimal Markov basis, then $\mathscr{B} \cup (-\mathscr{B})$, where $-\mathscr{B} = \{-z \mid z \in \mathscr{B}\}$ is a minimal sign-invariant Markov basis.

For each t, let $n = \deg t$ and consider the $\mathscr{B}_{n-1}$-equivalence classes of $\mathscr{F}_t$ in (5.3). Let $x_j \in \mathscr{F}_{t,j}$, $j = 1, \ldots, K_t$, be representative elements of the equivalence classes and let

$$z_{j_1,j_2} = x_{j_1} - x_{j_2}, \qquad j_1 \neq j_2$$

be a move connecting $\mathscr{F}_{t,j_1}$ and $\mathscr{F}_{t,j_2}$. Note that we can connect all equivalence classes with $K_t - 1$ moves of this type, by forming a tree, where the equivalence classes are interpreted as vertices and connecting moves are interpreted as edges of an undirected graph. Now we state the main theorem of this chapter. The following result was already known to algebraists in Theorem 2.5 of [28]. For the rest of this chapter we write $|t|$ for $\deg t$.

Theorem 5.1. *Let $\mathscr{B}$ be a minimal Markov basis. For each t, $\mathscr{B} \cap \mathscr{B}_t$ consists of $K_t - 1$ moves connecting different $\mathscr{B}_{|t|-1}$-equivalence classes of $\mathscr{F}_t$, in such a way that the equivalence classes are connected into a tree by these moves.*

Conversely choose any $K_t - 1$ moves $z_{t,1}, \ldots, z_{t,K_t-1}$ connecting different $\mathscr{B}_{|t|-1}$-equivalence classes of $\mathscr{F}_t$, in such a way that the equivalence classes are connected into a tree by these moves. Then

$$\mathscr{B} = \bigcup_{t:K_t \geq 2} \{z_{t,1}, \ldots, z_{t,K_t-1}\} \tag{5.5}$$

is a minimal Markov basis.

Proofs of this theorem and the following corollaries are given at the end of this section. Note that no move is needed from $\mathscr{F}_t$ with $K_t = 1$, including the case where $\mathscr{F}_t$ is a one-element set. If $\mathscr{F}_t = \{x\}$ is a one-element set, no nonzero move is applicable to x, but at the same time we do not need to move from x at all for such an $\mathscr{F}_t$.

In principle this theorem can be used to construct a minimal Markov basis from below as follows. As the initial step we consider t with the sample size $n = |t| = 1$. Because $\mathscr{B}_0$ consists only of the zero move $\mathscr{B}_0 = \{0\}$, each point $x \in \mathscr{F}_t$, $|t| = 1$, is isolated and forms an equivalence class by itself. For each t with $|t| = 1$, we choose

$K_t - 1$ degree 1 moves to connect K_t points of $\mathscr{F}_t$ into a tree. Let $\tilde{\mathscr{B}}_1$ be the set of chosen moves. $\tilde{\mathscr{B}}_1$ is a subset of the set $\mathscr{B}_1$ of all degree 1 moves. Every degree 1 move can be expressed by nonnegative integer combination of chosen degree 1 moves, thus it follows that $\tilde{\mathscr{B}}_1$ and $\mathscr{B}_1$ induce the same equivalence classes for each $\mathscr{F}_t$ with $|t| = 2$. Therefore as the second step we consider $\tilde{\mathscr{B}}_1$-equivalence classes of $\mathscr{F}_t$ for each t with $|t| = 2$ and choose representative elements from each equivalence class to form degree 2 moves connecting the equivalence classes into a tree. We add the chosen moves to $\tilde{\mathscr{B}}_1$ and form a set $\tilde{\mathscr{B}}_2$.

We can repeat this process for $n = |t| = 3, 4, \ldots$. By the Hilbert basis theorem there exists some n_0 such that for $n \geq n_0$ no new moves need to be added. Then a minimal Markov basis $\mathscr{B}$ of (5.5) is written as $\mathscr{B} = \tilde{\mathscr{B}}_{n_0}$. The difficulty with this approach is that the known theoretical upper bound for n_0 in Proposition 5.3 below is large.

Theorem 5.1 clarifies to what extent the minimal Markov basis is unique. If an equivalence class consists of more than one element, then any element can be chosen as the representative element of the equivalence class. Another indeterminacy is how to form a tree of the equivalence classes. In addition there exists a trivial indeterminacy of a Markov basis $\mathscr{B}$ in changing the signs of its elements.

We say that a minimal basis $\mathscr{B}$ is *unique* if all minimal bases differ only by sign changes of the elements; that is, if $\mathscr{B} \cup (-\mathscr{B})$ is the unique minimal sign-invariant Markov basis. In terms of binomials, a unique minimal Markov basis corresponds to a unique minimal system of binomial generators of a toric ideal. In view of Lemma 5.3 below, we have the following corollary to Theorem 5.1.

Corollary 5.1. *A minimal Markov basis is unique if and only if for each t, $\mathscr{F}_t$ itself constitutes one $\mathscr{B}_{|t|-1}$-equivalence class or $\mathscr{F}_t$ is a two-element set.*

In this corollary the importance of a two-element set $\mathscr{F}_t = \{x, y\}$ is suggested. Therefore we make the following definition.

Definition 5.1. A move $z = y - x$ is *indispensable* if $\mathscr{F}_t = \{x, y\}$ is a two-element set.

In this definition we are not assuming that the supports of x and y are disjoint. However $\min(x, y)$ is canceled in $z = y - x$. We also call a binomial $u^{z^+} - u^{z^-}$ indispensable if $z = z^+ - z^-$ is an indispensable move. The notion of the indispensable move was given in Takemura and Aoki [142]. Ohsugi and Hibi [110] proved some properties of indispensable moves, in particular for configurations arising from finite graphs. In Theorem 2.4 of [111], Ohsugi and Hibi showed that the set of indispensable binomials is characterized as the intersection of binomials in the reduced Gröbner bases with respect to all lexicographic term orders.

Using the notion of indispensability, we state another corollary, which is more convenient to use.

Corollary 5.2. *The unique minimal Markov basis exists if and only if the set of indispensable moves forms a Markov basis. In this case, the set of indispensable moves is the unique minimal Markov basis.*

From these corollaries it seems that the minimal Markov basis is unique only under special conditions. It is therefore of great interest whether the minimal Markov basis is unique for some standard problems in m-way ($m \geq 2$) contingency tables with fixed marginals. On the other hand for the simplest case of one-way contingency tables, the minimal Markov basis is not unique. These facts are confirmed in Sect. 5.4, Chaps. 8 and 9.

For the rest of this section we give a proof of Theorem 5.1 and its corollaries. We also state some lemmas, which are of some independent interest.

Lemma 5.1. *If a move $\boldsymbol{z}$ is applicable to at least one element of $\mathscr{F}_{\boldsymbol{t}}$, then*

$$\deg \boldsymbol{z} \leq |\boldsymbol{t}|, \tag{5.6}$$

where the equality holds if and only if $\boldsymbol{t} = A\boldsymbol{z}^+ = A\boldsymbol{z}^-$.

Proof. Let $\boldsymbol{z}$ be applicable to $\boldsymbol{x} \in \mathscr{F}_{\boldsymbol{t}}$. Then by (4.12), $x(\boldsymbol{i}) \geq z^-(\boldsymbol{i})$, $\forall \boldsymbol{i} \in \mathscr{I}$. Summing over $\mathscr{I}$ yields (5.6).

Concerning the equality, if $\boldsymbol{z}$ is applicable to $\boldsymbol{x} \in \mathscr{F}_{\boldsymbol{t}}$ and the equality holds in (5.6), then $x(\boldsymbol{i}) = z^-(\boldsymbol{i})$, $\forall \boldsymbol{i} \in \mathscr{I}$ and

$$\boldsymbol{t} = A\boldsymbol{x} = \sum_{\boldsymbol{i} \in \mathscr{I}} \boldsymbol{a}(\boldsymbol{i}) x(\boldsymbol{i}) = \sum_{\boldsymbol{i} \in \mathscr{I}} \boldsymbol{a}(\boldsymbol{i}) z^-(\boldsymbol{i}) = A\boldsymbol{z}^- .$$

Conversely if $\boldsymbol{t} = A\boldsymbol{z}^+ = A\boldsymbol{z}^-$, then $\deg \boldsymbol{z} = |\boldsymbol{t}|$ by the definition of $\deg \boldsymbol{z}$ and $|\boldsymbol{t}|$. □

Lemma 5.1 implies that in considering mutual accessibility between $\boldsymbol{x}, \boldsymbol{y} \in \mathscr{F}_{\boldsymbol{t}}$, we only need to consider moves of degree smaller than $|\boldsymbol{t}|$ or moves $\boldsymbol{z}$ with $\boldsymbol{t} = A\boldsymbol{z}^+ = A\boldsymbol{z}^-$. Lemma 5.1 also implies the following simple but useful fact.

Lemma 5.2. *Suppose that $\mathscr{F}_{\boldsymbol{t}} = \{\boldsymbol{x}, \boldsymbol{y}\}$ is a two-element set and suppose that the supports of $\boldsymbol{x}$ and $\boldsymbol{y}$ are disjoint. Then $K_{\boldsymbol{t}} = 2$ and $\boldsymbol{x}, \boldsymbol{y}$ are $\mathscr{B}_{|\boldsymbol{t}|-1}$-equivalence classes by themselves. Furthermore $\boldsymbol{z} = \boldsymbol{y} - \boldsymbol{x}$ belongs to each Markov basis.*

Proof. Suppose that $\boldsymbol{y}$ is accessible from $\boldsymbol{x}$ by $\mathscr{B}_{|\boldsymbol{t}|-1}$. Then there exists a nonzero move $\boldsymbol{z}$ with $\deg \boldsymbol{z} \leq |\boldsymbol{t}| - 1$ such that $\boldsymbol{z}$ is applicable to $\boldsymbol{x}$. If $\boldsymbol{x} + \boldsymbol{z} = \boldsymbol{y}$, then $\boldsymbol{z} = \boldsymbol{y} - \boldsymbol{x}$ and $\deg \boldsymbol{z} = |\boldsymbol{t}|$, because the supports of $\boldsymbol{x}$ and $\boldsymbol{y}$ are disjoint. Therefore $\boldsymbol{x} + \boldsymbol{z} \neq \boldsymbol{y}$ and $\mathscr{F}_{\boldsymbol{t}}$ contains a third element $\boldsymbol{x} + \boldsymbol{z}$, which is a contradiction. Therefore $\boldsymbol{y}$ and $\boldsymbol{x}$ are in different $\mathscr{B}_{|\boldsymbol{t}|-1}$-equivalence classes, implying that $\boldsymbol{y}$ and $\boldsymbol{x}$ are $\mathscr{B}_{|\boldsymbol{t}|-1}$-equivalence classes by themselves.

Now consider moving from $\boldsymbol{x}$ to $\boldsymbol{y}$. Because they are $\mathscr{B}_{|\boldsymbol{t}|-1}$-equivalence classes by themselves, no nonzero move $\boldsymbol{z}$ of degree $\deg \boldsymbol{z} < |\boldsymbol{t}|$ is applicable to $\boldsymbol{x}$. By Lemma 5.1, only moves $\boldsymbol{z}$ with $\boldsymbol{t} = A\boldsymbol{z}^+ = A\boldsymbol{z}^-$ are applicable to $\boldsymbol{x}$. If any such move is different from $\boldsymbol{y} - \boldsymbol{x}$, then as above $\mathscr{F}_{\boldsymbol{t}}$ contains a third element. It follows that in order to move from $\boldsymbol{x}$ to $\boldsymbol{y}$, we have to move by exactly one step using the move $\boldsymbol{z} = \boldsymbol{y} - \boldsymbol{x}$. Therefore $\boldsymbol{z}$ has to belong to any Markov basis. □

Lemma 5.2 can be slightly modified to yield the following result for the case where supports of $\boldsymbol{x}$ and $\boldsymbol{y}$ are not necessarily disjoint.

Lemma 5.3. *Suppose that $\mathscr{F}_{\boldsymbol{t}} = \{\boldsymbol{x}, \boldsymbol{y}\}$ is a two-element set. Then $\boldsymbol{z} = \boldsymbol{y} - \boldsymbol{x}$ belongs to each Markov basis.*

Proof. If the supports of $\boldsymbol{x}$ and $\boldsymbol{y}$ are disjoint, then the result is already contained in Lemma 5.2. Otherwise let $\boldsymbol{v} = \min(\boldsymbol{x}, \boldsymbol{y})$ and consider $\boldsymbol{y} - \boldsymbol{v}$ and $\boldsymbol{x} - \boldsymbol{v}$. Then the supports of $\boldsymbol{y} - \boldsymbol{v}$ and $\boldsymbol{x} - \boldsymbol{v}$ are disjoint and by Lemma 5.2 again

$$\boldsymbol{z} = (\boldsymbol{y} - \boldsymbol{v}) - (\boldsymbol{x} - \boldsymbol{v}) = \boldsymbol{y} - \boldsymbol{x}$$

belongs to each Markov basis. □

The following lemma concerns replacing a move by series of moves.

Lemma 5.4. *Let $\mathscr{B}$ be a set of moves and let $\boldsymbol{z}_0 \notin \mathscr{B}$ be another nonzero move. Assume that $\boldsymbol{z}_0^+$ is accessible from $\boldsymbol{z}_0^-$ by $\mathscr{B}$. Then for each $\boldsymbol{x}$, to which $\boldsymbol{z}_0$ is applicable, $\boldsymbol{x} + \boldsymbol{z}_0$ is accessible from $\boldsymbol{x}$ by $\mathscr{B}$.*

This lemma shows that if $\boldsymbol{z}_0^+$ is accessible from $\boldsymbol{z}_0^-$ by $\mathscr{B}$, then we can always replace $\boldsymbol{z}_0$ by a series of moves from $\mathscr{B}$.

Proof. Suppose that $\boldsymbol{z}_0$ is applicable to $\boldsymbol{x}$. Then we assume $\boldsymbol{x} \geq \boldsymbol{z}_0^-$ without loss of generality. By the definition of accessibility (cf. (5.1)), we can move from $\boldsymbol{z}_0^-$ to $\boldsymbol{z}_0^+$ by moves from $\mathscr{B}$ without causing negative elements on the way. Then the same sequence of moves can be applied to $\boldsymbol{x}$ without causing negative elements on the way, leading from $\boldsymbol{x}$ to $\boldsymbol{x} + \boldsymbol{z}_0$. □

Now we are ready to prove Theorem 5.1 and its corollaries.

Proof (Theorem 5.1). Let $\mathscr{B}$ be a minimal Markov basis. For each $\boldsymbol{z} \in \mathscr{B}_n \setminus (\mathscr{B} \cap \mathscr{B}_n)$, $\boldsymbol{z}^+$ is accessible from $\boldsymbol{z}^-$ by $\mathscr{B} \cap \mathscr{B}_n$, because no move of degree greater than n is applicable to $\boldsymbol{z}^+$ as stated in Lemma 5.1. Considering this fact and Lemma 5.4, it follows that $\mathscr{B}_n$ and $\mathscr{B} \cap \mathscr{B}_n$ induce the same equivalence classes in $\mathscr{F}_{\boldsymbol{t}}$, $|\boldsymbol{t}| = n+1$. Fix a particular $\boldsymbol{t}$. Write

$$\{\boldsymbol{z}_1, \ldots, \boldsymbol{z}_L\} = \mathscr{B} \cap \mathscr{B}_{\boldsymbol{t}},$$

where $\mathscr{B}_{\boldsymbol{t}}$ is the set of moves belonging to $\mathscr{F}_{\boldsymbol{t}}$ defined in (5.4). For any $j = 1, \ldots, L$, let

$$\boldsymbol{x} = \boldsymbol{z}_j^+, \qquad \boldsymbol{y} = \boldsymbol{z}_j^-.$$

If $\boldsymbol{x}$ and $\boldsymbol{y}$ are in the same $\mathscr{B}_{|\boldsymbol{t}|-1}$-equivalence class, then by Lemma 5.4, $\boldsymbol{z}_j$ can be replaced by a series of moves of lower degree from $\mathscr{B}$ and $\mathscr{B} \setminus \{\boldsymbol{z}_j\}$ remains to be a Markov basis. This contradicts the minimality of $\mathscr{B}$. Therefore $\boldsymbol{z}_j^+$ and $\boldsymbol{z}_j^-$ are in two different $\mathscr{B}_{|\boldsymbol{t}|-1}$-equivalence classes connecting them. Now we consider an undirected graph whose vertices are $\mathscr{B}_{|\boldsymbol{t}|-1}$-equivalence classes of $\mathscr{F}_{\boldsymbol{t}}$ and whose edges are moves $\boldsymbol{z}_1, \ldots, \boldsymbol{z}_L$. Considering that $\mathscr{B}$ is a Markov basis, and no move of degree greater than $|\boldsymbol{t}|$ is applicable to each element of $\mathscr{F}_{\boldsymbol{t}}$ as stated in Lemma 5.1, this graph is connected. On the other hand if the graph contains a cycle, then there exist z_j, such that z_j^+ and z_j^- are mutually accessible by $\mathscr{B} \setminus \{z_j\}$. By Lemma 5.4 again, this contradicts the minimality of $\mathscr{B}$. It follows that the graph is a tree. Because any tree with K_t vertices has $K_t - 1$ edges, $L = K_t - 1$.

Reversing the above argument, it is now easy to see that if $K_t - 1$ moves $z_{t,1}, \ldots, z_{t,K_t-1}$ connecting different $\mathscr{B}_{|t|-1}$-equivalence classes of $\mathscr{F}_t$ are chosen in such a way that the equivalence classes are connected into a tree by these moves, then

$$\mathscr{B} = \bigcup_{t:K_t \geq 2} \{z_{t,1}, \ldots, z_{t,K_t-1}\}$$

is a minimal Markov basis. □

Proof (Corollary 5.1). From our argument preceding Corollary 5.1, it follows that if the minimal Markov basis is unique, then for each t, $\mathscr{F}_t$ itself constitutes one $\mathscr{B}_{|t|-1}$-equivalence class or $\mathscr{F}_t$ is a two-element set $\{x_{t,1}, x_{t,2}\}$, such that $x_{t,1} \not\sim x_{t,2} \pmod{\mathscr{B}_{|t|-1}}$. Therefore we only need to prove the converse. Suppose that for each t, $\mathscr{F}_t$ itself constitutes one $\mathscr{B}_{|t|-1}$-equivalence class or $\mathscr{F}_t$ is a two-element set. By Lemma 5.3, for each two-element set $\mathscr{F}_t = \{x, y\}$, the move $z = y - x$ belongs to each Markov basis. However, by Theorem 5.1 each minimal Markov basis consists only of these moves. Therefore a minimal Markov basis is unique. □

Proof (Corollary 5.2). By Lemma 5.3, indispensable moves belong to each Markov basis. Therefore if the set of indispensable moves forms a Markov basis, then it is the unique minimal Markov basis.

On the other hand if the set of indispensable moves does not constitute a Markov basis, then there is a term with $K_t \geq 3$ in (5.5) and in this case a minimal Markov basis $\mathscr{B}$ is not unique as discussed after Theorem 5.1.

From these considerations it is obvious that if the unique minimal Markov basis exists, it coincides with the set of indispensable moves. □

5.3 Minimum Fiber Markov Basis

In this section we discuss the union of all minimal Markov bases and define the minimum fiber Markov basis. Let $z = z^+ - z^-$ be a move of degree n. We call z *nonreplaceable by lower degree moves* if

$$z^+ \not\sim z^- \pmod{\mathscr{B}_{n-1}}, \tag{5.7}$$

that is, if z connects different $\mathscr{B}_{n-1}$-equivalence classes of $\mathscr{F}_t \ni z^+$. Clearly an indispensable move is nonreplaceable by lower degree moves. Let

$$\mathscr{B}_{\mathrm{MF}} = \{z \mid z \text{ is nonreplaceable by lower degree moves}\}. \tag{5.8}$$

We call $\mathscr{B}_{\mathrm{MF}}$ the *minimum fiber Markov basis*. In actuality, we have the following fact.

Proposition 5.1. *$\mathscr{B}_{\mathrm{MF}}$ is the union of all minimal Markov bases.*

Proof. From the argument of the previous section, for all minimal Markov bases $\mathscr{B}$, the set of sufficient statistics

$$\{\boldsymbol{t} \mid \boldsymbol{t} = A\boldsymbol{z}^+ = A\boldsymbol{z}^-, \boldsymbol{z} \in \mathscr{B}\} = \{\boldsymbol{t} \mid \mathscr{F}_{\boldsymbol{t}} \text{ is not a single } \mathscr{B}_{|\boldsymbol{t}|-1}\text{-equivalence class}\}$$

is common and it is equal to the set of fibers of the moves in $\mathscr{B}_{\mathrm{MF}}$. A minimal Markov basis is constructed by arbitrarily choosing moves $\boldsymbol{z}$ connecting different $\mathscr{B}_{|\boldsymbol{t}|-1}$-equivalence classes of $\mathscr{F}_{\boldsymbol{t}}$ into a tree. Because $\mathscr{B}_{\mathrm{MF}}$ is the union of all these moves, it is the union of all minimal Markov bases. □

By construction the minimum fiber Markov basis is invariant in the sense of Chap. 7.

5.4 Examples of Minimal Markov Bases

5.4.1 One-Way Contingency Tables

We start with the simplest case of one-way contingency tables. Let $\boldsymbol{x} = (x_i)$ be an I-dimensional frequency vector and $A = (1,\dots,1) = \mathbf{1}'_I$. In this case, $\boldsymbol{t}$ is the sample size n. This situation corresponds to testing the homogeneity of mean parameters for I independent Poisson variables conditional on the total sample size n. In this case, a minimal Markov basis is formed as a set of $I-1$ degree 1 moves, but is not unique.

A minimal Markov basis is constructed as follows. First consider the case of $n = |\boldsymbol{t}| = 1$. There are I elements in $\mathscr{F}_{\boldsymbol{t}}$ as

$$\mathscr{F}_{\boldsymbol{t}} = \{\boldsymbol{e}_1,\dots,\boldsymbol{e}_I\}.$$

Each element $\boldsymbol{x} \in \mathscr{F}_{\boldsymbol{t}}$ forms an equivalence class by itself. To connect these points into a tree, there are I^{I-2} ways of choosing $I-1$ degree 1 moves by Cayley's theorem (see, e.g., Chap. 4 of Wilson [149]). One example is

$$\mathscr{B} = \{\boldsymbol{e}_1 - \boldsymbol{e}_2, \boldsymbol{e}_2 - \boldsymbol{e}_3, \dots, \boldsymbol{e}_{I-1} - \boldsymbol{e}_I\}.$$

It is easily verified that no move of degree larger than 1 is needed.

5.4.2 Independence Model of Two-Way Contingency Tables

The next example is the independence model of two-way contingency tables discussed in Chaps. 1 and 2. In Theorem 2.1 we have shown that the set of degree 2 moves displayed as

$$\boxed{\begin{array}{cc} +1 & -1 \\ -1 & +1 \end{array}} \tag{5.9}$$

is a Markov basis. In addition, this is the unique minimal Markov basis.

Indeed, consider marginal frequencies, where the ith row sum, the i'th row sum, the jth column sum, and the j'th column sum are ones and all other marginal frequencies are zeros. Then the relevant marginal frequencies are displayed as follows.

$$\begin{array}{c|c|c|c} & j & j' & \\ \hline i & & & 1 \\ \hline i' & & & 1 \\ \hline & 1 & 1 & 2 \end{array}.$$

Clearly there are only two elements of this fiber:

$$\boxed{\begin{array}{cc} 1 & 0 \\ 0 & 1 \end{array}}, \quad \boxed{\begin{array}{cc} 0 & 1 \\ 1 & 0 \end{array}}.$$

The move in (5.9) is the difference of these two elements and hence it is indispensable.

5.4.3 The Unique Minimal Markov Basis for the Lawrence Lifting

One remarkable example of the existence of the unique minimal Markov basis is the Lawrence lifting $\Lambda(A)$ in Sect. 4.6. In Proposition 4.3 we gave the Graver basis of $\Lambda(A)$. We now show that it is actually the unique minimal Markov basis of $\Lambda(A)$, by showing that each move in (4.25) is indispensable. This also shows that the Graver basis for the configuration A is finite, because a minimal Markov basis for $\Lambda(A)$ is finite.

Proposition 5.2. *For the Lawrence lifting $\Lambda(A)$, the Graver basis given by (4.25) is the unique minimal Markov basis*

Proof. Let $\boldsymbol{z} = \boldsymbol{z}^+ - \boldsymbol{z}^-$ be a conformally primitive move for A and let $\boldsymbol{t} = A\boldsymbol{z}^+$. Consider the fiber of

$$\begin{pmatrix} \boldsymbol{z} \\ -\boldsymbol{z} \end{pmatrix}$$

for $\Lambda(A)$. Note that

$$\begin{pmatrix} z \\ -z \end{pmatrix}^+ = \begin{pmatrix} z^+ \\ z^- \end{pmatrix}, \quad \begin{pmatrix} z \\ -z \end{pmatrix}^- = \begin{pmatrix} z^- \\ z^+ \end{pmatrix}.$$

Then

$$\begin{pmatrix} t \\ t \\ z^+ + z^- \end{pmatrix} = \Lambda(A) \begin{pmatrix} z^+ \\ z^- \end{pmatrix} = \Lambda(A) \begin{pmatrix} z^- \\ z^+ \end{pmatrix}, \qquad \text{where} \quad \Lambda(A) = \begin{pmatrix} A & 0 \\ 0 & A \\ E_\eta & E_\eta \end{pmatrix}.$$

Now suppose that

$$\begin{pmatrix} t \\ t \\ z^+ + z^- \end{pmatrix} = \Lambda(A) \begin{pmatrix} x \\ y \end{pmatrix} = \begin{pmatrix} Ax \\ Ay \\ x+y \end{pmatrix}.$$

Then $z^+, z^-, x, y \in \mathscr{F}_t$ for the configuration A. Note that $\mathrm{supp}(z^+) \cap \mathrm{supp}(z^-) = \emptyset$. Now decomposing x, y into these disjoint supports write

$$x = x_1 + x_2, \qquad y = y_1 + y_2, \quad \text{s.t. } \mathrm{supp}(x_1), \mathrm{supp}(y_1) \subset \mathrm{supp}(z^+),$$
$$\mathrm{supp}(x_2), \mathrm{supp}(y_2) \subset \mathrm{supp}(z^-).$$

Then by $z^+ + z^- = x + y$ we have

$$z^+ = x_1 + y_1, \qquad z^- = x_2 + y_2.$$

This implies

$$t = Ax_1 + Ay_1 = Ax_2 + Ay_2.$$

On the other hand $t = Ax = Ax_1 + Ax_2$. Similarly $t = Ay_1 + Ay_2$. Hence

$$\mathbf{0} = (Ax_1 + Ay_1) - (Ax_1 + Ax_2) = A(y_1 - x_2).$$

Similarly $\mathbf{0} = A(x_1 - y_2)$. Then

$$\begin{aligned} z = z^+ - z^- &= (x_1 + y_1) - (x_2 + y_2) \\ &= (y_1 - x_2) + (x_1 - y_2), \end{aligned}$$

which is a conformal sum. By primitiveness of z, either $y_1 - x_2 = 0$ or $x_1 - y_2 = 0$. It follows that

$$\left\{ \begin{pmatrix} z^+ \\ z^- \end{pmatrix}, \begin{pmatrix} z^- \\ z^+ \end{pmatrix} \right\}$$

is a two-element fiber for $\Lambda(A)$. □

Concerning the finiteness of the Graver basis, we cite the following important theoretical upper bound to the highest degree of elements in the Graver basis. In Definition 3.3 the toric ideal is defined in terms of $\ker_{\mathbb{Z}} A$. When we remove linearly dependent rows from the configuration A, $\ker_{\mathbb{Z}} A$ does not change. Hence we can assume without loss of generality that the rows of $A : \nu \times \eta$ are linearly independent.

Proposition 5.3 (Corollary 4.15 of Sturmfels [139]). *The degree of primitive moves for the configuration A with linearly independent rows is bounded from above by*

$$\frac{1}{2}(\nu+1)(\eta-\nu)D(A),$$

where

$$D(A) = \max\{|\det(\boldsymbol{a}_{i_1},\ldots,\boldsymbol{a}_{i_\nu})| \mid 1 \le i_1 < \cdots < i_\nu \le \eta\}$$

is the maximum of the absolute value of the determinants of $\nu \times \nu$ submatrices of A.

For the case that the semigroup generated by the columns of A is normal, the following much better upper bound on Markov bases is known.

Proposition 5.4 (Theorem 13.14 of Sturmfels [139]). *Let $A : \nu \times \eta$ be a configuration such that the semigroup generated by columns of A is normal. Then the toric ideal I_A is generated by binomials of degree at most ν.*

5.5 Indispensable Monomials

Extending the notion of indispensable moves, in this section we define indispensable monomials of a toric ideal and establish some of their properties. Indispensable monomials were introduced in [19]. They are useful for searching for indispensable binomials of a toric ideal and for proving the existence or nonexistence of a unique minimal system of binomial generators of a toric ideal. In this section we identify a frequency vector $\boldsymbol{x}$ with a monomial $\boldsymbol{u}^{\boldsymbol{x}}$ and use two notations interchangeably.

First we define an *indispensable monomial*. Hereafter, we say that a Markov basis $\mathscr{B}$ contains $\boldsymbol{x}$ if it contains a move $\boldsymbol{z}$ containing $\boldsymbol{x}$ (i.e., $\boldsymbol{z}^+ = \boldsymbol{x}$ or $\boldsymbol{z}^- = \boldsymbol{x}$ holds) by abusing the terminology.

Definition 5.2. A monomial $\boldsymbol{u}^{\boldsymbol{x}}$ is *indispensable* if every system of binomial generators of I_A contains a binomial f such that $\boldsymbol{u}^{\boldsymbol{x}}$ is a term of f.

We also call a frequency vector $\boldsymbol{x}$ indispensable if $\boldsymbol{u}^{\boldsymbol{x}}$ is an indispensable monomial. From this definition, any Markov basis contains all indispensable monomials. Therefore the set of indispensable monomials is finite. Note that both terms of an indispensable binomial $\boldsymbol{u}^{\boldsymbol{z}^+} - \boldsymbol{u}^{\boldsymbol{z}^-}$ are indispensable monomials, but the converse does not hold in general.

Now we present an alternative definition.

Definition 5.3. A frequency vector $\boldsymbol{x}$ is a minimal multielement if $|\mathscr{F}_{A\boldsymbol{x}}| \geq 2$ and $|\mathscr{F}_{A(\boldsymbol{x}-\boldsymbol{e}_i)}| = 1$ for every $\boldsymbol{i} \in \text{supp}(\boldsymbol{x})$.

Here $\boldsymbol{x} - \boldsymbol{e}_i$ is the frequency vector obtained by subtracting one frequency from $\boldsymbol{x}$ in the cell $\boldsymbol{i}$.

Proposition 5.5. *$\boldsymbol{x}$ is an indispensable monomial if and only if $\boldsymbol{x}$ is a minimal multielement.*

Proof. First, we suppose that $\boldsymbol{x}$ is a minimal multielement and want to show that it is an indispensable monomial. Let $n = \deg \boldsymbol{x}$ and consider the fiber $\mathscr{F}_{A\boldsymbol{x}}$. We claim that $\{\boldsymbol{x}\}$ forms a single $\mathscr{B}_{n-1}$-equivalence class. In order to show this, we argue by contradiction. If $\{\boldsymbol{x}\}$ does not form a single $\mathscr{B}_{n-1}$-equivalence class, then there exists a move $\boldsymbol{z}$ with degree less than or equal to $n-1$, such that

$$\boldsymbol{x}+\boldsymbol{z} = (\boldsymbol{x}-\boldsymbol{z}^-)+\boldsymbol{z}^+ \in \mathscr{F}_{A\boldsymbol{x}}.$$

Inasmuch as $\deg \boldsymbol{x} = n$, $\deg \boldsymbol{z} \leq n-1$, we have $\boldsymbol{0} \neq \boldsymbol{x} - \boldsymbol{z}^-$ and

$$\text{supp}(\boldsymbol{x}) \cap \text{supp}(\boldsymbol{x}+\boldsymbol{z}) \neq \emptyset.$$

Therefore we can choose $\boldsymbol{i} \in \text{supp}(\boldsymbol{x}) \cap \text{supp}(\boldsymbol{x}+\boldsymbol{z})$ such that

$$A(\boldsymbol{x}-\boldsymbol{e}_i) = A(\boldsymbol{x}+\boldsymbol{z}-\boldsymbol{e}_i), \quad \boldsymbol{x}-\boldsymbol{e}_i \neq \boldsymbol{x}+\boldsymbol{z}-\boldsymbol{e}_i.$$

This shows that $|\mathscr{F}_{A(\boldsymbol{x}-\boldsymbol{e}_i)}| \geq 2$, which contradicts the assumption that $\boldsymbol{x}$ is a minimal multielement. Therefore we have shown that $\{\boldsymbol{x}\}$ forms a single $\mathscr{B}_{n-1}$-equivalence class. Because we are assuming that $|\mathscr{F}_{A\boldsymbol{x}}| \geq 2$, there exists some other $\mathscr{B}_{n-1}$-equivalence class in $\mathscr{F}_{A\boldsymbol{x}}$. By Theorem 5.1 each Markov basis has to contain a move connecting $\{\boldsymbol{x}\}$ to another equivalence class of $\mathscr{F}_{A\boldsymbol{x}}$, which implies that each Markov basis has to contain a move $\boldsymbol{z}$ containing $\boldsymbol{x}$. We now have shown that each minimal multielement has to be contained in each Markov basis; that is, a minimal multielement is an indispensable monomial.

Now we show the converse. It suffices to show that if $\boldsymbol{x}$ is not a minimal multielement, then $\boldsymbol{x}$ is a dispensable monomial. Suppose that $\boldsymbol{x}$ is not a minimal multielement. If $\boldsymbol{x}$ is a 1-element ($|\mathscr{F}_{A\boldsymbol{x}}| = 1$), obviously it is dispensable. Hence assume $|\mathscr{F}_{A\boldsymbol{x}}| \geq 2$. In the case that $\mathscr{F}_{A\boldsymbol{x}}$ is a single $\mathscr{B}_{n-1}$-equivalence class, no move containing $\boldsymbol{x}$ is needed in a minimal Markov basis by Theorem 5.1. Therefore we only need to consider the case that $\mathscr{F}_{A\boldsymbol{x}}$ contains more than one $\mathscr{B}_{n-1}$-equivalence class. Because $\boldsymbol{x}$ is not a minimal multielement, there exists some $\boldsymbol{i} \in \text{supp}(\boldsymbol{x})$ such that $|\mathscr{F}_{A(\boldsymbol{x}-\boldsymbol{e}_i)}| \geq 2$. Then there exists $\boldsymbol{y} \neq \boldsymbol{x} - \boldsymbol{e}_i$, such that $A\boldsymbol{y} = A(\boldsymbol{x}-\boldsymbol{e}_i)$. Noting that $\deg \boldsymbol{y} = \deg(\boldsymbol{x}-\boldsymbol{e}_i) = n-1$, a move of the form

$$\boldsymbol{z} = \boldsymbol{y} - (\boldsymbol{x}-\boldsymbol{e}_i) = (\boldsymbol{y}+\boldsymbol{e}_i) - \boldsymbol{x}$$

satisfies $0 < \deg z \le n-1$. Then

$$y + e_i = x + z$$

and x and $y + e_i$ belong to the same $\mathscr{B}_{n-1}$-equivalence class of $\mathscr{F}_{Ax}$. Because $x \neq y + e_i$, Theorem 5.1 states that we can construct a minimal Markov basis containing $y + e_i$, but not containing x. Therefore x is a dispensable monomial. □

We give yet another definition.

Definition 5.4. x is a minimal i-lacking 1-element if $|\mathscr{F}_{Ax}| = 1$, $|\mathscr{F}_{A(x+e_i)}| \ge 2$, and $|\mathscr{F}_{A(x+e_i-e_j)}| = 1$ for each $j \in \mathrm{supp}(x)$.

We then have the following result.

Proposition 5.6. *The following three conditions are equivalent: (1) x is an indispensable monomial, (2) for each $i \in \mathrm{supp}(x)$, $x - e_i$ is a minimal i-lacking 1-element, (3) for some $i \in \mathrm{supp}(x)$, $x - e_i$ is a minimal i-lacking 1-element.*

By the previous theorem we can replace the condition (1) by the condition that x is a minimal multielement.

Proof. (1) ⇒ (2). Suppose that x is a minimal multielement. Then for any $i \in \mathrm{supp}(x)$, $x - e_i$ is a 1-element and $|\mathscr{F}_{A((x-e_i)+e_i)}| = |\mathscr{F}_{Ax}| \ge 2$. If $x - e_i$ is not a minimal i-lacking 1-element, then for some $j \in \mathrm{supp}(x - e_i)$, $|\mathscr{F}_{A(x-e_j)}| \ge 2$. However, $j \in \mathrm{supp}(x - e_i) \subset \mathrm{supp}(x)$ and $|\mathscr{F}_{A(x-e_j)}| \ge 2$ contradicts the assumption that x is a minimal multielement.

It is obvious that (2) ⇒ (3).

Finally we prove (3) ⇒ (1). Suppose that for some $i \in \mathrm{supp}(x)$, $x - e_i$ is a minimal i-lacking 1-element. Note that $|\mathscr{F}_{Ax}| = |\mathscr{F}_{A((x-e_i)+e_i)}| \ge 2$. Now consider $j \in \mathrm{supp}(x)$. If $j \in \mathrm{supp}(x - e_i)$ then $|\mathscr{F}_{A(x-e_j)}| = |\mathscr{F}_{A((x-e_i)+e_i-e_j)}| = 1$. On the other hand if $j \notin \mathrm{supp}(x - e_i)$, then $j = i$ because $j \in \mathrm{supp}(x)$. In this case $|\mathscr{F}_{A(x-e_j)}| = 1$. This shows that x is a minimal multielement. □

Proposition 5.6 suggests the following: Find any 1-element x. It is often the case that each e_i, $i = 1, \dots, \eta$, is a 1-element. Randomly choose $1 \le i \le \eta$ and check whether $x + e_i$ remains to be a 1-element. Once $|\mathscr{F}_{x+e_i}| \ge 2$, then subtract e_js, $j \neq i$, one by one from x such that it becomes a minimal i-lacking 1-element. We can apply this procedure for finding indispensable monomials of some actual statistical problem.

We illustrate this procedure with an example of a $2 \times 2 \times 2$ contingency table. Consider the following problem where $\eta = 8$, $\nu = 4$, and A is given as

$$A = \begin{pmatrix} 1\,1\,1\,1\,1\,1\,1\,1 \\ 1\,1\,1\,1\,0\,0\,0\,0 \\ 1\,1\,0\,0\,1\,1\,0\,0 \\ 1\,0\,1\,0\,1\,0\,1\,0 \end{pmatrix}.$$

We write indeterminates as

$$\boldsymbol{u} = (u_{111}, u_{112}, u_{121}, u_{122}, u_{211}, u_{212}, u_{221}, u_{222}).$$

To find indispensable monomials for this problem, we start with the monomial $\boldsymbol{u}^{\boldsymbol{x}} = u_{111}$ and consider $\boldsymbol{x} + \boldsymbol{e}_{\boldsymbol{i}}, \boldsymbol{i} \in \mathscr{I}$. Then we see that

- $u_{111}^2, u_{111}u_{112}, u_{111}u_{121}, u_{111}u_{211}$ are 1-element monomials
- $u_{111}u_{122}, u_{111}u_{212}, u_{111}u_{221}$ are 2-element monomials
- $u_{111}u_{222}$ is a 4-element monomial.

From this, we found four indispensable monomials, $u_{111}u_{122}, u_{111}u_{212}, u_{111}u_{221}$, and $u_{111}u_{222}$, because each of $u_{122}, u_{212}, u_{221}, u_{222}$ is a 1-element monomial.

Starting from the other monomials, similarly, we can find the following list of indispensable monomials:

- $u_{111}u_{122}, u_{111}u_{212}, u_{111}u_{221}, u_{112}u_{121}, u_{112}u_{211}, u_{112}u_{222}, u_{121}u_{222}, u_{121}u_{211}, u_{122}u_{221}, u_{122}u_{212}, u_{211}u_{222}, u_{212}u_{221}$, each of which is a 2-element monomial.
- $u_{111}u_{222}, u_{112}u_{221}, u_{121}u_{212}, u_{122}u_{211}$, each of which is a 4-element monomial.

The next step is to consider the newly produced 1-element monomials, $u_{111}^2, u_{111}u_{112}, u_{111}u_{121}, u_{111}u_{211}$, and so on. For each of these monomials, consider adding $\boldsymbol{e}_{\boldsymbol{i}}, \boldsymbol{i} \in \mathscr{I}$ one by one, checking whether they are multielement. For example, we see that the monomials such as

$$u_{111}^3, u_{111}^2u_{112}, u_{111}u_{112}^2, \ldots$$

are again 1-element monomials (and we have to consider these 1-element monomials in the next step). On the other hand, monomials such as

$$u_{111}^2u_{122}, u_{111}u_{112}u_{122}, u_{111}^2u_{222}, u_{111}u_{112}u_{221}, \ldots$$

are multielement monomials. However, it is seen that they are not minimal multielement, inasmuch as

$$u_{111}u_{122}, u_{112}u_{122}, u_{111}u_{222}, u_{112}u_{221}, \ldots$$

are not 1-element monomials. To find all indispensable monomials for this problem, we have to repeat the above procedure for monomials of degree $4, 5, \ldots$. Indispensable monomials belong to any Markov basis, in particular to the Graver basis, therefore Proposition 5.3 again gives an upper bound for the degree of indispensable monomials and we can stop at this bound.

We mention that analogous to Theorem 2.4 in [111], the set of indispensable monomials is characterized as the intersection of monomials in reduced Gröbner bases with respect to all lexicographic term orders. Further characterizations of indispensable binomials and indispensable monomials are given in [31, 63, 115].

Chapter 6
Method of Distance Reduction

6.1 Distance Reducing Markov Bases

Throughout this book, we use the method of distance reduction (or a distance-reducing argument) of Takemura and Aoki [143] for finding a Markov basis for a given configuration. We have already seen a typical example in Sect. 2.1 for proving that the set of basic moves in (2.4) forms a Markov basis for the independence model of $I \times J$ contingency tables. In this section we first formalize the idea of distance reduction by a set of moves.

Consider a metric $d(\boldsymbol{x},\boldsymbol{y})$ on a fiber $\mathscr{F}_{\boldsymbol{t}}$. Although we are mainly concerned with the 1-norm in the following, here we consider a general metric. A metric $d = d_{\boldsymbol{t}}$ on $\mathscr{F}_{\boldsymbol{t}}$ can be defined in various ways. If a metric d is defined on the whole space of frequency vectors $\mathscr{X} = \mathbb{N}^{|\mathscr{I}|}$, we can consider the restriction of d to each $\mathscr{F}_{\boldsymbol{t}}$

$$d_{\boldsymbol{t}}(\boldsymbol{x},\boldsymbol{y}) = d(\boldsymbol{x},\boldsymbol{y}), \quad \boldsymbol{x},\boldsymbol{y} \in \mathscr{F}_{\boldsymbol{t}}.$$

If $d(\cdot)$ is a norm on the set $\mathbb{Z}^{|\mathscr{I}|}$ of integer vectors, such as the 1-norm $|\boldsymbol{z}| = \sum_{\boldsymbol{i}\in\mathscr{I}} |z(\boldsymbol{i})|$, $d_{\boldsymbol{t}}$ is defined by $d_{\boldsymbol{t}}(\boldsymbol{x},\boldsymbol{y}) = d(\boldsymbol{x}-\boldsymbol{y})$. For notational simplicity we suppress the subscript $\boldsymbol{t}$ in $d_{\boldsymbol{t}}$ hereafter.

Now we introduce the notion of a distance reduction by a set of moves. Let $\mathscr{B}$ be a set of moves. We call $\mathscr{B}$ *d-reducing* for $\boldsymbol{x},\boldsymbol{y} \in \mathscr{F}_{\boldsymbol{t}}$ if there exists an element $\boldsymbol{z} \in \mathscr{B}$ and $\varepsilon = \pm 1$ such that $\varepsilon\boldsymbol{z}$ is applicable to $\boldsymbol{x}$ or $\boldsymbol{y}$ and we can decrease the distance; that is,

$$\begin{aligned} &\boldsymbol{x}+\varepsilon\boldsymbol{z} \in \mathscr{F}_{\boldsymbol{t}} \quad \text{and} \quad d(\boldsymbol{x}+\varepsilon\boldsymbol{z},\boldsymbol{y}) < d(\boldsymbol{x},\boldsymbol{y}), \qquad \text{or} \\ &\boldsymbol{y}+\varepsilon\boldsymbol{z} \in \mathscr{F}_{\boldsymbol{t}} \quad \text{and} \quad d(\boldsymbol{x},\boldsymbol{y}+\varepsilon\boldsymbol{z}) < d(\boldsymbol{x},\boldsymbol{y}). \end{aligned} \tag{6.1}$$

We simply call $\mathscr{B}$ *d-reducing* if $\mathscr{B}$ is d-reducing for every $\boldsymbol{x},\boldsymbol{y}(\neq \boldsymbol{x}) \in \mathscr{F}_{\boldsymbol{t}}$ and for every $\boldsymbol{t}$. Alternatively we say that $\mathscr{B}$ is *norm-reducing* if it is clear which metric d is used in the context.

S. Aoki et al., *Markov Bases in Algebraic Statistics*, Springer Series in Statistics 199, DOI 10.1007/978-1-4614-3719-2_6,

We call $\mathscr{B}$ *strongly d-reducing* for $\boldsymbol{x}, \boldsymbol{y} \in \mathscr{F}_{\boldsymbol{t}}$ if there exist elements $z_1, z_2 \in \mathscr{B}$ and $\varepsilon_1, \varepsilon_2 = \pm 1$ such that $\boldsymbol{x} + \varepsilon_1 z_1 \in \mathscr{F}_{\boldsymbol{t}}$, $\boldsymbol{y} + \varepsilon_2 z_2 \in \mathscr{F}_{\boldsymbol{t}}$ and

$$d(\boldsymbol{x} + \varepsilon_1 z_1, \boldsymbol{y}) < d(\boldsymbol{x}, \boldsymbol{y}) \quad \text{and} \quad d(\boldsymbol{x}, \boldsymbol{y} + \varepsilon_2 z_2) < d(\boldsymbol{x}, \boldsymbol{y}). \tag{6.2}$$

We call $\mathscr{B}$ *strongly d-reducing* if $\mathscr{B}$ is strongly d-reducing for every $\boldsymbol{x}, \boldsymbol{y}(\neq \boldsymbol{x}) \in \mathscr{F}_{\boldsymbol{t}}$ and for every $\boldsymbol{t}$. Clearly if $\mathscr{B}$ is strongly d-reducing, then $\mathscr{B}$ is d-reducing.

The following fact on d-reducing set of moves $\mathscr{B}$ is obvious, but very useful.

Proposition 6.1. *Let a metric d be given on each fiber $\mathscr{F}_{\boldsymbol{t}}$. A set of finite moves $\mathscr{B}$ is a Markov basis if it is d-reducing.*

Instead of a formal proof, we give the following argument on how two states are connected by a set of moves from $\mathscr{B}$.

If $\mathscr{B}$ is d-reducing, then for every $\boldsymbol{x}, \boldsymbol{y}(\neq \boldsymbol{x}) \in \mathscr{F}_{\boldsymbol{t}}$, there exist $k > 0$, $\varepsilon_l = \pm 1$, $z_l \in \mathscr{B}$, $\boldsymbol{x}_l \in \mathscr{F}_{\boldsymbol{t}}$, $\boldsymbol{y}_l \in \mathscr{F}_{\boldsymbol{t}}$, $l = 1, \ldots, k$, with the following properties.

(i) $\boldsymbol{x}_k = \boldsymbol{y}_k$.
(ii) $d(\boldsymbol{x}_l, \boldsymbol{y}_l) < d(\boldsymbol{x}_{l-1}, \boldsymbol{y}_{l-1})$, $l = 1, \ldots, k$, where $\boldsymbol{x}_0 = \boldsymbol{x}$ and $\boldsymbol{y}_0 = \boldsymbol{y}$.
(iii) $(\boldsymbol{x}_l, \boldsymbol{y}_l) = (\boldsymbol{x}_{l-1} + \varepsilon_l z_l, \boldsymbol{y}_{l-1})$ or $(\boldsymbol{x}_l, \boldsymbol{y}_l) = (\boldsymbol{x}_{l-1}, \boldsymbol{y}_{l-1} + \varepsilon_l z_l)$, $l = 1, \ldots, k$.

Given the above sequence of frequency vectors, we can move from $\boldsymbol{x}$ to $\boldsymbol{x}_k = \boldsymbol{y}_k$ and then reversing the moves we can move from $\boldsymbol{y}_k$ to $\boldsymbol{y}$. Thus $\boldsymbol{y}$ is accessible from $\boldsymbol{x}$ by $\mathscr{B}$. Note that in this sequence of moves the distances

$$d(\boldsymbol{x}, \boldsymbol{x}_1), \ldots, d(\boldsymbol{x}, \boldsymbol{x}_k), d(\boldsymbol{x}, \boldsymbol{y}_{k-1}) \ldots, d(\boldsymbol{x}, \boldsymbol{y})$$

might not be monotone increasing.

On the other hand, if $\mathscr{B}$ is strongly d-reducing, then starting from $\boldsymbol{y}$, we can always decrease the distance by moving from the side of $\boldsymbol{y}$; that is, we can find $k > 0$ and $\boldsymbol{y} = \boldsymbol{y}_0, \boldsymbol{y}_1, \ldots, \boldsymbol{y}_{k-1}, \boldsymbol{y}_k = \boldsymbol{x}$ in $\mathscr{F}_{\boldsymbol{t}}$ such that $\boldsymbol{y}_l = \boldsymbol{y}_{l-1} + \varepsilon_l z_l$, $\varepsilon_l = \pm 1$, $z_l \in \mathscr{B}$, $l = 1, \ldots, k$, and

$$d(\boldsymbol{x}, \boldsymbol{y}_{k-1}), d(\boldsymbol{x}, \boldsymbol{y}_{k-2}), \ldots, d(\boldsymbol{x}, \boldsymbol{y})$$

are strictly increasing.

Note that a Markov basis is not necessarily d-reducing. By a Markov basis, we can connect any two states $\boldsymbol{x}, \boldsymbol{y}$ in the same fiber as

$$\boldsymbol{x} = \boldsymbol{x}_0 \to \boldsymbol{x}_1 \to \cdots \to \boldsymbol{x}_{k-1} \to \boldsymbol{x}_k = \boldsymbol{y}$$

by the moves in a Markov basis. Hence by moving from $\boldsymbol{x}$ to $\boldsymbol{y}$, we can eventually decrease the distance between $\boldsymbol{x}$ and $\boldsymbol{y}$ to 0. However, in difficult cases, we might need to make a detour, so that initially the distance increases or stays the same as

$$d(\boldsymbol{x}, \boldsymbol{y}) \le d(\boldsymbol{x}_1, \boldsymbol{y}) \quad \text{and} \quad d(\boldsymbol{x}, \boldsymbol{y}) \le d(\boldsymbol{x}, \boldsymbol{x}_{k-1}).$$

6.2 Examples of Distance-Reducing Proofs

In this section we give two examples of distance-reducing arguments. The first example is a minimal Markov basis for a complete independence model of three-way contingency tables. The second example is a minimal Markov basis for the Hardy–Weinberg model. These examples are treated again in Sect. 7.2 from the viewpoint of symmetry of Markov bases.

6.2.1 The Complete Independence Model of Three-Way Contingency Tables

Consider $I \times J \times K$ contingency tables. Under the complete independence model the probability of the cell (i, j, k) is written as

$$p_{ijk} = p_{i++}p_{+j+}p_{++k},$$

where $p_{i++}, p_{+j+}, p_{++k}$ denote one-dimensional marginal probabilities. With lexicographic ordering of indices, the configuration A for the complete independence model of three-way tables is written as

$$A = \begin{pmatrix} \mathbf{1}'_I \otimes \mathbf{1}'_J \otimes E_K \\ \mathbf{1}'_I \otimes E_J \otimes \mathbf{1}'_K \\ E_I \otimes \mathbf{1}'_J \otimes \mathbf{1}'_K \end{pmatrix}.$$

A sufficient statistic consists of one-dimensional marginal frequencies. In this case, we construct a minimal Markov basis as follows.

There are two obvious patterns of moves of degree 2. An example of moves of type I is

$$z_{111} = z_{222} = 1, \quad z_{211} = z_{122} = -1,$$

with the other elements being 0. For the case of a $2 \times 2 \times 2$ table, this move can be displayed as follows.

$$\boxed{\begin{matrix} +1 & 0 \\ 0 & -1 \end{matrix}} \quad \boxed{\begin{matrix} -1 & 0 \\ 0 & +1 \end{matrix}}\ .$$

All the other moves of type I are obtained by permutation of indices or axes.

An example of moves of type II is

$$z_{111} = z_{122} = 1, \quad z_{112} = z_{121} = -1,$$

with the other elements being 0. For the case of a $2 \times 2 \times 2$ table, this move can be displayed as follows.

$$\begin{array}{|cc|}\hline +1 & -1 \\ -1 & +1 \\ \hline\end{array} \quad \begin{array}{|cc|}\hline 0 & 0 \\ 0 & 0 \\ \hline\end{array}\ .$$

All the other moves of type II are obtained by permutation of indices or axes. Let $\mathscr{B}^*$ be the set of type I and type II degree 2 moves. Then we have the following proposition.

Proposition 6.2. *$\mathscr{B}^*$ is a Markov basis for three-way contingency tables with fixed one-dimensional marginals.*

Proof. In this problem it is obvious that no degree 1 move is applicable to any frequency vector. Furthermore it is easy to verify that every degree 2 move is either of type I or type II. It remains to verify that for $\deg \boldsymbol{t} \geq 3$, $\mathscr{F}_{\boldsymbol{t}}$ itself constitutes one $\mathscr{B}^*$-equivalence class. Suppose that for some $\boldsymbol{t}$, $\mathscr{F}_{\boldsymbol{t}}$ consists of more than one $\mathscr{B}^*$-equivalence class. Let $\mathscr{F}_1, \mathscr{F}_2$ denote two different $\mathscr{B}^*$-equivalence classes. Choose $\boldsymbol{x} \in \mathscr{F}_1, \boldsymbol{y} \in \mathscr{F}_2$ such that

$$|\boldsymbol{z}| = |\boldsymbol{y} - \boldsymbol{x}| = \sum_{i,j,k} |y_{ijk} - x_{ijk}|$$

is minimized. Because $\boldsymbol{x}$ and $\boldsymbol{y}$ are chosen from different $\mathscr{B}^*$-equivalence classes, this minimum has to be positive. In the following we let $z_{111} > 0$ without loss of generality.

Case 1. Suppose that there exists a negative cell $z_{i_0 11} < 0, i_0 \geq 2$. Then because $\sum_{j,k} z_{i_0 jk} = 0$, there exists $(j,k), j+k > 2$, with $z_{i_0 jk} > 0$. Then the four cells

$$(1,1,1),\ (i_0,1,1),\ (i_0,j,k),\ (1,j,k)$$

are in the positions of either a type I move or a type II move. In either case we can apply a type I move or a type II move to $\boldsymbol{x}$ or $\boldsymbol{y}$ and make $|\boldsymbol{z}| = |\boldsymbol{y} - \boldsymbol{x}|$ smaller, which is a contradiction. This argument shows that $\boldsymbol{z}$ cannot contain both positive and negative elements in any one-dimensional slice.

Case 2. Now we consider the remaining case, where no one-dimensional slice of $\boldsymbol{z}$ contains both positive and negative elements. Because $\sum_{j,k} z_{1jk} = 0$, there exists (j_1,k_1), $j_1,k_1 \geq 2$, such that $z_{1j_1k_1} < 0$. Similarly there exists (i_1,k_2), $i_1,k_2 \geq 2$, such that $z_{i_1 1 k_2} < 0$. Then the four cells

$$(1,j_1,k_1),(1,1,k_1),(i_1,1,k_2),(i_1,j_1,k_2)$$

are in the positions of a type II move (if $k_1 = k_2$) or a type I move (if $k_1 \neq k_2$) and we can apply a degree 2 move. By doing this, $|\boldsymbol{z}| = |\boldsymbol{y} - \boldsymbol{x}|$ may remain the same, but now z_{11k_1} becomes negative and this case reduces to Case 1. Therefore Case 2 itself is a contradiction. □

We show in the following that $\mathscr{B}^*$ is not a minimal Markov basis. Let $\boldsymbol{z}$ be a degree 2 move and let $\boldsymbol{t} = A\boldsymbol{z}^+$. If $\boldsymbol{z}$ is a type II move, it is easy to verify that $\mathscr{F}_{\boldsymbol{t}}$ is a two-element set $\{z^+, z^-\}$. Therefore degree 2 moves of type II are indispensable. On the other hand, if z is a type I move, $\mathscr{F}_t$ is a four-element set. For the $2\times2\times2$ case, let $t = (z_{1++}, z_{2++}, z_{+1+}, z_{+2+}, z_{++1}, z_{++2})' = (1,1,1,1,1,1)'$. Then we have

$$\mathscr{F}_{(1,1,1,1,1,1)'} = \left\{ \boxed{\begin{matrix}1&0\\0&0\end{matrix}}\ \boxed{\begin{matrix}0&0\\0&1\end{matrix}},\ \boxed{\begin{matrix}0&1\\0&0\end{matrix}}\ \boxed{\begin{matrix}0&0\\1&0\end{matrix}},\ \boxed{\begin{matrix}0&0\\1&0\end{matrix}}\ \boxed{\begin{matrix}0&1\\0&0\end{matrix}},\ \boxed{\begin{matrix}0&0\\0&1\end{matrix}}\ \boxed{\begin{matrix}1&0\\0&0\end{matrix}} \right\}.$$

To connect these elements to a tree, only three moves of type I are needed. In the $2\times2\times2$ case, there are $4^{4-2} = 16$ possibilities, such as

$$\left\{ \boxed{\begin{matrix}+1&-1\\0&0\end{matrix}}\ \boxed{\begin{matrix}0&0\\-1&+1\end{matrix}},\ \boxed{\begin{matrix}0&+1\\-1&0\end{matrix}}\ \boxed{\begin{matrix}0&-1\\+1&0\end{matrix}},\ \boxed{\begin{matrix}0&0\\+1&-1\end{matrix}}\ \boxed{\begin{matrix}-1&+1\\0&0\end{matrix}} \right\}$$

or

$$\left\{ \boxed{\begin{matrix}+1&-1\\0&0\end{matrix}}\ \boxed{\begin{matrix}0&0\\-1&+1\end{matrix}},\ \boxed{\begin{matrix}+1&0\\-1&0\end{matrix}}\ \boxed{\begin{matrix}0&-1\\0&+1\end{matrix}},\ \boxed{\begin{matrix}+1&0\\0&-1\end{matrix}}\ \boxed{\begin{matrix}-1&0\\0&+1\end{matrix}} \right\}$$

and so on. From these considerations, a minimal Markov basis for $I\times J\times K$ tables consists of

$$3\binom{I}{2}\binom{J}{2}\binom{K}{2}$$

degree 2 moves of type I and

$$I\binom{J}{2}\binom{K}{2} + J\binom{I}{2}\binom{K}{2} + K\binom{I}{2}\binom{J}{2}$$

degree 2 moves of type II.

6.2.2 Hardy–Weinberg Model

We next discuss the Hardy–Weinberg model. It is a standard model in population genetics. Consider a multiallele locus with alleles $A_1, A_2, \ldots, A_I$. The allele frequency data are usually given as the genotype frequency. The probability of the genotype A_iA_j in an individual from a random breeding population is given by

$$P(A_iA_j) = \begin{cases} q_i^2 & (i = j) \\ 2q_iq_j & (i \neq j), \end{cases}$$

where q_i is the proportion of the allele A_i, $i = 1,\ldots,I$. Because the Hardy–Weinberg law plays an important role in the field of population genetics and often serves as a basis for genetic inference, much attention has been paid to tests of the hypothesis that a population being sampled is in the Hardy–Weinberg equilibrium against the hypothesis that disturbing forces cause some deviation from the Hardy–Weinberg ratio. See [43, 67, 90].

For the Hardy–Weinberg model, the frequency vector is written as

$$\boldsymbol{x} = (x_{11}, x_{12}, \dots, x_{1I}, x_{22}, x_{23}, \dots, x_{2I}, x_{33}, \dots, x_{II})'.$$

If the frequencies are written in a matrix, only the upper triangular part has the frequencies. A sufficient statistic $\boldsymbol{t} = (t_1, \dots, t_I)'$ is the frequencies of the alleles A_i, $i = 1, \dots, I$, and is given as

$$t_i = 2x_{ii} + \sum_{j \neq i} x_{ij}, \quad i = 1, \dots, I,$$

where we write $x_{ij} = x_{ji}$ for $i > j$. The configuration A is an $I \times \big(I(I+1)/2\big)$ matrix. In terms of the standard basis vectors, the columns of A are written as $2\boldsymbol{e}_i$, $i = 1, \dots, I$ and $\boldsymbol{e}_i + \boldsymbol{e}_j$, $1 \leq i < j \leq I$.

Guo and Thompson [67] constructed a connected Markov chain over any fiber. Their basis consists of three types of degree 2 moves, namely, type 0, type 1, and type 2. Here "type" refers to the number of nonzero diagonal cells in the move. The examples of the moves are displayed as

$$\text{type 0:} \begin{array}{|cccc|} \hline 0 & +1 & -1 & 0 \\ & 0 & 0 & -1 \\ & & 0 & +1 \\ & & & 0 \\ \hline \end{array}, \quad \text{type 1:} \begin{array}{|cccc|} \hline +1 & -1 & -1 & 0 \\ & 0 & +1 & 0 \\ & & 0 & 0 \\ & & & 0 \\ \hline \end{array}, \quad \text{type 2:} \begin{array}{|cccc|} \hline +1 & -2 & 0 & 0 \\ & +1 & 0 & 0 \\ & & 0 & 0 \\ & & & 0 \\ \hline \end{array}.$$

By the distance-reducing argument we first show that these moves form a Markov basis.

Proposition 6.3. *The above three types of moves form a Markov basis for the Hardy–Weinberg model.*

Proof. Suppose that $\boldsymbol{x}$ and $\boldsymbol{y}$ $(\boldsymbol{y} \neq \boldsymbol{x})$ are in the same fiber.

First consider the case that $x_{ii} = y_{ii}$, $i = 1, \dots, I$. We look at the type 0 move above. Because $\boldsymbol{y} \neq \boldsymbol{x}$, there exist some $i < j$ such that $x_{ij} > y_{ij}$. By relabeling the levels we can assume that $i = 1, j = 2$. Then because $2x_{11} + \sum_{j=2}^{I} x_{1j} = 2y_{11} + \sum_{j=2}^{I} y_{1j}$, there exists some $j' > 2$ such that $x_{1j'} < y_{1j'}$. We can again assume that $j' = 3$. Also because $x_{12} + 2x_{22} + \sum_{j=3}^{I} x_{2j} = y_{12} + 2y_{22} + \sum_{j=3}^{I} y_{2j}$, there exists some $j'' > 2$ such that $x_{2j''} < y_{2j''}$. If $j'' \neq 3$, then we can add a type 0 move to $\boldsymbol{y}$ and reduce the distance $|\boldsymbol{y} - \boldsymbol{x}|$. In the case $j'' = 3$, we can then find $j''' > 3$ such that $x_{3j'''} > y_{3j'''}$. We can put $j''' = 4$. In this case we can subtract a type 0 move from $\boldsymbol{x}$ and reduce the distance $|\boldsymbol{y} - \boldsymbol{x}|$.

Now consider the case that there exists some i, such that $x_{ii} > y_{ii}$. We can assume $i = 1$. If there are $1 < j < j'$ such that $x_{1j} < y_{1j}$, $x_{1j'} < y_{1j'}$, then letting $j = 2, j' = 3$, we can add a type 1 move to $\boldsymbol{y}$ and reduce $|\boldsymbol{y}-\boldsymbol{x}|$. On the other hand, if there is only one $j>1$ satisfying $x_{1j} \le y_{1j}$, then $y_{1j} \ge 2$ holds and we can add a type 2 move to $\boldsymbol{y}$ and reduce $|\boldsymbol{y}-\boldsymbol{x}|$. □

We now show the above basis is not minimal and a minimal basis is not unique. Consider $\mathscr{F}_{\boldsymbol{t}}$ with $\deg \boldsymbol{t}=2$ for the above three types of moves. If $\boldsymbol{t}=A\boldsymbol{z}^+=A\boldsymbol{z}^-$ for moves $\boldsymbol{z}$ of type 1 or type 2, there are two elements in $\mathscr{F}_{\boldsymbol{t}}$ and the move of type 1 or type 2 is the difference of these two elements. Hence type 1 and type 2 moves are indispensable.

But if $\boldsymbol{t}=A\boldsymbol{z}^+=A\boldsymbol{z}^-$ for a move $\boldsymbol{z}$ of type 0, there are three elements in $\mathscr{F}_{\boldsymbol{t}}$. Then to connect these three elements to form a tree, we can choose two moves to construct a minimal Markov basis. (There are three ways of doing this.) For example, consider the case of $I=4$ and $\boldsymbol{t}=(1,1,1,1)'$. $\mathscr{F}_{(1,1,1,1)'}$ is written as

$$\mathscr{F}_{(1,1,1,1)'} = \left\{ \begin{array}{|cccc|} 0&1&0&0\\ &0&0&0\\ &&0&1\\ &&&0 \end{array}, \begin{array}{|cccc|} 0&0&1&0\\ &0&0&1\\ &&0&0\\ &&&0 \end{array}, \begin{array}{|cccc|} 0&0&0&1\\ &0&1&0\\ &&0&0\\ &&&0 \end{array} \right\}.$$

To connect these three elements to a tree, any two of the following type 0 moves of degree 2,

$$\begin{array}{|cccc|} 0&+1&-1&0\\ &0&0&-1\\ &&0&+1\\ &&&0 \end{array}, \quad \begin{array}{|cccc|} 0&+1&0&-1\\ &0&-1&0\\ &&0&+1\\ &&&0 \end{array}, \quad \begin{array}{|cccc|} 0&0&-1&+1\\ &0&+1&-1\\ &&0&0\\ &&&0 \end{array},$$

can be included in a minimal Markov basis. Accordingly, $I(I-1)(I-2)(I-3)/12$ moves of type 0, $I(I-1)(I-2)/2$ moves of type 1 and $I(I-1)/2$ moves of type 2 constitute a minimal Markov basis.

6.3 Graver Basis and 1-Norm Reducing Markov Bases

The 1-norm $|\boldsymbol{z}| = \sum_{i\in\mathscr{I}} |z_i| = 2\deg \boldsymbol{z}$ on the set $\mathbb{Z}^{|\mathscr{I}|}$ of integer vectors is a natural norm to consider for Markov bases. In this section we discuss the relation between the Graver basis and the 1-norm reducing Markov bases. We show that the Graver basis is always 1-norm reducing.

We have the following proposition.

Proposition 6.4. *The Graver basis is strongly* 1*-norm reducing.*

Proof. Let $\boldsymbol{x},\boldsymbol{y} \in \mathscr{F}_{\boldsymbol{t}}$, $\boldsymbol{y} \neq \boldsymbol{x}$, be in the same fiber. Express $\boldsymbol{y}-\boldsymbol{x}$ as a conformal sum of nonzero elements of the Graver basis:

$$\boldsymbol{y}-\boldsymbol{x} = \boldsymbol{z}_1 + \cdots + \boldsymbol{z}_m.$$

Then $|\mathbf{y}-\mathbf{x}| = |\mathbf{z}_1| + \cdots + |\mathbf{z}_m|$. Now $\mathbf{z}_1$ can be subtracted from $\mathbf{y}$ and at the same time $\mathbf{z}_1$ can be added to $\mathbf{x}$ to give

$$|(\mathbf{y}-\mathbf{z}_1)-\mathbf{x}| = |\mathbf{y}-(\mathbf{x}+\mathbf{z}_1)| = |\mathbf{z}_2| + \cdots + |\mathbf{z}_m| < |\mathbf{y}-\mathbf{x}|. \qquad \square$$

Note that the Graver basis is rich enough that we can take $\mathbf{z}_1 = \mathbf{z}_2$ in the definition of strong distance reduction in (6.2).

Proposition 6.5. *A set of moves $\mathscr{B}$ is 1-norm reducing if and only if for every element $\mathbf{z} = \mathbf{z}^+ - \mathbf{z}^-$ of the Graver basis, $\mathscr{B}$ is 1-norm reducing for $\mathbf{z}^+, \mathbf{z}^-$.*

Proof. We only have to prove sufficiency. Let $\mathbf{x}, \mathbf{y} \in \mathscr{F}_\mathbf{t}$ be arbitrarily given and let $\mathbf{y}-\mathbf{x} = \mathbf{z}_1 + \cdots + \mathbf{z}_m$ be a conformal sum of elements of the Graver basis. By assumption $\mathscr{B}$ is 1-norm reducing for $\mathbf{z}_1^+, \mathbf{z}_1^-$. Among four possible cases, without loss of generality, consider the case that $\mathbf{z} \in \mathscr{B}$ is applicable to $\mathbf{z}_1^+$ and

$$|(\mathbf{z}_1^+ + \mathbf{z}) - \mathbf{z}_1^-| < |\mathbf{z}_1^+ - \mathbf{z}_1^-| = |\mathbf{z}_1|. \tag{6.3}$$

Because $\mathbf{z}$ is applicable to $\mathbf{z}_1^+$, $\mathbf{z}^- \le \mathbf{z}_1^+ \le (\mathbf{y}-\mathbf{x})^+$. Furthermore (6.3) implies that

$$\emptyset \neq \mathrm{supp}(\mathbf{z}^+) \cap \mathrm{supp}(\mathbf{z}_1^-) \subset \mathrm{supp}(\mathbf{z}^+) \cap \mathrm{supp}((\mathbf{y}-\mathbf{x})^-).$$

It follows that $\mathbf{z}$ is applicable to $\mathbf{y}$ and $|(\mathbf{y}+\mathbf{z})-\mathbf{x}| < |\mathbf{y}-\mathbf{x}|$. $\square$

Note that the same statement holds for strong 1-norm reduction with exactly the same proof.

Proposition 6.6. *A set of moves $\mathscr{B}$ is strongly 1-norm reducing if and only if for every element $\mathbf{z} = \mathbf{z}^+ - \mathbf{z}^-$ of the Graver basis, $\mathscr{B}$ is strongly 1-norm reducing for $\mathbf{z}^+, \mathbf{z}^-$.*

6.4 Some Results on Minimality of 1-Norm Reducing Markov Bases

In this section we discuss minimality of 1-norm reducing Markov bases. Inasmuch as the material in this section is not used in other parts of this book, the reader can skip this section.

Suppose that $\mathscr{B}$ is a 1-norm reducing Markov basis. Then any $\mathscr{B}' \supset \mathscr{B}$ is a 1-norm reducing Markov basis as well. In view of this, it is of interest to consider minimality of 1-norm reducing Markov bases. A 1-norm reducing Markov basis $\mathscr{B}$ is minimal if no proper subset of $\mathscr{B}$ is a 1-norm reducing Markov basis. For a 1-norm reducing Markov basis $\mathscr{B}$, we can examine each element $\mathbf{z}$ of $\mathscr{B}$ one by one, and see whether $\mathscr{B} \setminus \{\mathbf{z}\}$ remains to be a 1-norm reducing Markov basis. If $\mathscr{B} \setminus \{\mathbf{z}\}$ remains to be 1-norm reducing, we remove $\mathbf{z}$, recursively, until none of the remaining elements can be removed any further. Then we arrive at a minimal 1-norm reducing Markov basis. Therefore every 1-norm reducing Markov basis $\mathscr{B}$ contains a minimal 1-norm reducing Markov basis.

Exactly the same argument holds concerning minimality of strongly 1-norm reducing Markov bases. Every strongly 1-norm reducing Markov basis contains a minimal strongly 1-norm reducing Markov basis.

In Chap. 5 we considered minimality of Markov bases. A similar argument can be applied to the question of minimality of 1-norm reducing Markov bases.

In order to study this minimality question we introduce three closely related notions of degree reduction of a move z by other moves. We say that a move $z = z^+ - z^-$ is *1-norm reducible by another move* $z' \neq \pm z$ if z' is applicable to z^+ and $|z+z'| < |z|$ or z' is applicable to z^- and $|-z+z'| = |z-z'| < |z|$. We say that a move $z = z^+ - z^-$ is *strongly 1-norm reducible by a pair of (other) moves* $z_1, z_2 \neq \pm z$ if z_1 is applicable to z^+ and $|z+z_1| < |z|$ and furthermore z_2 is applicable to z^- and $|z-z_2| < |z|$. Finally we say that z is *1-norm reducible by a lower degree move* z' if $|z'| < |z|$ and z is 1-norm reducible by z'.

Consider the implications among these notions. If z is strongly 1-norm reducible by z_1, z_2, then z is clearly 1-norm reducible by z_1 (or z_2). Now we show that if z is 1-norm reducible by a lower degree move z', then z is strongly 1-norm reducible either by the pair $z', z+z'$ or by the pair $z'-z, z'$. To show this, first consider the case that z' is applicable to z^+ and $|z+z'| < |z|$. Let $z'' = z+z'$. Then $|z-z''| = |z'| < |z|$ and we only need to check that z'' is applicable to z^-. In fact

$$\begin{aligned} z'' &= z^+ - z^- + (z')^+ - (z')^- = (z^+ - (z')^-) + (z')^+ - z^- \\ &\geq (z')^+ - z^-. \end{aligned}$$

This implies that $(z'')^- \leq z^-$ and z'' is applicable to z^-. Similarly if z' is applicable to z^-, we can check that z is strongly 1-norm reducible by the pair $z'-z, z'$.

Based on the above observation, we define three notions of irreducibility of a move. We call z *1-norm irreducible* if it is not 1-norm reducible by any other move $z' \neq z$. We call z *strongly 1-norm irreducible* if it is not strongly 1-norm reducible by any pair of other moves. Finally we call z *1-norm lower degree irreducible* if it is not 1-norm reducible by any lower degree move. We state the above implications of the properties of moves, as well as further implications among indispensability and conformal primitiveness, in the following proposition.

Proposition 6.7. *For a move z, the following implications hold.*

$$\begin{aligned} \textit{indispensable} &\Rightarrow \textit{1-norm irreducible} \\ &\Rightarrow \textit{strongly 1-norm irreducible} \\ &\Rightarrow \textit{1-norm lower degree irreducible} \\ &\Rightarrow \textit{conformally primitive.} \end{aligned} \tag{6.4}$$

Proof. If z is not conformally primitive, then z is clearly 1-norm reducible by a lower degree move. This proves the last implication.

If z is 1-norm reducible by $z' \neq \pm z$, then $z^- \neq z^+ + z' \in \mathscr{F}_t$ or $z^+ \neq z^- + z' \in \mathscr{F}_t$, where $t = Az^+$. Therefore $\mathscr{F}_t$ is not a two-element fiber. Therefore z is not indispensable. This proves the first implication. Other implications hold by definition. □

We now state two lemmas.

Lemma 6.1. *If z is 1-norm reducible by another move $z' \neq \pm z$, then there exists a conformally primitive move $z'' \neq z$, $|z''| \leq |z'|$, such that z is 1-norm reducible by z''.*

Proof. If z' is itself conformally primitive, just let $z'' = z'$. If z' is not conformally primitive, write z' as a conformal sum $z' = z_1 + \cdots + z_m$ of nonzero elements of the Graver basis. Among two possible cases, without loss of generality, consider the case that z' is applicable to z^+ and $|z + z'| < |z|$. In this case $(z')^- \leq z^+$ and $\mathrm{supp}((z')^+) \cap \mathrm{supp}(z^-) \neq \emptyset$. Because $\mathrm{supp}((z')^+) = \mathrm{supp}(z_1^+) \cup \cdots \cup \mathrm{supp}(z_m^+)$, there exists some l such that $\mathrm{supp}((z_l)^+) \cap \mathrm{supp}(z^-) \neq \emptyset$. Furthermore $z_l^- \leq (z')^- \leq z^+$ and $|z_l| < |z'| \leq |z|$. This implies that z is 1-norm reducible by $z'' = z_l \neq z$. □

Lemma 6.2. *Let z be a 1-norm irreducible move. Then either z or $-z$ belongs to every 1-norm reducing Markov basis.*

Proof. We argue by contradiction. Let $z = z^+ - z^-$ be 1-norm irreducible and let $\mathscr{B}$ be a 1-norm reducing Markov basis containing neither z nor $-z$. Because $\mathscr{B}$ is 1-norm reducing, $\mathscr{B}$ is 1-norm reducing for z^+, z^-. But this contradicts the 1-norm irreducibility of z in view of (6.1). □

We say that there exists a unique minimal 1-norm reducing Markov basis if all minimal 1-norm reducing Markov bases coincide except for sign changes of their elements. We now state the following proposition.

Proposition 6.8. *There exists a unique minimal 1-norm reducing Markov basis if and only if 1-norm irreducible moves form a 1-norm reducing Markov basis.*

Proof. Every 1-norm irreducible move (or its sign change) belongs to every 1-norm reducing Markov basis, thus if the set of 1-norm irreducible moves is a 1-norm reducing Markov basis, then it is clearly the unique minimal 1-norm reducing Markov basis ignoring the sign of each move.

Conversely suppose that 1-norm irreducible moves do not form a 1-norm reducing Markov basis. Then every 1-norm reducing Markov basis contains a 1-norm reducible move. Let $\mathscr{B}$ be a minimal 1-norm reducing Markov basis and let $z_0 \in \mathscr{B}$ be 1-norm reducible. Consider

$$\tilde{\mathscr{B}} = (\mathscr{B} \cup \mathscr{B}_{\mathrm{Graver}}) \setminus \{z_0, -z_0\},$$

where $\mathscr{B}_{\text{Graver}}$ is the Graver basis. We show that $\tilde{\mathscr{B}}$ is a 1-norm reducing Markov basis. If this is the case, $\tilde{\mathscr{B}}$ contains a minimal 1-norm reducing Markov basis different from $\mathscr{B}$ even if we change the signs of the elements.

Now by Propositions 6.1 and 6.5, it suffices to show that for every $\boldsymbol{z} = \boldsymbol{z}^+ - \boldsymbol{z}^- \in \mathscr{B}_{\text{Graver}}$, $\tilde{\mathscr{B}}$ is 1-norm reducing for $\boldsymbol{z}^+, \boldsymbol{z}^-$. If $\boldsymbol{z}_0$ is not conformally primitive, $\tilde{\mathscr{B}} \supset \mathscr{B}_{\text{Graver}}$ and $\tilde{\mathscr{B}}$ is 1-norm reducing. Therefore let $\boldsymbol{z}_0$ be conformally primitive. Each conformally primitive $\boldsymbol{z} = \boldsymbol{z}^+ - \boldsymbol{z}^- \neq \boldsymbol{z}_0$ is already in $\tilde{\mathscr{B}}$ and $\tilde{\mathscr{B}}$ is 1-norm reducing for $\boldsymbol{z}^+, \boldsymbol{z}^-$. The only remaining case is $\boldsymbol{z} = \boldsymbol{z}_0$ itself, but by Lemma 6.1, $\boldsymbol{z}_0$ is 1-norm reducible by a conformally primitive $\boldsymbol{z}' \neq \pm \boldsymbol{z}_0$, $\boldsymbol{z}' \in \tilde{\mathscr{B}}$. □

Chapter 7
Symmetry of Markov Bases

7.1 Motivations for Invariance of Markov Bases

In this chapter we study properties of Markov bases from the viewpoint of invariance. This is partly motivated by the fact that Gröbner bases depend on a given term order and a reduced Gröbner basis does not preserve the symmetry inherent in a given statistical model. For example, hierarchical models for multiway contingency tables (cf. Sect. 1.5) are symmetric with respect to permutations of the levels of each axis of the table. In group-theoretic terminology, the direct product of symmetric groups acts on the set of multiway tables and hierarchical models are invariant with respect to this group action.

By utilizing invariance we can give a concise description of Markov bases by orbit lists. To illustrate this, we consider the no-three-factor interaction model for three-way tables, which is treated in Chap. 9. In Table 7.1 we list the numbers of the elements of the unique minimal Markov basis, along with the numbers of the reduced Gröbner basis elements calculated by 4ti2 [1] and the numbers of the orbits with respect to the action of the direct product of symmetric groups for the problem of $3 \times 3 \times K$ $(K \leq 7)$ contingency tables with fixed two-dimensional marginals. As we show in Sect. 7.6, a set of moves is partitioned into orbits that are equivalence classes by the action of the group. As we show in Chap. 9, there are at most six orbits of indispensable moves for these problems.

In these examples, a minimal Markov basis is unique. Furthermore it is minimal invariant in the sense of Sect. 7.6. Therefore the representative basis elements for each orbit contain all the information of the minimal Markov basis. To perform the Markov chain Monte Carlo simulations using these orbit lists, users can first randomly choose an orbit, and then apply a random group action to the representative basis element for each step of the chain. Another interesting consideration is how to choose a minimal Markov basis if it is not unique. For such cases, different minimal Markov bases contain different numbers of orbits in general, and some basis elements in these orbits are not necessarily needed for connectivity.

S. Aoki et al., *Markov Bases in Algebraic Statistics*, Springer Series in Statistics 199, DOI 10.1007/978-1-4614-3719-2_7,

Table 7.1 Number of the elements of the unique minimal Markov bases, the reduced Gröbner bases, and orbits for $3 \times 3 \times K, K \leq 7$, tables with fixed two-dimensional marginals

K	3	4	5	6	7
Number of the elements in the unique minimal Markov basis	81	450	2,670	10,665	31,815
Number of the elements in the reduced Gröbner basis	110	622	3,240	12,085	34,790
Number of orbits in the unique minimal Markov basis	4	5	6	6	6

In Table 7.1 we have considered permutation of the levels for each axis. If the number of levels of the axes is common and if in addition the hierarchical log-linear model considered is symmetric with respect to permutations of axes, we can further consider the permutation of the axes themselves. For example in the case of the $3 \times 3 \times 3$ contingency tables with no three-factor interactions, we can consider the permutation of the axes. As we show in Chap. 9, if this additional symmetry of axes is considered, there are only two orbits corresponding to moves of degree 4 and degree 6, whereas if this additional symmetry is not considered there are four orbits as indicated in Table 7.1. This question leads to the notion of the largest symmetry in a given model, which we define in Sect. 7.4.

7.2 Examples of Invariant Markov Bases

In this section we consider two simple examples of invariant Markov bases. They are the $2 \times 2 \times 2$ contingency tables with fixed one-dimensional marginals and the Hardy–Weinberg model, which were already treated in Sect. 6.2.

We use the following notation for our moves. Moves in minimal bases contain many zero cells. Furthermore, often the nonzero elements of a move contain either 1 or -1. Therefore a move can be concisely denoted by locations of its nonzero cells. We express a move $\boldsymbol{z}$ of degree n as $\boldsymbol{z} = \boldsymbol{i}_1 \cdots \boldsymbol{i}_n - \boldsymbol{j}_1 \cdots \boldsymbol{j}_n$, where $\boldsymbol{i}_1, \ldots, \boldsymbol{i}_n$ are the cells of positive frequencies of $\boldsymbol{z}$ and $\boldsymbol{j}_1, \ldots, \boldsymbol{j}_n$ are the cells of negative frequencies of $\boldsymbol{z}$. In the case $z(\boldsymbol{i}) > 1$, $\boldsymbol{i}$ is repeated $z(\boldsymbol{i})$ times. Similarly $\boldsymbol{j}$ is repeated $-z(\boldsymbol{j})$ times if $z(\boldsymbol{j}) < -1$. We use a similar notation for contingency tables as well. $\boldsymbol{x}$ with $\deg \boldsymbol{x} = n$ is simply denoted as $\boldsymbol{x} = \boldsymbol{i}_1 \cdots \boldsymbol{i}_n$.

First consider the $2 \times 2 \times 2$ contingency tables with fixed one-dimensional marginals. As shown in Sect. 6.2, the minimal Markov basis for this problem is not unique. Each minimal Markov basis contains six indispensable elements and three dispensable elements. Consider dispensable elements. The reduced Gröbner basis with respect to the graded reverse lexicographic order contains three dispensable moves (binomials) such as

$$(121)(212) - (111)(222),\ (122)(211) - (111)(222),\ (112)(221) - (111)(222).$$

It is seen that these three dispensable basis elements are in different orbits with respect to permutation of levels. On the other hand, another minimal basis is constructed from three dispensable basis elements such as

$$(121)(212)-(111)(222),\ (122)(211)-(111)(222),\ (112)(221)-(121)(212).$$

In this basis, the second and the third binomials are in the same orbit. In fact, we see that $(112)(221)-(121)(212)$ can be produced from $(122)(211)-(111)(222)$ by interchanging the cell indices $1,2$ in the second axis. Accordingly, if we consider an action of the direct product of symmetric groups, only two basis elements such as

$$(121)(212)-(111)(222),\ (122)(211)-(111)(222)$$

have to be included in our list, because the third basis element can be produced by permuting the second axis.

Furthermore, because the number of levels is common for three axes in the $2 \times 2 \times 2$ case, we can also permute the axes. If we consider invariance with respect to this larger group, then a single representative element among dispensable ones such as

$$(112)(221)-(111)(222)$$

is sufficient to describe an invariant Markov basis.

We now consider the Hardy–Weinberg model of Sect. 6.2.2 for I alleles. The direct product of symmetric groups is not appropriate in this case, because the contingency table $\boldsymbol{x} = \{x_{ij}\}_{1 \le i \le j \le I}$ is of an upper triangular form. However, it is clear that this problem has the symmetry with respect to a simultaneous permutation of the levels (i.e., alleles). It can be checked (see [12]) that a minimal invariant Markov basis with respect to this group action consists of three orbits, with the representative moves given as

$$(11)(22)-(12)(12),\ (11)(23)-(12)(13),\ (12)(34)-(13)(24). \tag{7.1}$$

In this case, the unique minimal Markov basis does not exist as we have seen in Sect. 6.2.2. However, the minimal invariant Markov basis given in (7.1) can be shown to be the unique minimal invariant Markov basis.

7.3 Action of Symmetric Group on the Set of Cells

In this section we formulate the symmetry of a given toric model in terms of the action of a group on the set of cells. As the most important example we consider the direct product of symmetric groups acting on the cells of multiway contingency tables by permutations of levels for each axis. Decomposable models (Chap. 8) and more general hierarchical models (Chap. 9) are invariant with respect to this group.

First we give a brief list of definitions and notations of a group action. Basic facts on group action in statistical problems are found in Chap. 6 of [98] or Chap. 4 of [58]. More comprehensive treatment is given in [56]. Let a group G act (from the left) on a set $\mathscr{I}$. This means that each element $g \in G$ is a map $\mathscr{I} \to \mathscr{I}$ sending $\boldsymbol{i}$ to $g\boldsymbol{i}$, and the following conditions are satisfied

$$e\boldsymbol{i} = \boldsymbol{i}, \quad \forall \boldsymbol{i} \in \mathscr{I}, \tag{7.2}$$

$$(g_1 g_2)\boldsymbol{i} = g_1(g_2\boldsymbol{i}), \quad \forall g_1, g_2 \in G, \forall \boldsymbol{i} \in \mathscr{I}, \tag{7.3}$$

where e is the identity element of G. Under these conditions, the inverse element g^{-1} of g in G is also the inverse of g as a map from $\mathscr{I}$ to $\mathscr{I}$. This implies that each g is a bijection from $\mathscr{I}$ to $\mathscr{I}$. In this book as $\mathscr{I}$ we are considering the set of cells, which is a finite set. Hence each $g \in G$ is just a permutation of the cells of $\mathscr{I}$ and G is a subgroup of the symmetric group S_η, $\eta = |\mathscr{I}|$, which is the group of all permutations of the cells of $\mathscr{I}$.

Define $G(\boldsymbol{i}) = \{g\boldsymbol{i} \mid g \in G\}$ as the *orbit* through $\boldsymbol{i}$. Let $\mathscr{I}/G$ denote the *orbit space*, that is, the set of orbits. The action of G is called transitive, if the whole G is one orbit. Let $G_{\boldsymbol{i}} = \{g \mid g\boldsymbol{i} = \boldsymbol{i}\}$ denote the isotropy subgroup (pointwise stabilizer) of $\boldsymbol{i}$ in G. If G acts on $\mathscr{I}$, the action of G on the set of functions f on $\mathscr{I}$ is induced by $(gf)(\boldsymbol{i}) = f(g^{-1}\boldsymbol{i})$.

Because the frequency vector $\boldsymbol{x}$ is considered as a function $\mathscr{I} \to \mathbb{N}$, the action of G on the set $\mathscr{X} = \mathbb{N}^\eta$ of frequency vectors is defined as

$$(g\boldsymbol{x})(\boldsymbol{i}) = x(g^{-1}\boldsymbol{i}).$$

This is again just a permutation of elements of $\boldsymbol{x}$.

Let us write out the permutation matrix for g. $(g\boldsymbol{x})(\boldsymbol{i}) = x(g^{-1}\boldsymbol{i})$ means that the $\boldsymbol{i}$th element of $g\boldsymbol{x}$ is the $(g^{-1}\boldsymbol{i})$th element of $\boldsymbol{x}$. Let

$$P_g = \{p_{\boldsymbol{i}\boldsymbol{j}}\} = \{\delta_{\boldsymbol{i},g\boldsymbol{j}}\} \tag{7.4}$$

denote an $\eta \times \eta$ permutation matrix, where δ is Kronecker's delta. Then the $\boldsymbol{i}$th row of P_g has 1 at the column $\boldsymbol{j} = g^{-1}\boldsymbol{i}$ and hence the $\boldsymbol{i}$th element of $P_g\boldsymbol{x}$ is the $(g^{-1}\boldsymbol{i})$th element of $\boldsymbol{x}$. Equivalently, the $(g\boldsymbol{i})$th element of $P_g\boldsymbol{x}$ is the $\boldsymbol{i}$th element of $\boldsymbol{x}$. Therefore we have

$$g\boldsymbol{x} = P_g\boldsymbol{x}.$$

From this it follows that

$$P_{g_1 g_2} = P_{g_1} P_{g_2}, \quad \forall g_1, g_2 \in S_\eta \ \text{ and } \ P_{g^{-1}} = P'_g, \quad \forall g \in S_\eta. \tag{7.5}$$

Similarly G acts on a move $\boldsymbol{z}$ by $(g\boldsymbol{z})(\boldsymbol{i}) = z(g^{-1}\boldsymbol{i})$. If we write $\boldsymbol{z} = \boldsymbol{z}^+ - \boldsymbol{z}^-$, then

$$g\boldsymbol{z} = P_g\boldsymbol{z} = P_g\boldsymbol{z}^+ - P_g\boldsymbol{z}^- = g\boldsymbol{z}^+ - g\boldsymbol{z}^-.$$

We also call a move $\boldsymbol{z} = \boldsymbol{z}^+ - \boldsymbol{z}^-$ *symmetric* with respect to G if $\boldsymbol{z}^+ = g\boldsymbol{z}^-$ for some $g \in G$. Conversely, a move $\boldsymbol{z}$ is *asymmetric* if $G(\boldsymbol{z}^+) \cap G(\boldsymbol{z}^-) = \emptyset$.

As G consider the direct product of symmetric groups, which is our main example. Let $G^\ell = S_{I_\ell}$ denote the symmetric group on $\{1,\dots,I_\ell\}$ for $\ell = 1,\dots,m$ and let

$$G = G^1 \times G^2 \times \cdots \times G^m \tag{7.6}$$

be the direct product. We write an element of $g \in G$ as

$$g = g_1 \times \cdots \times g_m = \begin{pmatrix} 1 & \cdots & I_1 \\ \sigma_1(1) & \cdots & \sigma_1(I_1) \end{pmatrix} \times \cdots \times \begin{pmatrix} 1 & \cdots & I_m \\ \sigma_m(1) & \cdots & \sigma_m(I_m) \end{pmatrix}.$$

G acts on the set of cells $\mathscr{I}$ of m-way contingency tables by

$$\begin{aligned} \boldsymbol{i}' &= g\boldsymbol{i} \\ &= (g_1 i_1, \dots, g_m i_m) \\ &= (\sigma_1(i_1), \dots, \sigma_m(i_m)) . \end{aligned} \tag{7.7}$$

Now we consider the action of G on the set of sufficient statistics $\mathscr{T}$ in (4.6). We go back to the general definition of a group action. Let $h : \mathscr{I} \to \mathscr{T}$ be a surjection. If the following condition holds,

$$h(\boldsymbol{i}) = h(\boldsymbol{i}') \Rightarrow h(g\boldsymbol{i}) = h(g\boldsymbol{i}'), \quad \forall g \in G, \tag{7.8}$$

then the action of G on $\mathscr{T}$ is *induced* by defining

$$g\boldsymbol{t} = h(g\boldsymbol{i}), \text{ where } \boldsymbol{t} = h(\boldsymbol{i}). \tag{7.9}$$

Indeed $g\boldsymbol{t}$ is well defined, because by (7.8) $g\boldsymbol{t}$ does not depend on $\boldsymbol{i}$ such that $\boldsymbol{t} = h(\boldsymbol{i})$. Then by choosing $\boldsymbol{i}$ for each $\boldsymbol{t} \in \mathscr{T}$, (7.2) and (7.3) for $\boldsymbol{t}$ are easily verified as

$$\begin{aligned} &e\boldsymbol{t} = h(e\boldsymbol{i}) = h(\boldsymbol{i}) = \boldsymbol{t}, \\ &(g_1 g_2)\boldsymbol{t} = h((g_1 g_2)\boldsymbol{i}) = h(g_1(g_2\boldsymbol{i})) = g_1 h(g_2\boldsymbol{i}) = g_1(g_2 h(\boldsymbol{i})) = g_1(g_2\boldsymbol{t}). \end{aligned}$$

Note that (7.9) is written as

$$h(g\boldsymbol{i}) = gh(\boldsymbol{i}), \quad \forall g \in G, \forall \boldsymbol{i} \in \mathscr{I}. \tag{7.10}$$

We call h satisfying (7.10) *equivariant*. We can also say that h and the group action commute (i.e., $hg = gh$).

Conversely suppose that G acts on both $\mathscr{I}$ and $\mathscr{T}$ and the surjection $h : \mathscr{I} \to \mathscr{T}$ is equivariant. Then

$$h(\boldsymbol{i}) = h(\boldsymbol{i}') \;\Rightarrow\; gh(\boldsymbol{i}) = gh(\boldsymbol{i}') \;\Rightarrow\; h(g\boldsymbol{i}) = h(g\boldsymbol{i}')$$

and (7.8) holds.

Consider again the direct product of symmetric groups in (7.6). A sufficient statistic for hierarchical models for contingency tables consists of various marginal frequencies. Note that G acts on the marginal cells $\boldsymbol{i}_D$, $D = \{s_1,\dots,s_k\} \subset [m] = \{1,\dots,m\}$, by

$$\begin{aligned}\boldsymbol{i}'_D &= g\boldsymbol{i}_D \\ &= (g_{s_1}i_{s_1},\dots,g_{s_k}i_{s_k}) \\ &= (\sigma_{s_1}(i_{s_1}),\dots,\sigma_{s_k}(i_{s_k})).\end{aligned}$$

Hence G acts on marginal tables by

$$\boldsymbol{x}'_D = g\boldsymbol{x}_D = \{x_D(g^{-1}\boldsymbol{i}_D)\}_{\boldsymbol{i}_D \in \mathscr{I}_D}.$$

Considering this action simultaneously for various marginals $D_1,\dots,D_r \subset [m]$, the action of G on the sufficient statistic $\boldsymbol{t} = (\boldsymbol{x}_{D_1},\dots,\boldsymbol{x}_{D_r})$ of a hierarchical model is defined by

$$g\boldsymbol{t} = (g\boldsymbol{x}_{D_1},\dots,g\boldsymbol{x}_{D_r}).$$

An important point here is that the map of taking marginal frequencies is equivariant; that is, we have the following lemma.

Lemma 7.1. $(g\boldsymbol{x})_D = g\boldsymbol{x}_D$ *for all* $g \in G$ *and* $\boldsymbol{x} \in \mathbb{N}^\eta$.

This lemma clearly holds, because taking the marginal sums after permutation of levels of axes is the same as first taking the marginal sums and permuting the axes in the marginal cells. By Lemma 7.1 and the above argument, the action of G on the set $\mathscr{T}$ of marginal frequencies $\boldsymbol{t}$ is induced from the action of G on $\boldsymbol{x}$.

7.4 Symmetry of a Toric Model and the Largest Group of Invariance

In the previous section we considered the direct product of symmetric groups acting on the set of multiway contingency tables as our main example. In the case of the Hardy–Weinberg model, the set of cells was an upper triangular matrix and the symmetry was not described by the direct product of symmetric groups. Now we consider how to define a symmetry of a given toric model or a configuration A.

Consider the toric model in (4.5). The probability distribution of $\boldsymbol{x}$ depends on $\boldsymbol{\theta}'A$, and $\boldsymbol{\theta}'A$ is an element of the row space of A, rowspan(A). Assuming that $\boldsymbol{\theta} \in \mathbb{R}^\nu$ is a free parameter vector, the set of probability distributions of the toric model is identified with rowspan(A). In this sense, when we consider the symmetry of a given toric model, it is reasonable to require that the symmetry be defined in terms of rowspan(A).

Multiplying A from the right by P'_g results in a matrix AP'_g whose columns are permutations of columns of A by $g \in S_\eta$. The reason we take the transpose P'_g is to preserve the action of G "from the left." By defining $gA = AP'_g$, we have

$$(g_1g_2)A = AP'_{g_1g_2} = A(P_{g_1}P_{g_2})' = AP'_{g_2}P'_{g_1} = g_1(g_2A). \tag{7.11}$$

In $gA = AP'_g$, we can think of P'_g as multiplying each row of A from the right, namely G acts on the set of η-dimensional row vectors $\boldsymbol{\alpha}$ by $g\boldsymbol{\alpha} = \boldsymbol{\alpha}P'_g$. Note that the space of row vectors can be considered as the dual vector space of the space of column vectors.

In Sect. 7.2, the complete independence model of $2\times 2\times 2$ contingency tables is clearly invariant with respect to the direct product of symmetric groups $S_2 \times S_2 \times S_2$. The Hardy–Weinberg model for I alleles is invariant with respect to permutation of alleles S_I. In view of these examples we make the following definition.

Definition 7.1. Let $G \subset S_\eta$ be a subgroup of S_η. A configuration A is *invariant* with respect to G if $\mathrm{rowspan}(A) = \mathrm{rowspan}(AP'_g)$ for all $g \in G$.

In Sect. 4.2 we discussed the relation $\mathrm{rowspan}(A)^\perp = \ker A$. Then we have

$$\mathrm{rowspan}(A) = \mathrm{rowspan}(AP'_g) \;\Leftrightarrow\; \ker A = \ker(AP'_g).$$

Also note that

$$\ker(AP'_g) = \{\boldsymbol{z} \mid AP'_g\boldsymbol{z} = 0\} = \{P_g\boldsymbol{z} \mid A\boldsymbol{z} = 0\} = P_g \ker A, \qquad (7.12)$$

where on the right-hand side P_g is now multiplying column vectors from the left. Hence an equivalent definition of invariance with respect to G is given as follows.

Definition 7.2. A configuration A is *invariant* with respect to G if $\ker A = P_g \ker A$ for all $g \in G$.

So far we have defined the invariance of the configuration A with respect to a given G. When A is given first, it is natural to consider all $g \in S_{\mathscr{I}}$ such that $\mathrm{rowspan}(A) = \mathrm{rowspan}(AP'_g)$, or equivalently $\ker A = P_g \ker A$. Here the notion of setwise stabilizer [136] is useful. Let a group G act on a set $\mathscr{X}$ from the left. For a subset $\mathscr{V}$ of $\mathscr{X}$, let

$$G_{\mathscr{V}} = \{g \mid g\mathscr{V} = \mathscr{V}\}$$

denote the setwise stabilizer of $\mathscr{V}$. (Note that $G_{\mathscr{V}}$ forms a subgroup of G.) As we have discussed already, G acts on the set of η-dimensional column vectors by $g : \boldsymbol{x} \mapsto P_g\boldsymbol{x}$ and the set of η-dimensional row vectors by $g : \boldsymbol{\alpha} \mapsto \boldsymbol{\alpha}P'_g$. From this viewpoint, the set of g such that $\mathrm{rowspan}(A) = \mathrm{rowspan}(AP'_g)$ is the setwise stabilizer $G_{\mathrm{rowspan}(A)}$. Equivalently it is the setwise stabilizer $G_{\ker A}$. Therefore we are led to the following definition.

Definition 7.3. For a given configuration A, the *largest group of invariance* is the setwise stabilizer $G_{\ker A}$ of $\ker A$ in the symmetric group S_η, where G acts on the set of η-dimensional column vectors. Alternatively, it is the setwise stabilizer $G_{\mathrm{rowspan}(A)}$, where G acts on the set of η-dimensional row vectors.

From now on, among two equivalent definitions, we mainly consider $G_{\ker A}$. The notion of the largest group of invariance was introduced in [12].

In Sect. 7.3 we considered the induced action on the set of sufficient statistics. We want to check that the induced action on the set of sufficient statistics $\mathscr{T}$ is defined also for the largest group of invariance. We try to define $g\boldsymbol{t}$, $\boldsymbol{t} = A\boldsymbol{x}$, by

$$g\boldsymbol{t} = Ag\boldsymbol{x}.$$

If $\boldsymbol{t} = A\boldsymbol{x} = A\tilde{\boldsymbol{x}}$, then $\boldsymbol{x} - \tilde{\boldsymbol{x}} \in \ker A$. For $g \in G_{\ker A}$, $g(\boldsymbol{x} - \tilde{\boldsymbol{x}}) \in \ker A$ and hence $Ag(\boldsymbol{x} - \tilde{\boldsymbol{x}}) = 0$. Therefore $g\boldsymbol{t}$ does not depend on the choice of $\boldsymbol{x}$ in $\boldsymbol{t} = A\boldsymbol{x}$. Hence the induced action of $G_{\ker A}$ on $\mathscr{T}$ is well defined.

It should be noted that a configuration A is invariant with respect to a group H if and only if H is a subgroup of the largest group of invariance $G_{\ker A}$. Also because $G_{\ker A}$ acts on $\mathscr{T}$, any subgroup H of $G_{\ker A}$ also acts on $\mathscr{T}$.

We have given the definition of the largest group of invariance for a general configuration A. For many configurations A it is often surprisingly hard to determine the largest group of invariance $G_{\ker A}$, although some obvious subgroup of $G_{\ker A}$ is easy to find. We discuss one simple example in the next section.

7.5 The Largest Group of Invariance for the Independence Model of Two-Way Tables

As we have stated above, it is often surprisingly hard to determine the largest group of invariance G for a given A. The symmetry in the independence model of two-way tables seems to be trivial. However, to prove that the obvious symmetry is the largest, we need some careful arguments. For showing that a given candidate group is the largest group of invariance, in [135], we developed a "perturbation method." Here we illustrate the perturbation method with the independence model of two-way tables.

Consider $I \times J$ contingency tables with fixed row sums and column sums. The configuration is clearly invariant with respect to the direct product $S_I \times S_J$, which seems to be the largest group of invariance if $I \neq J$.

In the case of square tables $I = J$, there is an additional symmetry of interchanging the two axes. Although this is again a symmetric group S_2, for clarity we denote the group of interchanging the axes by H_2. Then the largest group of invariance for the square case seems to be the subgroup of S_{I^2} generated by $S_I \times S_I$ and H_2. In the square case, we can first decide whether to flip the axes, and then we can arbitrarily and independently permute the levels of two axes. This is called the *wreath product* (e.g., [129]) of groups and written as

$$S_I \text{ wr } H_2.$$

Now we have the following result.

Proposition 7.1. *The largest group of invariance for $I \times J$ contingency tables with fixed row sums and column sums is $S_I \times S_J$ if $I \neq J$ and S_I wr H_2 if $I = J$.*

The rest of this section is devoted to a sketch of a proof of this proposition. We mainly consider the case $I \neq J$. In the process of proving the case $I \neq J$, it will become clear that the additional symmetry in the square case is given by H_2. We now assume $I > J$ without loss of generality.

By the form of the configuration A in (1.20), an element of rowspan(A) is an $I \times J$-dimensional (row) vector with components of the form

$$\{\alpha_i + \beta_j\}_{1 \le i \le I, 1 \le j \le J}, \quad \alpha_i, \beta_j \in \mathbb{R}.$$

Write $\gamma_{ij} = \alpha_i + \beta_j$. For four distinct pairs $(i_1, j_1), (i_2, j_2), (i_3, j_3), (i_4, j_4)$ consider the following linear combination

$$\delta = \gamma_{i_1 j_1} + \gamma_{i_2 j_2} - \gamma_{i_3 j_3} - \gamma_{i_4 j_4}.$$

Inasmuch as $\ker A$ is spanned by the basic moves in (2.4), $\delta = 0$ if $(i_1, j_1), (i_2, j_2), (i_3, j_3), (i_4, j_4)$ are the four cells in a rectangular position as in (2.4). By taking α_i, β_j sufficiently "generic" it is clear that $\delta \neq 0$ unless the four cells are in a rectangular position.

This can be made explicit as follows. Let $b > 0$ be a large positive integer. For our case $b = 5$ is good enough. Let

$$\alpha_i = b^i, \quad \beta_j = b^{I+j}. \tag{7.13}$$

Then by the uniqueness of the base b expansion of a positive integer δ, for this choice of α_i and β_j, $\delta \neq 0$ unless the four cells are in a rectangular position. Let $\boldsymbol{\gamma} = \{b^i + b^{I+j}\}$ denote the $I \times J$-dimensional vector with these elements. In the following we consider $\boldsymbol{\gamma}$ as an $I \times J$ table.

Now consider $g \in G_{\text{rowspan}(A)}$ and let $\tilde{\boldsymbol{\gamma}} = g\boldsymbol{\gamma}$. By the consideration immediately following (7.4), the $g(i,j)$ element of $\tilde{\boldsymbol{\gamma}}$ is the (i,j) element of $\boldsymbol{\gamma}$. Consider the lower-right cell (I,J) of $\boldsymbol{\gamma}$. Then the value $b^I + b^{I+J}$ is in the $g(i,j)$ cell of $\tilde{\boldsymbol{\gamma}}$. Write $(i^*, j^*) = g(I,J)$ and consider the following linear combination of elements of $\tilde{\boldsymbol{\gamma}}$:

$$\begin{aligned} \tilde{\delta}(i_2, j_2) &= \tilde{\gamma}_{i^* j^*} + \tilde{\gamma}_{i_2 j_2} - \tilde{\gamma}_{i^* j_2} - \tilde{\gamma}_{i_2 j^*} \\ &= \gamma_{IJ} + \gamma_{g^{-1}(i_2, j_2)} - \gamma_{g^{-1}(i^*, j_2)} - \gamma_{g^{-1}(i_2, j^*)}, \end{aligned}$$

where the four cells $(i^*, j^*), (i_2, j_2), (i^*, j_2), (i_2, j^*)$ are distinct; that is, $i_2 \neq i^*$ and $j_2 \neq j^*$. Because g and g^{-1} are bijections, the four cells $(I,J), g^{-1}(i_2, j_2), g^{-1}(i^*, j_2), g^{-1}(i_2, j^*)$ are distinct as well. Furthermore, because $\tilde{\gamma} \in \text{rowspan}(AP_g') = \text{rowspan}(A)$, $\tilde{\delta}(i_2, j_2) = 0$ for all choices of (i_2, j_2). By our particular choice (7.13) of $\boldsymbol{\gamma}$, we see that $(I,J), g^{-1}(i_2, j_2), g^{-1}(i^*, j_2), g^{-1}(i_2, j^*)$ have to be in a rectangular position. This means that, either

1. $g^{-1}(i^*, j_2)$ is in the last row and $g^{-1}(i_2, j^*)$ is in the last column, or
2. $g^{-1}(i^*, j_2)$ is in the last column and $g^{-1}(i_2, j^*)$ is in the last row.

We note that these two cases cannot mix. In fact, if $g^{-1}(i_2, j^*)$ and $g^{-1}(i^*, j_2)$ are both in the last column for some i_2 and j_2, then it can be easily seen that $\tilde{\delta}(i_2, j_2)$ cannot be zero for this i_2 and j_2. It follows that either $g^{-1}(i_2, j^*)$, $i_2 \neq i^*$, are all in the last column, or they are all in the last row. Because we have assumed $I > J$, the second case is impossible and we have $\{g^{-1}(i_2, j^*)\}_{i_2=1,\dots,I} = \{(i,J)\}_{i=1,\dots,I}$. This means that the last column of $\boldsymbol{\gamma}$ is moved to the j^*th column of $\tilde{\boldsymbol{\gamma}}$.

Now we can similarly argue for other columns and other rows of $\boldsymbol{\gamma}$. Then it follows that g moves columns of $\boldsymbol{\gamma}$ to columns of $\tilde{\boldsymbol{\gamma}}$ and rows of $\boldsymbol{\gamma}$ to rows of $\tilde{\boldsymbol{\gamma}}$. Hence g is an element of $S_I \times S_J$.

7.6 Characterizations of a Minimal Invariant Markov Basis

Now we go back to invariant Markov bases. Let $\mathscr{B} \subset \ker_{\mathbb{Z}} A$ be a set of moves. For convenience in this section we assume that $\mathscr{B}$ is sign invariant (see Sect. 5.2). Let the configuration A be invariant with respect to a group G. We call $\mathscr{B}$ *G-invariant* if $G(\mathscr{B}) = \mathscr{B}$. Note that $\mathscr{B}$ is G-invariant if and only if

$$g \in G, \boldsymbol{z} \in \mathscr{B} \implies g\boldsymbol{z} \in \mathscr{B}.$$

In other words, $\mathscr{B}$ is G-invariant if and only if it is a union of orbits $\mathscr{B} = \bigcup_{\boldsymbol{z} \in \mathscr{B}^*} G(\boldsymbol{z})$ for some subset $\mathscr{B}^* \subset \ker_{\mathbb{Z}} A$ of moves.

A finite sign invariant set $\mathscr{B} \subset \ker_{\mathbb{Z}} A$ is an *invariant Markov basis* if it is a Markov basis and it is G-invariant. An invariant Markov basis is *minimal* if no proper sign invariant and G-invariant subset of $\mathscr{B}$ is a Markov basis. A minimal invariant Markov basis always exists, because from any invariant Markov basis, we can remove orbits one by one, until none of the remaining orbits can be removed any further.

Partition $\mathscr{X} = \mathbb{N}^\eta$ by the degree (total sample size) of the frequency vectors as

$$\mathscr{X} = \cup_{n=1}^{\infty} \mathscr{X}_n, \quad \mathscr{X}_n = \{\boldsymbol{x} \in \mathscr{X} \mid \deg \boldsymbol{x} = n\}.$$

Similarly partition the set of sufficient statistics as

$$\mathscr{T} = \cup_{n=1}^{\infty} \mathscr{T}_n.$$

In considering the orbits of G acting on $\mathscr{X}$, we note that $\deg \boldsymbol{x} = \deg(g\boldsymbol{x}), \forall g \in G$, and hence $G(\mathscr{X}_n) = \mathscr{X}_n$ for all n. Therefore we can consider the action of G on each $\mathscr{X}_n$ separately. Similarly we can consider the action of G on each $\mathscr{T}_n$ separately because $\deg \boldsymbol{t} = \deg(g\boldsymbol{t}), \forall g \in G$.

Consider a particular sufficient statistic $\boldsymbol{t} \in \mathscr{T}_n$. As in (5.4) let

$$\mathscr{B}_{\boldsymbol{t}} = \{\boldsymbol{z} \in \ker_{\mathbb{Z}} A \mid \boldsymbol{z}^+, \boldsymbol{z}^- \in \mathscr{F}_{\boldsymbol{t}}\}$$

be the set of moves whose positive and negative parts belong to the fiber $\mathscr{F}_{\boldsymbol{t}}$. Let $G(\boldsymbol{t}) \in \mathscr{T}_n / G$ be the orbit through $\boldsymbol{t}$. Let

$$\mathscr{B}_{G(\boldsymbol{t})} = \bigcup_{\boldsymbol{t}' \in G(\boldsymbol{t})} \mathscr{B}_{\boldsymbol{t}'}$$

denote the union of the set of moves $\mathscr{B}_{\boldsymbol{t}'}$ over the orbit $G(\boldsymbol{t})$ through $\boldsymbol{t}$.

Let $\mathscr{B} \subset \ker_{\mathbb{Z}} A$ be a finite set of moves. An important observation is that $\mathscr{B}$ is partitioned as

$$\mathscr{B} = \bigcup_n \bigcup_{\alpha \in \mathscr{T}_n/G} \mathscr{B}_{n,\alpha}, \tag{7.14}$$

where we define

$$\mathscr{B}_{n,\alpha} = \mathscr{B} \cap \mathscr{B}_{\alpha}, \quad \alpha \in \mathscr{T}_n/G.$$

Inasmuch as $\mathscr{B}$ is invariant if and only if it is a union of orbits $G(\boldsymbol{z})$, the following lemma holds.

Lemma 7.2. *$\mathscr{B}$ is invariant if and only if $\mathscr{B}_{n,\alpha}$ is invariant for each n and $\alpha \in \mathscr{T}_n/G$.*

Proof. Let $\boldsymbol{z} \in \mathscr{B}_{n,\alpha}$ and $\boldsymbol{t} = A\boldsymbol{z}^+ \in \alpha$. Then it follows that $g\boldsymbol{z} \in \mathscr{B}_{g\boldsymbol{t}} \subset \mathscr{B}_{\alpha}$ and the lemma is proved. □

This lemma shows that we can restrict our attention to each $\mathscr{B}_{n,\alpha}$ in studying the invariance of a Markov basis.

In characterizing a Markov basis and its minimality, in Chap. 5 we argued that it is essential to consider $\mathscr{B}_{|\boldsymbol{t}|-1}$-equivalence classes of $\mathscr{F}_{\boldsymbol{t}}$, where $\mathscr{B}_n$ is the set of moves of degree less than or equal to n defined in (5.2). As in Chap. 5 we write $|\boldsymbol{t}|$ for $\deg \boldsymbol{t}$.

Considering group actions on the set of moves and each fiber, we characterize the structure of a minimal invariant Markov basis. As we show in the following, the relation between the action of the isotropy subgroup $G_{\boldsymbol{t}}$ and $\mathscr{B}_{|\boldsymbol{t}|-1}$-equivalence classes of $\mathscr{F}_{\boldsymbol{t}}$ is important. For the rest of this section, we write the set of $\mathscr{B}_{|\boldsymbol{t}|-1}$-equivalence classes of $\mathscr{F}_{\boldsymbol{t}}$ as $\mathscr{H}_{\boldsymbol{t}}$ for simplicity; that is, $\mathscr{H}_{\boldsymbol{t}} = \mathscr{F}_{\boldsymbol{t}}/\mathscr{B}_{|\boldsymbol{t}|-1}$.

Now we state the following theorem.

Theorem 7.1. *Let $\mathscr{B}$ be a minimal G-invariant Markov basis and let $\mathscr{B} = \bigcup_n \bigcup_{\alpha \in \mathscr{T}_n/G} \mathscr{B}_{n,\alpha}$ be the partition in (7.14). Then each $\mathscr{B}_{n,\alpha}$, $\alpha \in \mathscr{T}_n/G$, is a minimal invariant set of moves, where $\mathscr{B}_{n,\alpha} \cap \mathscr{B}_{\boldsymbol{t}}$, $\boldsymbol{t} \in \alpha$, connects $\mathscr{B}_{|\boldsymbol{t}|-1}$-equivalence classes of $\mathscr{F}_{\boldsymbol{t}}$ and*

$$\mathscr{B}_{n,\alpha} = G(\mathscr{B}_{n,\alpha} \cap \mathscr{B}_{\boldsymbol{t}}) \tag{7.15}$$

for any $\boldsymbol{t} \in \alpha$.

Conversely, from each $\alpha \in \mathscr{T}_n/G$ with $|\mathscr{H}_{\boldsymbol{t}}| \geq 2$, where $\boldsymbol{t} \in \alpha$ is a representative sufficient statistic, choose a minimal $G_{\boldsymbol{t}}$-invariant set of moves $\mathscr{B}^* \subset \mathscr{B}_{\boldsymbol{t}}$ connecting $\mathscr{B}_{|\boldsymbol{t}|-1}$-equivalence classes of $\mathscr{F}_{\boldsymbol{t}}$, where $G_{\boldsymbol{t}} \subset G$ is the isotropy subgroup of $\boldsymbol{t}$, and extend $\mathscr{B}^*$ to $G(\mathscr{B}^*)$. Then

$$\mathscr{B} = \bigcup_n \bigcup_{\substack{\alpha \in \mathscr{T}_n/G \\ |\mathscr{H}_{\boldsymbol{t}}| \geq 2, \boldsymbol{t} \in \alpha}} G(\mathscr{B}^*)$$

is a minimal G-invariant Markov basis.

This theorem only adds a statement of minimal G-invariance to the structure of a minimal Markov basis considered in Chap. 5

In principle this theorem can be used to construct a minimal invariant Markov basis by considering $\bigcup_{\alpha\in\mathscr{T}_n/G}\mathscr{B}_{n,\alpha}, n=1,2,3,\dots$ step by step. By the Hilbert basis theorem, there exists some n_0 (cf. Proposition 5.3), such that for $n\geq n_0$ no new moves need to be added. Then a minimal invariant Markov basis is written as $\bigcup_{n=1}^{n_0}\bigcup_{\alpha\in\mathscr{T}_n/G}\mathscr{B}_{n,\alpha}$.

To prove Theorem 7.1, we prepare some lemmas in the following.

First, we derive some basic properties of orbits of G acting on each fiber. As we stated before, we consider the action of G on each $\mathscr{X}_n$ separately. Let

$$\mathscr{F}_{G(\boldsymbol{t})}=\bigcup_{\boldsymbol{t}'\in G(\boldsymbol{t})}\mathscr{F}_{\boldsymbol{t}'}$$

denote the union of fibers over the orbit $G(\boldsymbol{t})$ through $\boldsymbol{t}$. Let $\boldsymbol{x}\in\mathscr{F}_{\boldsymbol{t}}$. Because $\boldsymbol{t}(g\boldsymbol{x})=g\boldsymbol{t}$, it follows that

$$g\boldsymbol{x}\in\mathscr{F}_{g\boldsymbol{t}}\subset\mathscr{F}_{G(\boldsymbol{t})}.$$

Therefore $G(\mathscr{F}_{G(\boldsymbol{t})})=\mathscr{F}_{G(\boldsymbol{t})}$. This implies that $\mathscr{X}_n$ is partitioned as

$$\mathscr{X}_n=\bigcup_{\alpha\in\mathscr{T}_n/G}\mathscr{F}_\alpha, \tag{7.16}$$

where α runs over the set of different orbits and we can consider the action of G on each $\mathscr{F}_{G(\boldsymbol{t})}$ separately.

Consider a particular $\mathscr{F}_{G(\boldsymbol{t})}$. An important observation is that there is a direct product structure in $\mathscr{F}_{G(\boldsymbol{t})}$. Write

$$G(\boldsymbol{t})=\{\boldsymbol{t}_1,\dots,\boldsymbol{t}_a\}, \tag{7.17}$$

where $a=a(\boldsymbol{t})=|G(\boldsymbol{t})|$ is the number of elements of the orbit $G(\boldsymbol{t})\subset\mathscr{T}_n$. Let $b=b(\boldsymbol{t})=|\mathscr{F}_{G(\boldsymbol{t})}/G|$ be the number of orbits of G acting on $\mathscr{F}_{G(\boldsymbol{t})}$ and let $\boldsymbol{x}_1,\dots,\boldsymbol{x}_b$ be representative elements of different orbits; that is,

$$\mathscr{F}_{G(\boldsymbol{t})}=G(\boldsymbol{x}_1)\cup\cdots\cup G(\boldsymbol{x}_b) \tag{7.18}$$

gives a partition of $\mathscr{F}_{G(\boldsymbol{t})}$. Then we have the following lemma.

Lemma 7.3. *We use the notations* (7.17) *and* (7.18). *Then $\mathscr{F}_{G(\boldsymbol{t})}$ is partitioned as*

$$\mathscr{F}_{G(\boldsymbol{t})}=\bigcup_{i=1}^{a}\bigcup_{j=1}^{b}\mathscr{F}_{\boldsymbol{t}_i}\cap G(\boldsymbol{x}_j), \tag{7.19}$$

where each $\mathscr{F}_{\boldsymbol{t}_i}\cap G(\boldsymbol{x}_j)$ is nonempty. Furthermore if $\boldsymbol{t}'_i=g\boldsymbol{t}_i$, then $\mathscr{F}_{\boldsymbol{t}_i}\ni\boldsymbol{x}\mapsto g\boldsymbol{x}\in\mathscr{F}_{\boldsymbol{t}'_i}$ gives a bijection between $\mathscr{F}_{\boldsymbol{t}_i}\cap G(\boldsymbol{x})$ and $\mathscr{F}_{\boldsymbol{t}'_i}\cap G(\boldsymbol{x})$.

Proof. Let $\mathscr{F}_{G(\boldsymbol{t})} = \mathscr{F}_{\boldsymbol{t}_1} \cup \cdots \cup \mathscr{F}_{\boldsymbol{t}_a}$ be a partition. Intersecting this partition with $\mathscr{F}_{G(\boldsymbol{t})} = \bigcup_{j=1}^{b} G(\boldsymbol{x}_j)$ gives the partition of (7.19). Let $\boldsymbol{x} \in \mathscr{F}_{\boldsymbol{t}}$. Then the orbit $G(\boldsymbol{x})$ intersects each fiber; that is, $G(\boldsymbol{x}) \cap \mathscr{F}_{\boldsymbol{t}_i} \neq \emptyset$ for $i = 1, \ldots, a$. Because every $g \in G$ is a bijection of $\mathscr{F}_{G(\boldsymbol{t})}$ to itself and

$$g(\mathscr{F}_{\boldsymbol{t}} \cap G(\boldsymbol{x})) = \mathscr{F}_{g\boldsymbol{t}} \cap G(\boldsymbol{x}),$$

g gives a bijection between $\mathscr{F}_{\boldsymbol{t}_i} \cap G(\boldsymbol{x})$ and $\mathscr{F}_{\boldsymbol{t}'_i} \cap G(\boldsymbol{x})$. □

In particular for each j, $\mathscr{F}_{\boldsymbol{t}_i} \cap G(\boldsymbol{x}_j)$, $i = 1, \ldots, a$, have the same number of elements

$$|\mathscr{F}_{\boldsymbol{t}_1} \cap G(\boldsymbol{x}_j)| = \cdots = |\mathscr{F}_{\boldsymbol{t}_a} \cap G(\boldsymbol{x}_j)|.$$

In addition, for $\boldsymbol{t}_i, \boldsymbol{t}'_i \in G(\boldsymbol{t})$ such that $\boldsymbol{t}'_i = g\boldsymbol{t}_i$, the map $g : G_{\boldsymbol{t}_i} \to gG_{\boldsymbol{t}_i}g^{-1}$ gives an isomorphism between $G_{\boldsymbol{t}_i}$ and $G_{\boldsymbol{t}'_i} = gG_{\boldsymbol{t}_i}g^{-1}$, where $G_{\boldsymbol{t}_i}$ and $G_{\boldsymbol{t}'_i}$ are the isotropy subgroup of $\boldsymbol{t}_i$ and $\boldsymbol{t}'_i$ in G, respectively. Therefore there exists the following isomorphic structures in $\mathscr{F}_{\boldsymbol{t}_i}$,

$$(G_{\boldsymbol{t}_i}, \mathscr{F}_{\boldsymbol{t}_i}) \simeq (G_{\boldsymbol{t}'_i}, \mathscr{F}_{\boldsymbol{t}'_i}). \tag{7.20}$$

Considering the isomorphic structure of (7.20), now we can focus our attention on each fiber. Consider a particular fiber $\mathscr{F}_{\boldsymbol{t}}$. Here we can restrict our attention to the action of $G_{\boldsymbol{t}}$ on $\mathscr{F}_{\boldsymbol{t}}$. As we have stated before, the relation between the action of $G_{\boldsymbol{t}}$ and $\mathscr{H}_{\boldsymbol{t}} = \mathscr{F}_{\boldsymbol{t}}/\mathscr{B}_{|\boldsymbol{t}|-1}$ (the $\mathscr{B}_{|\boldsymbol{t}|-1}$-equivalence classes of $\mathscr{F}_{\boldsymbol{t}}$) is essential. First we show the following lemma.

Lemma 7.4. *For any integer n, if $\boldsymbol{x}'$ is accessible from $\boldsymbol{x}$ by $\mathscr{B}_n$, then $g\boldsymbol{x}'$ is accessible from $g\boldsymbol{x}$ by $\mathscr{B}_n$.*

Proof. Note that $\deg \boldsymbol{z} \leq n$ if and only if $\deg(g\boldsymbol{z}) \leq n$. If $\boldsymbol{x}'$ is accessible from $\boldsymbol{x}$ by $\mathscr{B}_n$, then there exist $L > 0$, $\boldsymbol{z}_1, \ldots \boldsymbol{z}_L \in \mathscr{B}_n$, $\varepsilon_1, \ldots, \varepsilon_L \in \{-1, 1\}$, satisfying

$$\boldsymbol{x}' = \boldsymbol{x} + \sum_{s=1}^{L} \varepsilon_s \boldsymbol{z}_s, \quad \boldsymbol{x} + \sum_{s=1}^{l} \varepsilon_s \boldsymbol{z}_s \in \mathscr{F}_{\boldsymbol{t}} \quad \text{for } 1 \leq l \leq L.$$

Applying g to both sides of the equations we get

$$g\boldsymbol{x}' = g\boldsymbol{x} + \sum_{s=1}^{L} \varepsilon_s g\boldsymbol{z}_s, \quad g\boldsymbol{x} + \sum_{s=1}^{l} \varepsilon_s g\boldsymbol{z}_s \in \mathscr{F}_{g\boldsymbol{t}} \quad \text{for } 1 \leq l \leq L\,.$$

Because $g\boldsymbol{z}_s \in \mathscr{B}_n$ for $s = 1, \ldots, L$, the lemma is proved. □

This lemma holds for all $g \in G$. In particular, $g\boldsymbol{x} \in \mathscr{F}_{\boldsymbol{t}(\boldsymbol{x})}$ if $g \in G_{\boldsymbol{t}}$. This implies that an action of $G_{\boldsymbol{t}}$ is induced on $\mathscr{H}_{\boldsymbol{t}}$. In the sequel let $X_\gamma \in \mathscr{H}_{\boldsymbol{t}}$ denote each equivalence class:

$$\mathscr{H}_{\boldsymbol{t}} = \{X_\gamma\}_{1 \leq \gamma \leq |\mathscr{H}_{\boldsymbol{t}}|}.$$

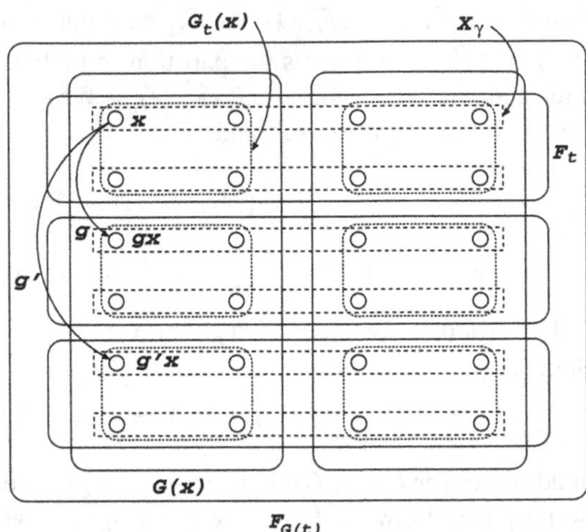

Fig. 7.1 A direct product structure of $\mathscr{F}_{G(t)}$ ($a = 3$, $b = 2, p = 1, q_i = 2, r_i = 2$)

Let $\pi : \boldsymbol{x} \mapsto X_\gamma$ denote the natural projection of $\boldsymbol{x}$ to its equivalence class. Then Lemma 7.4 states

$$\pi(\boldsymbol{x}) = \pi(\boldsymbol{x}') \Rightarrow \pi(g\boldsymbol{x}) = \pi(g\boldsymbol{x}').$$

Let $\boldsymbol{x} \in X_\gamma$ and $g \in G_{\boldsymbol{t}}$. Then $g\boldsymbol{x}$ belongs to some $\mathscr{B}_{n-1}$-equivalence class $X_{\gamma'}$. By Lemma 7.4, this γ' does not depend on the choice of $\boldsymbol{x} \in X_\gamma$ and we may write $\gamma' = g\gamma$. By definition a group action is bijective, therefore the following lemma holds.

Lemma 7.5. *$g \in G_{\boldsymbol{t}} : X_\gamma \mapsto X_{\gamma'}$ is a bijection of $\mathscr{H}_{\boldsymbol{t}}$ to itself.*

Now we give a proof of Theorem 7.1.

Proof (Theorem 7.1). Let $\mathscr{B}$ be a minimal invariant Markov basis and consider the partition (7.14). Then each $\mathscr{B}_{n,\alpha}, \alpha \in \mathscr{T}_n/G$, is G-invariant from Lemma 7.2. Moreover, from the argument of Chap. 5, each $\boldsymbol{z} = \boldsymbol{z}^+ - \boldsymbol{z}^- \in \mathscr{B}_{n,\alpha}$ is a move connecting $X_\gamma \in \mathscr{H}_{\boldsymbol{t}}$ and $X_{\gamma'} \in \mathscr{H}_{\boldsymbol{t}}$, $\gamma \neq \gamma'$, that is, $\boldsymbol{z}^+ \in X_\gamma$ and $\boldsymbol{z}^- \in X_{\gamma'}$, from the minimality of $\mathscr{B}$. In this case, $g\boldsymbol{z} = g\boldsymbol{z}^+ - g\boldsymbol{z}^-$ is a move connecting $X_{g\gamma}$ and $X_{g\gamma'}$. Applying g^{-1} the converse is also true. This implies that the way $\mathscr{B}_{n,\alpha} \cap \mathscr{B}_{\boldsymbol{t}}$ connects the $\mathscr{B}_{n-1}$-equivalence classes $\mathscr{H}_{\boldsymbol{t}}$ is the same for all $\boldsymbol{t} \in \alpha$ and hence the relation (7.15) holds.

Conversely, to construct a minimal invariant Markov basis, we only have to consider sets of moves connecting $\mathscr{B}_{|\boldsymbol{t}|-1}$-equivalence classes of each $\mathscr{F}_{\boldsymbol{t}}$ from the argument of Chap. 5. Considering the isomorphic structure (7.20) of Lemma 7.3 and Lemma 7.5, we see that the structure of $\mathscr{H}_{\boldsymbol{t}'}$ is common for all $\boldsymbol{t}' \in G(\boldsymbol{t})$. Therefore it suffices to consider the $G_{\boldsymbol{t}}$-invariant set of moves $\mathscr{B}_{\boldsymbol{t}}$ for some representative sufficient statistic $\boldsymbol{t} \in \alpha$ satisfying $|\mathscr{H}_{\boldsymbol{t}}| \geq 2$ for each $\alpha \in \mathscr{T}_n/G$. □

Here we give an illustration of a direct product structure of $\mathscr{F}_{G(\boldsymbol{t})}$. Figure 7.1 shows a structure of $\mathscr{F}_{G(\boldsymbol{t})}$ where $a = a(\boldsymbol{t}) = |G(\boldsymbol{t})| = 3$ and $b = b(\boldsymbol{t}) = |\mathscr{F}_{G(\boldsymbol{t})}/G| = 2$. In each $\mathscr{F}_{\boldsymbol{t}} \subset \mathscr{F}_{G(\boldsymbol{t})}$, there are two $\mathscr{B}_{|\boldsymbol{t}|-1}$-equivalence classes: $|\mathscr{H}_{\boldsymbol{t}}| = 2$.

Figure 7.1 also shows G_t orbits in each $\mathscr{F}_t$. In fact, Fig. 7.1 is derived from an example of $2 \times 2 \times 2 \times 2$ contingency tables, where the following marginals are fixed:

$$D_1 = \{1,2\}, \quad D_2 = \{1,3\}, \quad D_3 = \{2,3\}, \quad D_4 = \{3,4\}.$$

We see the above structure by considering $\boldsymbol{x} = (1111)(1221)(2122)(2212)$, for example. In this case, $\mathscr{F}_{\boldsymbol{t}(\boldsymbol{x})}$ is an eight-element set as follows.

$$\begin{array}{l} \left.\begin{array}{l}(1111)(1221)(2122)(2212),\ (1111)(1222)(2121)(2212),\\ (1112)(1222)(2121)(2211),\ (1112)(1221)(2122)(2211),\end{array}\right\} X_\gamma(\ni \boldsymbol{x}) \\ (1121)(1211)(2112)(2222),\ (1121)(1212)(2111)(2222), \\ \underbrace{(1122)(1212)(2111)(2221)}_{G_t(\boldsymbol{x})},\ (1122)(1211)(2112)(2221). \end{array}$$

In this chapter we have discussed properties of minimal invariant Markov bases. Then a natural question is to seek some conditions for the uniqueness of a minimal invariant Markov basis. In [13] we gave some characterizations of the uniqueness of a minimal invariant Markov basis. However, the characterizations are not simple and the argument is rather long. Therefore we omit discussion of uniqueness of a minimal invariant Markov basis.

Part III
Markov Bases for Specific Models

In Part III of this book, we present results on Markov bases for some specific models, which are important for applications. We give many numerical examples to illustrate the application of Markov basis methodology to practical statistical problems.

In Chap. 8 we give a thorough discussion of Markov bases for decomposable models of contingency tables. For decomposable models we have a complete description of minimal Markov bases and minimal invariant Markov bases.

In Chap. 9 we discuss Markov bases for no-three-factor interaction models of three-way contingency tables and some other hierarchical models. We see that for general hierarchical models the structure of Markov bases is very complicated.

In Chap. 10 we discuss two-way tables with structural zeros and fixed subtable sums. We give explicit forms of Markov bases and give some numerical examples of a running Markov chain with the obtained Markov bases.

In Chap. 11 we explain applications of the Markov basis approach to experimental designs, where the response variables are discrete. In standard textbooks on experimental design, the response variables are usually assumed to be normally distributed. When response variables are discrete it is more appropriate to use exact tests. We give many numerical examples, because this topic is of practical importance.

In Chap. 12 we introduce groupwise selection models, where the Gröbner basis approach works particularly well and testing these models can be performed easily. We illustrate the use of these models by analyzing educational and allele frequency data.

Finally in Chap. 13 we study the problem of connecting some specific fibers by a subset of a Markov basis. In some problems, when we consider connectivity of specific fibers, it is possible to describe a subset of a Markov basis that connects these fibers. A typical example is the logistic regression model with positive sample size for each level of a covariate.

Chapter 8
Decomposable Models of Contingency Tables

8.1 Chordal Graphs and Decomposable Models

In this section we summarize some properties of the decomposable model and chordal graphs according to Lauritzen [97] and Hara and Takemura [74, 75, 76].

We use the notation of hierarchical models introduced in Sect. 1.5. Let $\Delta = [m] = \{1,\dots,m\}$ denote the set of variables of an m-way contingency table $\boldsymbol{x} = \{x(\boldsymbol{i}) \mid \boldsymbol{i} \in \mathscr{I}\}$. Let $\mathscr{D} = \{D_1,\dots,D_r\}$ be the set of facets of a simplicial complex $\mathscr{K}$ such that $\Delta = \cup_{j=1}^r D_j$. Let $p(\boldsymbol{i})$ denote the cell probability for $\boldsymbol{i}$. Then the hierarchical model for $\mathscr{D}$ is defined as

$$\log p(\boldsymbol{i}) = \sum_{D \in \mathscr{D}} \mu_D(\boldsymbol{i}),$$

where μ_D depends only on $\boldsymbol{i}_D$. $\mathscr{D}$ is called a generating class for the model. In this chapter, we often identify a hierarchical model with its generating class $\mathscr{D}$.

As defined in Sect. 1.4, for a subset of the variables $V \subset \Delta$, let $\boldsymbol{x}_V$ and $\boldsymbol{z}_V$ denote the V-marginal sums of $\boldsymbol{x}$ and $\boldsymbol{z}$ with entries given by

$$x_V(\boldsymbol{i}_V) = \sum_{\boldsymbol{i}_{V^C} \in \mathscr{I}_{V^C}} x(\boldsymbol{i}_V, \boldsymbol{i}_{V^C}), \quad z_V(\boldsymbol{i}_V) = \sum_{\boldsymbol{i}_{V^C} \in \mathscr{I}_{V^C}} z(\boldsymbol{i}_V, \boldsymbol{i}_{V^C})$$

for $\boldsymbol{i}_V \in \mathscr{I}_V = \prod_{\delta \in V} \mathscr{I}_\delta$. We often denote $\boldsymbol{i} = (\boldsymbol{i}_V, \boldsymbol{i}_{V^C})$ by appropriately reordering indices. A sufficient statistic $\boldsymbol{t}$ for $\mathscr{D}$ is the set of marginal sums for all $D \in \mathscr{D}$,

$$\boldsymbol{t} = \{\boldsymbol{x}_D \mid D \in \mathscr{D}\}.$$

Hence a move $\boldsymbol{z}$ for the generating class $\mathscr{D}$ satisfies $\boldsymbol{z}_D = \boldsymbol{0}$ for all $D \in \mathscr{D}$.

Marginal tables $\boldsymbol{x}_{D_1},\dots,\boldsymbol{x}_{D_r}$ are called *consistent* if, for any r_1, r_2, $(D_{r_1} \cap D_{r_2})$-marginal of $\boldsymbol{x}_{D_{r_1}}$ is equal to the $(D_{r_1} \cap D_{r_2})$-marginal of $\boldsymbol{x}_{D_{r_2}}$ ([52]). The consistency of the marginal tables is obviously a necessary condition for the existence of $\boldsymbol{x}$. However, it does not necessarily guarantee the existence of $\boldsymbol{x}$ in general (e.g., [46, 91, 148]). This is closely related to the notion of normality of semigroups given in Sect. 4.3. We again discuss normality in Sect. 9.5.

S. Aoki et al., *Markov Bases in Algebraic Statistics*, Springer Series in Statistics 199, DOI 10.1007/978-1-4614-3719-2_8,

Let $\mathscr{G}$ be a graph with the vertex set Δ and an edge between $\delta, \delta' \in \Delta$ if and only if there exists $D \in \mathscr{D}$ such that $\delta, \delta' \in D$. $\mathscr{G}$ is called the *independence graph* of $\mathscr{D}$ (e.g., Dobra and Sullivant [54]). For $V \subset \Delta$, denote by $\mathscr{G}(V)$ the subgraph induced by V; that is, V is the set of vertices of $\mathscr{G}(V)$ and the edges of $\mathscr{G}(V)$ are those in $\mathscr{G}$ restricted to V. $V \subset \Delta$ is called a *clique* if $\mathscr{G}(V)$ is complete; that is, every pair of vertices in V is an edge of $\mathscr{G}$. A clique V of $\mathscr{G}$ is called maximal if every proper superset of V is not a clique of $\mathscr{G}$. A hierarchical model for $\mathscr{D}$ is called *graphical* if there exists a graph whose set of maximal cliques is given by $\mathscr{D}$ (e.g., Edwards [57]).

A graphical model is called *decomposable* if $\mathscr{G}$ is chordal; that is, every cycle of $\mathscr{G}$ with length greater than three has a chord. A *clique tree* $\mathscr{T} = (\mathscr{D}, \mathscr{E})$ of $\mathscr{G}$ is a tree with the vertex set $\mathscr{D}$ satisfying

$$D \cap D'' \subset D' \quad \text{for all } D' \text{ on the path between } D \text{ and } D'' \text{ in } \mathscr{T}.$$

A graph is chordal if and only if there exists a clique tree of it ([30, 64]). When $(D, D') \in \mathscr{E}$, $S = D \cap D'$ is called a *minimal vertex separator* of $\mathscr{G}$. Let $\mathscr{S}$ be the *multiset*

$$\mathscr{S} := \{D \cap D' \mid (D, D') \in \mathscr{E}\},$$

where the same minimal vertex separator may be included several times (e.g., [97]). Denote a marginal probability for $\boldsymbol{i}_D$ by $p_D(\boldsymbol{i}_D)$. Then $p(\boldsymbol{i})$ and its maximum likelihood estimator $\hat{p}(\boldsymbol{i})$ are written by

$$p(\boldsymbol{i}) = \frac{\prod_{D \in \mathscr{D}} p_D(\boldsymbol{i}_D)}{\prod_{S \in \mathscr{S}} p_S(\boldsymbol{i}_S)}, \quad \hat{p}(\boldsymbol{i}) = \frac{\prod_{D \in \mathscr{D}} x_D(\boldsymbol{i}_D)}{n \prod_{S \in \mathscr{S}} x_S(\boldsymbol{i}_S)},$$

respectively.

A vertex is called *simplicial* if its adjacent vertices form a clique of $\mathscr{G}$. Any chordal graph with at least two vertices has at least two simplicial vertices and if the graph is not complete, these can be chosen to be nonadjacent ([51]). For $D \in \mathscr{D}$ of a decomposable model, let Simp(D) denote the set of simplicial vertices in D and let Sep(D) denote the set of nonsimplicial vertices in D. If Simp$(D) \neq \emptyset$, D is called a *simplicial clique*. A simplicial clique D is called a *boundary clique* if there exists another clique $D' \in \mathscr{D}$ such that Sep$(D) = D \cap D'$ ([137]). Simplicial vertices in boundary cliques are called *simply separated vertices*. A maximal clique D is a boundary clique if and only if there exists a clique tree such that D is its endpoint ([74]).

Dobra [52] showed that decomposable models have a Markov basis consisting of only square-free moves of degree 2. In the next section, we give a proof of this fact. For convenience, denote a square-free move $\boldsymbol{z}$ of degree 2 with $z(\boldsymbol{i}) = z(\boldsymbol{j}) = 1$ and $z(\boldsymbol{i}') = z(\boldsymbol{j}') = -1$ by

$$\boldsymbol{z} = \boldsymbol{i}\boldsymbol{j} - \boldsymbol{i}'\boldsymbol{j}'.$$

This notation was already used in Sect. 7.2. Similarly, $\boldsymbol{x} = \boldsymbol{i}\boldsymbol{j}$ denotes a frequency vector with one frequency at cells $\boldsymbol{i}$ and $\boldsymbol{j}$.

8.2 Markov Bases for Decomposable Models

The simplest decomposable model is the two-way complete independence model $\mathscr{D} = \{\{1\},\{2\}\}$. Consider an $R \times C$ table. A sufficient statistics for this model is the set of row sums and column sums, thus every move $\boldsymbol{z} = \{z_{ij}\}$ satisfies

$$z_{i+} = \sum_{j=1}^{C} z_{ij} = 0, \qquad z_{+j} = \sum_{i=1}^{R} z_{ij} = 0.$$

In Theorem 2.1 we saw that the set of the following degree 2 moves

$$\boldsymbol{z}(i,i';j,j') := (ij)(i'j') - (ij')(i'j), \quad 1 \le i < i' \le R, \quad 1 \le j < j' \le C \tag{8.1}$$

forms a Markov basis for $R \times C$ two-way complete independence models.

Consider a decomposable model consisting of two maximal cliques $\mathscr{D} = \{D, D'\}$. Denote $A := D \setminus D'$, $B := D' \setminus D$, and $S := D \cap D'$. When $A = \{1\}$, $B = \{2\}$, and $S = \emptyset$, the model coincides with the two-way complete independence model. When $D = \{1,2\}, D' = \{2,3\}, A = \{1\}, B = \{3\}$, and $S = \{2\}$, the model coincides with the conditional independence model of three-way contingency tables in Sect. 1.4.

Proposition 8.1. *Define $\mathscr{B}(D,D')$ by the following set of square-free moves of degree* 2,

$$\mathscr{B}(D,D') = \{(\boldsymbol{i}_A\boldsymbol{i}_S\boldsymbol{i}_B)(\boldsymbol{i}'_A\boldsymbol{i}_S\boldsymbol{i}'_B) - (\boldsymbol{i}_A\boldsymbol{i}_S\boldsymbol{i}'_B)(\boldsymbol{i}'_A\boldsymbol{i}_S\boldsymbol{i}_B) \mid \boldsymbol{i}_A, \boldsymbol{i}'_A \in \mathscr{I}_A, \boldsymbol{i}_B, \boldsymbol{i}'_B \in \mathscr{I}_B, \boldsymbol{i}_S \in \mathscr{I}_S\}.$$

Then $\mathscr{B}(D,D')$ forms a Markov basis for $\mathscr{D} = \{D,D'\}$.

Proof. Let $\boldsymbol{x}, \boldsymbol{y}$ $(\boldsymbol{y} \neq \boldsymbol{x})$ be two tables in the same fiber of the model $\mathscr{D}$ and let $\boldsymbol{z} = \boldsymbol{y} - \boldsymbol{x}$. Denote by $\boldsymbol{z}^{\boldsymbol{i}_S}$ the $\boldsymbol{i}_S$-slice of $\boldsymbol{z}$:

$$\boldsymbol{z}^{\boldsymbol{i}_S} = \{z(\boldsymbol{i}_A\boldsymbol{i}_S\boldsymbol{i}_B) \mid \boldsymbol{i}_A \in \mathscr{I}_A, \boldsymbol{i}_B \in \mathscr{I}_B\}.$$

Assume $\boldsymbol{z}^{\boldsymbol{i}_S} \neq \boldsymbol{0}$ without loss of generality. Consider $\boldsymbol{z}^{\boldsymbol{i}_S}$ as a two-way integer array with the set of levels $\mathscr{I}_A \times \mathscr{I}_B$. Let $\boldsymbol{z}_A^{\boldsymbol{i}_S}$ and $\boldsymbol{z}_B^{\boldsymbol{i}_S}$ denote the A-marginal table and the B-marginal table of $\boldsymbol{z}^{\boldsymbol{i}_S}$, respectively. The assumption that $\boldsymbol{z}_D = \boldsymbol{0}$ and $\boldsymbol{z}_{D'} = \boldsymbol{0}$ implies that $\boldsymbol{z}_A^{\boldsymbol{i}_S} = \boldsymbol{0}$ and $\boldsymbol{z}_B^{\boldsymbol{i}_S} = \boldsymbol{0}$. Therefore $\boldsymbol{z}^{\boldsymbol{i}_S}$ is regarded as a move of a two-way complete independence model. Hence from Theorem 2.1 we can reduce $|\boldsymbol{z}|_1$ by a square-free move of degree 2 of the form $(\boldsymbol{i}_A\boldsymbol{i}_S\boldsymbol{i}_B)(\boldsymbol{i}'_A\boldsymbol{i}_S\boldsymbol{i}'_B) - (\boldsymbol{i}_A\boldsymbol{i}_S\boldsymbol{i}'_B)(\boldsymbol{i}'_A\boldsymbol{i}_S\boldsymbol{i}_B)$. □

The above arguments are generalized to general decomposable models. Let $\mathscr{T}$ be a clique tree of $\mathscr{G}$. Denote by $\mathscr{T}_e = (\mathscr{D}_e, \mathscr{E}_e)$ and $\mathscr{T}'_e = (\mathscr{D}'_e, \mathscr{E}'_e)$ the two induced subtrees of $\mathscr{T}$ obtained by removing an edge $e \in \mathscr{E}$ from $\mathscr{T}$. Let V_e and V'_e be

$$V_e = \bigcup_{D \in \mathscr{D}_e} D, \qquad V'_e = \bigcup_{D \in \mathscr{D}'_e} D.$$

Then $\mathscr{T}_e$ and $\mathscr{T}_e'$ are clique trees of chordal graphs $\mathscr{G}(V_e)$ and $\mathscr{G}(V_e')$, respectively. Define the set $\mathscr{B}^{\mathscr{T}}$ of square-free moves of degree 2 as

$$\mathscr{B}^{\mathscr{T}} = \bigcup_{e\in\mathscr{E}} \mathscr{B}(V_e, V_e'). \tag{8.2}$$

Denote $S_e := V_e \cap V_e'$, $R_e := V_e \setminus S_e$ and $R_e' := V_e' \setminus S_e$.

Lemma 8.1. *Suppose that*

$$\begin{gathered} \boldsymbol{z}^* = (\boldsymbol{i}_{R_e}\boldsymbol{i}_{S_e})(\boldsymbol{i}'_{R_e}\boldsymbol{i}'_{S_e}) - (\boldsymbol{j}_{R_e}\boldsymbol{i}_{S_e})(\boldsymbol{j}'_{R_e}\boldsymbol{i}'_{S_e}) \in \mathscr{B}^{\mathscr{T}_e}, \\ \boldsymbol{i}_{R_e}, \boldsymbol{i}'_{R_e}, \boldsymbol{j}_{R_e}, \boldsymbol{j}'_{R_e} \in \mathscr{I}_{R_e}, \quad \boldsymbol{i}_{S_e}, \boldsymbol{i}'_{S_e} \in \mathscr{I}_{S_e} \end{gathered}$$

is a move of $\mathscr{D}_e$. Then

$$\boldsymbol{z} = (\boldsymbol{i}_{R_e}\boldsymbol{i}_{S_e}\boldsymbol{i}_{R_e'})(\boldsymbol{i}'_{R_e}\boldsymbol{i}'_{S_e}\boldsymbol{i}'_{R_e'}) - (\boldsymbol{j}_{R_e}\boldsymbol{i}_{S_e}\boldsymbol{i}_{R_e'})(\boldsymbol{j}'_{R_e}\boldsymbol{i}'_{S_e}\boldsymbol{i}'_{R_e'}) \in \mathscr{B}^{\mathscr{T}}$$

for any $\boldsymbol{i}_{R_e'}, \boldsymbol{i}'_{R_e'} \in \mathscr{I}_{R_e'}$.

Proof. Obviously $\boldsymbol{z}_{V_e'} = \boldsymbol{0}$ and hence $\boldsymbol{z}_D = \boldsymbol{0}$ for all $D \in \mathscr{D}_e'$. Because $\boldsymbol{z}_{V_e} = \boldsymbol{z}^*$, we also have $\boldsymbol{z}_D = \boldsymbol{0}$ for all $D \in \mathscr{D}_e$. Hence $\boldsymbol{z}$ is a move for $\mathscr{D} = \mathscr{D}_e \cup \mathscr{D}_e'$. Because $\boldsymbol{z}_{V_e'} = \boldsymbol{0}$, there exists an edge $e^* \in \mathscr{E}_e$ such that $\boldsymbol{z} \in \mathscr{B}(V_{e^*}, V_{e^*}')$. □

Theorem 8.1 (Dobra [52]). *$\mathscr{B}^{\mathscr{T}}$ forms a Markov basis of the decomposable model $\mathscr{D}$.*

Proof. The proof is by induction on the number of maximal cliques r. When $r = 2$, $\mathscr{B}^{\mathscr{T}}$ coincides with the Markov basis in Proposition 8.1. Suppose that the theorem holds for any decomposable models with $r-1$ maximal cliques.

Let $\boldsymbol{x}, \boldsymbol{y}$ ($\boldsymbol{y} \neq \boldsymbol{x}$) be two tables in the same fiber $\mathscr{F}$ of the decomposable model $\mathscr{D}$. Let $D \in \mathscr{D}$ be an endpoint of $\mathscr{T}$ and suppose that $e := (D, D') \in \mathscr{E}$. Then we can set

$$V_e = \Delta \setminus (D \setminus D'), \quad V_e' = D, \quad \mathscr{D}_e = \mathscr{D} \setminus \{D\}, \quad \mathscr{D}_e' = \{D\}$$

and define $\mathscr{T}_e$ as above. Then the marginal tables $\boldsymbol{x}_{V_e}$ and $\boldsymbol{y}_{V_e}$ lie in the same fiber $\mathscr{F}'$ of $\mathscr{D}_e$. From the inductive assumption, $\mathscr{B}^{\mathscr{T}_e}$ is a Markov basis for $\mathscr{D}_e$. Hence there exists a sequence of moves $\boldsymbol{z}^1_{V_e}, \ldots, \boldsymbol{z}^l_{V_e}$ such that

$$\boldsymbol{y}_{V_e} = \boldsymbol{x}_{V_e} + \sum_{k=1}^{l} \boldsymbol{z}^k_{V_e}, \quad \boldsymbol{x}_{V_e} + \sum_{k=1}^{l'} \boldsymbol{z}^k_{V_e} \in \mathscr{F}'$$

for $1 \leq l' \leq l$. From Lemma 8.1, there exists a sequence of moves $\boldsymbol{z}^1, \ldots, \boldsymbol{z}^l$ of $\mathscr{D}'$ such that

$$\boldsymbol{x} + \sum_{k=1}^{l'} \boldsymbol{z}^k \in \mathscr{F}$$

for $1 \le l' \le l$. Define $\boldsymbol{y}' = \boldsymbol{x} + \sum_{k=1}^{l} \boldsymbol{z}^k$. Then $\boldsymbol{y}$ and $\boldsymbol{y}'$ satisfy $\boldsymbol{y}_{V_e} = \boldsymbol{y}'_{V_e}$ and $\boldsymbol{y}_{V'_e} = \boldsymbol{y}'_{V'_e}$. Hence $\boldsymbol{y}$ and $\boldsymbol{y}'$ are accessible by moves in $\mathscr{B}(V_e, V'_e)$. □

Dobra [52] proposed the following algorithm for generating moves from $\mathscr{B}^{\mathscr{T}}$.

Algorithm 8.1 (Dobra [52])

1. For each edge $e \in \mathscr{E}$ of $\mathscr{T}$:

   a. Define V_e, V'_e and S_e as above.
   b. Calculate the weights w_e representing the number of degree 2 moves:

$$w_e \leftarrow \left[2 \cdot \prod_{\delta \in V_e \setminus S_e} \binom{I_\delta}{2} \cdot \prod_{\delta \in V'_e \setminus S_e} \binom{I_\delta}{2} \right]^{\prod_{\delta \in S_e} I_\delta}.$$

2. Normalize the weights $w_2, \ldots, w_r$.
3. Randomly select an edge $e \in \mathscr{E}$ with probability w_e.
4. Uniformly pick up a move in $\mathscr{B}(V_e, V'_e)$.

8.3 Structure of Degree 2 Fibers

In the previous section we showed that every decomposable model has a Markov basis consisting of square-free moves of degree 2. As discussed in Sect. 5.3, the set of fibers of the minimum fiber Markov basis for a decomposable model coincides with the set of degree 2 fibers with more than one element. Therefore the structure of degree 2 moves is equivalent to that of degree 2 fibers. In this section, we discuss the structure of such fibers in detail this section is mainly based on [71].

Let $\mathscr{F}_{\boldsymbol{t}}$ be a fiber with $\deg \boldsymbol{t} = 2$. For a given $\boldsymbol{t}$ we say that a variable $\delta \in \Delta$ is *degenerate* if there exists a unique level i_δ such that $x_{\{\delta\}}(i_\delta) = 2$. Otherwise, if there exist two levels $i_\delta \ne i'_\delta$ such that $x_{\{\delta\}}(i_\delta) = x_{\{\delta\}}(i'_\delta) = 1$, then we say that δ is *nondegenerate*. Degeneracy or nondegeneracy of δ does not depend on a particular $\boldsymbol{x} \in \mathscr{F}_{\boldsymbol{t}}$, because one-dimensional marginals are determined from marginals of the facets $\boldsymbol{x}_D$, $D \in \mathscr{D}$. If all the variables $\delta \in \Delta$ are degenerate, then $\mathscr{F}_{\boldsymbol{t}} = \{\boldsymbol{x}\}$ is a one-element fiber with frequency $x(\boldsymbol{i}) = 2$ at a particular cell $\boldsymbol{i}$. This case is trivial, therefore below we consider the case that at least one variable is nondegenerate.

From the fact that there exist at most two levels with positive one-dimensional marginals for each variable, it follows that we only need to consider $2 \times \cdots \times 2$ tables for studying degree 2 fibers. Therefore we set $I_1 = \cdots = I_m = 2$, $\mathscr{I} = \{0,1\}^m$ without loss of generality.

For a given $\boldsymbol{t}$ of degree 2, let $\bar{\Delta}_{\boldsymbol{t}}$ denote the set of nondegenerate variables. As noted above, we assume that $\bar{\Delta}_{\boldsymbol{t}} \ne \emptyset$. Each $\boldsymbol{x} \in \mathscr{F}_{\boldsymbol{t}}$ is of the form $x(\boldsymbol{i}) = x(\boldsymbol{i}') = 1$ for $\boldsymbol{i} \ne \boldsymbol{i}'$ and remaining entries are 0. For nondegenerate $\delta \in \bar{\Delta}_{\boldsymbol{t}}$ the levels of the variable δ in $\boldsymbol{i}$ and $\boldsymbol{i}'$ are different:

$$\{i_\delta, i'_\delta\} = \{0,1\}, \quad \forall \delta \in \bar{\Delta}_{\boldsymbol{t}},$$

or equivalently $i'_\delta = 1 - i_\delta$, $\forall \delta \in \bar{\Delta}_t$. In the following, we use the notation $i^*_\delta = 1 - i_\delta$. More generally, for a subset of variables $V = \{\delta_1, \ldots, \delta_k\}$ and a marginal cell $\boldsymbol{i}_V = (i_{\delta_1}, \ldots, i_{\delta_k})$, we write

$$\boldsymbol{i}^*_V \equiv (i^*_{\delta_1}, \ldots, i^*_{\delta_k}) = (1 - i_{\delta_1}, \ldots, 1 - i_{\delta_k}).$$

Let us identify $\boldsymbol{x} = \boldsymbol{i}\boldsymbol{i}' \in \mathscr{F}_{\boldsymbol{t}}$ with the set $\{\boldsymbol{i}, \boldsymbol{i}'\}$ of its two cells of frequency one. Then we see that the number of elements of fibers $|\mathscr{F}_{\boldsymbol{t}}|$ is at most $2^{|\bar{\Delta}_t|-1}$. Let $\mathscr{G}(\bar{\Delta}_t)$ be the subgraph of $\mathscr{G}$ induced by $\bar{\Delta}_t \subset \Delta$.

Lemma 8.2. *Suppose that $\boldsymbol{t}$ is a set of consistent marginal frequencies of a contingency table with* $\deg \boldsymbol{t} = 2$. *Let Γ be any subset of a connected component in $\mathscr{G}(\bar{\Delta}_t)$. Then the marginal table $\boldsymbol{x}_\Gamma = \{x_\Gamma(\boldsymbol{i}_\Gamma) \mid i_\Gamma \in \mathscr{I}_\Gamma\}$ is uniquely determined.*

Proof. Let $r(\Gamma)$ be the number of generating sets $D \in \mathscr{D}$ satisfying $\Gamma \cap D \neq \emptyset$. We prove this lemma by induction on $r(\Gamma)$. When $r(\Gamma) = 1$, the lemma obviously holds. Suppose that the lemma holds for all $r(\Gamma) < r'$ and we now assume that $r(\Gamma) = r'$. Let $\Gamma_1 \subset \Gamma$ and $\Gamma_2 \subset \Gamma$ satisfy

$$\Gamma_1 \cup \Gamma_2 = \Gamma, \quad \Gamma_1 \cap \Gamma_2 \neq \emptyset, \quad r(\Gamma_1) < r', \quad r(\Gamma_2) < r'.$$

Because $r(\Gamma_1) < r'$ and $r(\Gamma_2) < r'$, both $\boldsymbol{x}_{\Gamma_1}$ and $\boldsymbol{x}_{\Gamma_2}$ are uniquely determined. Suppose that

$$x_{\Gamma_1}(\boldsymbol{i}_{\Gamma_1 \setminus \Gamma_2}, \boldsymbol{i}_{\Gamma_1 \cap \Gamma_2}) = 1, \quad x_{\Gamma_1}(\boldsymbol{i}^*_{\Gamma_1 \setminus \Gamma_2}, \boldsymbol{i}^*_{\Gamma_1 \cap \Gamma_2}) = 1. \tag{8.3}$$

Then from the consistency of $\boldsymbol{t}$, there uniquely exists $\boldsymbol{i}_{\Gamma_2 \setminus \Gamma_1} \in \mathscr{I}_{\Gamma_2 \setminus \Gamma_1}$, such that

$$x_{\Gamma_2}(\boldsymbol{i}_{\Gamma_2 \setminus \Gamma_1}, \boldsymbol{i}_{\Gamma_1 \cap \Gamma_2}) = 1, \quad x_{\Gamma_2}(\boldsymbol{i}^*_{\Gamma_2 \setminus \Gamma_1}, \boldsymbol{i}^*_{\Gamma_1 \cap \Gamma_2}) = 1. \tag{8.4}$$

Hence the table $\boldsymbol{x}_\Gamma = \{x(\boldsymbol{j}_\Gamma) \mid \boldsymbol{j}_\Gamma \in \mathscr{I}_\Gamma\}$ with entries

$$x(\boldsymbol{j}_\Gamma) = \begin{cases} 1, & \text{if} \quad \boldsymbol{j}_\Gamma = (\boldsymbol{i}_{\Gamma_1 \setminus \Gamma_2}, \boldsymbol{i}_{\Gamma_1 \cap \Gamma_2}, \boldsymbol{i}_{\Gamma_2 \setminus \Gamma_1}) \text{ or } \boldsymbol{j}_\Gamma = (\boldsymbol{i}^*_{\Gamma_1 \setminus \Gamma_2}, \boldsymbol{i}^*_{\Gamma_1 \cap \Gamma_2}, \boldsymbol{i}^*_{\Gamma_2 \setminus \Gamma_1}), \\ 0, & \text{otherwise} \end{cases}$$

is consistent with $\boldsymbol{t}$.

Suppose that there exists another marginal table $\boldsymbol{x}'_\Gamma$ which is consistent with $\boldsymbol{t}$ such that $x_\Gamma(\boldsymbol{j}_\Gamma) = x_\Gamma(\boldsymbol{j}^*_\Gamma) = 1$ and $\boldsymbol{j}_\Gamma \neq (\boldsymbol{i}_{\Gamma_1 \setminus \Gamma_2}, \boldsymbol{i}_{\Gamma_1 \cap \Gamma_2}, \boldsymbol{i}_{\Gamma_2 \setminus \Gamma_1})$. Then we have at least one of

$$x_{\Gamma_1}(\boldsymbol{i}_{\Gamma_1}) = 0 \quad \text{or} \quad x_{\Gamma_2}(\boldsymbol{i}_{\Gamma_2}) = 0.$$

This contradicts (8.3) and (8.4). □

Theorem 8.2. *Let $\mathscr{F}_{\boldsymbol{t}}$ be a degree 2 fiber such that $\bar{\Delta}_t \neq \emptyset$ and let $c(\boldsymbol{t})$ be the number of connected components of $\mathscr{G}(\bar{\Delta}_t)$. Then*

$$|\mathscr{F}_{\boldsymbol{t}}| = 2^{c(\boldsymbol{t})-1}.$$

Proof. Denote by $\Gamma_1,\dots,\Gamma_c$, $c = c(\boldsymbol{t})$, the connected components of $\mathscr{G}(\bar{\Delta}_{\boldsymbol{t}})$. Define $\Gamma_{c+1} := \Delta \setminus \bar{\Delta}_{\boldsymbol{t}}$. By definition, there exists $\boldsymbol{i}_{\Gamma_{c+1}}$ such that

$$\boldsymbol{i}_{\Gamma_{c+1}} = \{\boldsymbol{i}_\delta \mid \delta \in \Gamma_{c+1},\ x_{\{\delta\}}(\boldsymbol{i}_\delta) = 2\}.$$

From Lemma 8.2, the marginal cells $\boldsymbol{i}_{\Gamma_k}$, $k = 1,\dots,c$, satisfying $x_{\Gamma_k}(\boldsymbol{i}_{\Gamma_k}) = x_{\Gamma_k}(\boldsymbol{i}^*_{\Gamma_k}) = 1$ uniquely exist. Now define $\mathscr{I}_{\boldsymbol{t}}$ by

$$\mathscr{I}_{\boldsymbol{t}} = \{\boldsymbol{i}_{\Gamma_1}, \boldsymbol{i}^*_{\Gamma_1}\} \times \{\boldsymbol{i}_{\Gamma_2}, \boldsymbol{i}^*_{\Gamma_2}\} \times \cdots \times \{\boldsymbol{i}_{\Gamma_c}, \boldsymbol{i}^*_{\Gamma_c}\} \times \{\boldsymbol{i}_{\Gamma_{c+1}}\},$$

where $\times$ denotes the direct product of sets. Suppose that $\boldsymbol{j} \in \mathscr{I}_{\boldsymbol{t}}$. Define

$$\boldsymbol{x}^{\boldsymbol{j}} = \{x^{\boldsymbol{j}}(\boldsymbol{i}) \mid \boldsymbol{i} \in \mathscr{I}\}, \quad x^{\boldsymbol{j}}(\boldsymbol{i}) = \begin{cases} 1, & \text{if } \ \boldsymbol{i} = \boldsymbol{j} \ \text{ or } \ \boldsymbol{i} = \boldsymbol{j}^*, \\ 0, & \text{otherwise.} \end{cases}$$

Then we have $\mathscr{F}(\mathscr{I}_{\boldsymbol{t}}) := \{\boldsymbol{x}^{\boldsymbol{j}} \mid \boldsymbol{j} \in \mathscr{I}_{\boldsymbol{t}}\} \subset \mathscr{F}_{\boldsymbol{t}}$ and $|\mathscr{F}(\mathscr{I}_{\boldsymbol{t}})| = 2^{c-1}$.

Suppose that $\boldsymbol{x} \in \mathscr{F}(\mathscr{I}_{\boldsymbol{t}})$. If there exists $\boldsymbol{x}' = \{x'(\boldsymbol{i}) \mid \boldsymbol{i} \in \mathscr{I}\}$ such that $\boldsymbol{x}' \in \mathscr{F}_{\boldsymbol{t}}$, $\boldsymbol{x}' \notin \mathscr{F}(\mathscr{I}_{\boldsymbol{t}})$, there exists a cell $\boldsymbol{j} \in \mathscr{I}$ and $1 \le k \le c+1$ such that $x'(\boldsymbol{j}) = 1$ and $\boldsymbol{j}_{\Gamma_k} \neq \boldsymbol{i}_{\Gamma_k}$. This implies that there exists $D_l \in \mathscr{D}$ such that $\boldsymbol{x}'_{D_l}(\boldsymbol{i}_{D_l}) \neq \boldsymbol{x}_{D_l}(\boldsymbol{i}_{D_l})$. Hence we have $|\mathscr{F}_{\boldsymbol{t}}| = 2^{c-1}$. □

As mentioned in Sect. 8.1, for a consistent $\boldsymbol{t}$ such that $\deg \boldsymbol{t} > 2$, the fiber $\mathscr{F}_{\boldsymbol{t}}$ may be empty in general. However Theorem 8.2 shows that, in the case of $\deg \boldsymbol{t} = 2$, if a consistent $\boldsymbol{t}$ such that $\bar{\Delta}_{\boldsymbol{t}} \neq \emptyset$ is given, then $\mathscr{F}_{\boldsymbol{t}} \neq \emptyset$ for any hierarchical model. We also note that Theorem 8.2 holds for general hierarchical models.

8.4 Minimal Markov Bases for Decomposable Models

In this section we discuss Markov bases for decomposable models from a viewpoint of minimality (cf. Chap. 5).

Let $\deg \boldsymbol{t} = 2$. Let $\mathscr{T}_{\boldsymbol{t}}$ be any tree whose nodes are elements of $\mathscr{F}_{\boldsymbol{t}}$. Denote the set of edges in $\mathscr{T}_{\boldsymbol{t}}$ by $\mathscr{B}_{\mathscr{T}_{\boldsymbol{t}}}$. We can identify each edge $(\boldsymbol{x}, \boldsymbol{x}') \in \mathscr{B}_{\mathscr{T}_{\boldsymbol{t}}}$ with a move $\boldsymbol{z} = \boldsymbol{x} - \boldsymbol{x}'$. So we identity $\mathscr{B}_{\mathscr{T}_{\boldsymbol{t}}}$ with a set of moves for $\mathscr{F}_{\boldsymbol{t}}$. In this section, we consider only sign invariant Markov bases. Hence identify $\boldsymbol{z} = \boldsymbol{x} - \boldsymbol{x}'$ with $-\boldsymbol{z} = \boldsymbol{x}' - \boldsymbol{x}$ and consider the edges in $\mathscr{T}_{\boldsymbol{t}}$ as undirected.

Let $\mathscr{B}_{\rm nd}$ be

$$\mathscr{B}_{\rm nd} = \{\boldsymbol{t} \mid \deg \boldsymbol{t} = 2,\ |\mathscr{F}_{\boldsymbol{t}}| \ge 2\}. \tag{8.5}$$

As mentioned above, the set of fibers of the minimum fiber Markov basis for decomposable models coincides with the set of degree 2 fibers with more than one element. Hence we can provide the complete description of minimal Markov bases for decomposable models as follows.

Theorem 8.3. *Define* $\mathscr{B}^0$ *by*

$$\mathscr{B}^0 = \bigcup_{\boldsymbol{t} \in \mathscr{B}_{\rm nd}} \mathscr{B}_{\mathscr{T}_{\boldsymbol{t}}}. \tag{8.6}$$

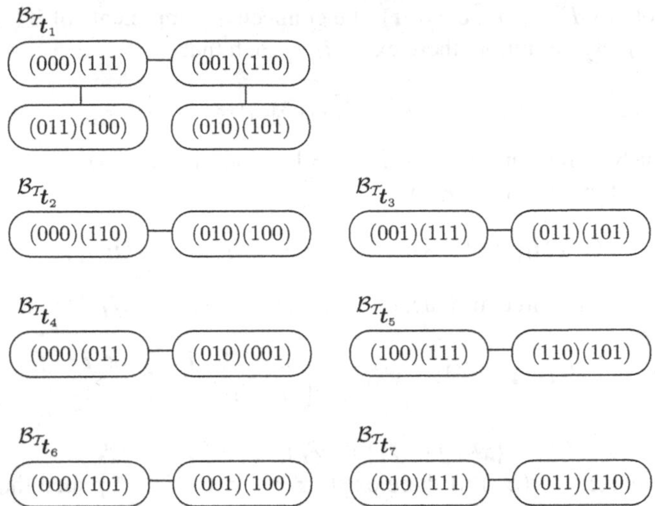

Fig. 8.1 $\mathscr{B}_{\mathscr{T}_{t_l}}$ in the complete independence model of three-way contingency tables

Then $\mathscr{B}^0$ *is a minimal Markov basis and (8.6) is a disjoint union. Conversely every minimal Markov basis can be written as in (8.6).*

Example 8.1 (The complete independence model of $2 \times 2 \times 2$ *contingency tables).* Consider the model $\mathscr{D} = \{\{1\},\{2\},\{3\}\}$. $\mathscr{B}_{\rm nd}$ for the model has seven elements. Denote them by $\boldsymbol{t}_1,\ldots,\boldsymbol{t}_7$. Figure 8.1 shows an example of $\mathscr{B}_{\mathscr{T}_{t_l}}$ for $t = 1,\ldots,7$. $\boldsymbol{t}_1,\ldots,\boldsymbol{t}_7$ satisfy

$$\begin{aligned} \bar{\Delta}_{t_1} &= \{1,2,3\}, \qquad \bar{\Delta}_{t_2} = \bar{\Delta}_{t_3} = \{1,2\}, \\ \bar{\Delta}_{t_4} &= \bar{\Delta}_{t_5} = \{2,3\}, \ \bar{\Delta}_{t_6} = \bar{\Delta}_{t_7} = \{1,3\}. \end{aligned} \tag{8.7}$$

The union of all these moves is a minimal Markov basis for the model. Inasmuch as $\mathscr{F}_{t_1}$ is a four-element fiber, $\mathscr{T}_{t_1}$ is not uniquely determined. Hence minimal Markov bases are not unique for this model.

As seen from this example, minimal Markov bases are not necessarily uniquely determined. The following corollary provides a necessary and sufficient condition on decomposable models to have the unique minimal Markov basis.

Corollary 8.1. *There exists the unique minimal Markov basis for a decomposable model if and only if the number of connected components in any induced subgraph of* $\mathscr{G}$ *is less than three.*

Proof. Suppose that $\mathscr{G}(\bar{\Delta}_t)$ has more than two connected components. Then because $|\mathscr{F}_t| \geq 4$ from Theorem 8.2, $\mathscr{T}_t$ is not uniquely determined. For a different tree $\mathscr{T}_t'$, $\mathscr{B}_{\mathscr{T}_t} \neq \mathscr{B}_{\mathscr{T}_t'}$. Hence minimal Markov bases are not unique either.

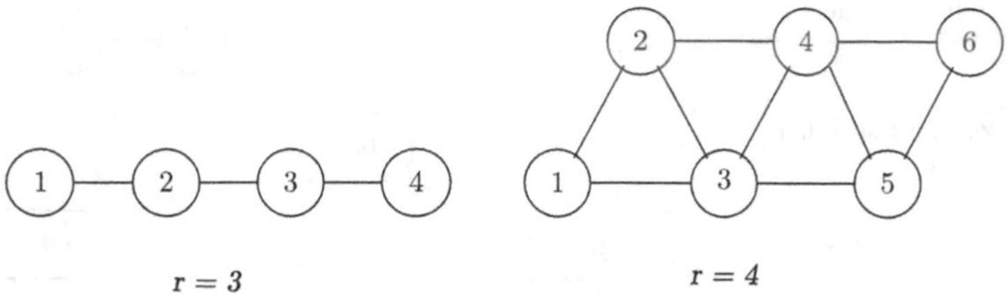

Fig. 8.2 Examples of chordal graphs satisfying the condition of Corollary 8.2

Conversely assume that the number of connected components of $\mathscr{G}(\bar{\Delta}_{\boldsymbol{t}})$ for any $\boldsymbol{t} \in \mathscr{B}_{\text{nd}}$ is two. Then $\mathscr{T}_{\boldsymbol{t}}$ for any $\boldsymbol{t} \in \mathscr{B}_{\text{nd}}$ is uniquely determined. Hence the minimal Markov basis is also unique. □

Corollary 8.2. *For a decomposable model, there exists the unique minimal Markov basis if and only if $\mathscr{G}$ has only two boundary cliques D and D' such that $D'' \subset D \cup D'$ for all $D'' \in \mathscr{D}$.*

Proof. Suppose that $\mathscr{G}$ has two boundary cliques D and D' such that $D'' \subset D \cup D'$ for all $D'' \in \mathscr{D}$. Then any vertex in D'' is adjacent to D or D'. Hence the number of connected components for any induced subgraph of $\mathscr{G}$ is at most two.

Conversely suppose that there exists $D'' \in \mathscr{D}$ such that $D'' \not\subset D \cup D'$. Then the subgraph induced by the union of $D'' \setminus (D \cup D')$, $\text{Simp}(D)$ and $\text{Simp}(D')$ has three connected components. □

The graphs with $r = 2$ always satisfy the conditions of the corollary. For $r \geq 3$ the graph with

$$\mathscr{D} = \{\{1,\ldots,r-1\},\{2,\ldots,r\},\ldots,\{r,\ldots,2r-2\}\} \tag{8.8}$$

satisfies the conditions of the corollary. Figure 8.2 shows the graphs satisfying (8.8) for $r = 3,4$. We can easily see that any induced subgraph of the graphs in the figure has at most two connected components.

From a viewpoint of minimality, Dobra's Markov basis $\mathscr{B}^{\mathscr{T}}$ is characterized as follows.

Theorem 8.4. *A decomposable model has a clique tree $\mathscr{T}$ such that $\mathscr{B}^{\mathscr{T}}$ is a minimal Markov basis if and only if the model has the unique minimal Markov basis.*

Proof. When a decomposable model has a unique minimal Markov basis, $\mathscr{B}^{\mathscr{T}}$ coincides with it.

Suppose that there exist three vertices in $\mathscr{G}$ which are not adjacent to one another. Let $1,2$, and 3 be three such vertices and assume that $l \in D_l$, $D_l \in \mathscr{D}$, for $l = 1,2,3$. Define $\{1,2,3\}^C = \Delta \setminus \{1,2,3\}$. Consider a degree 2 fiber $\mathscr{F}_{\boldsymbol{t}}$ such that $\bar{\Delta}_{\boldsymbol{t}} = \{1,2,3\}$ and $x_{\{1,2,3\}^C}(\boldsymbol{i}_{\{1,2,3\}^C}) = 2$ for some $\boldsymbol{i}_{\{1,2,3\}^C}$. Then $|\mathscr{F}_{\boldsymbol{t}}| = 4$ from Theorem 8.2 and we can denote the four elements by

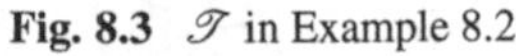

Fig. 8.3 $\mathscr{T}$ in Example 8.2

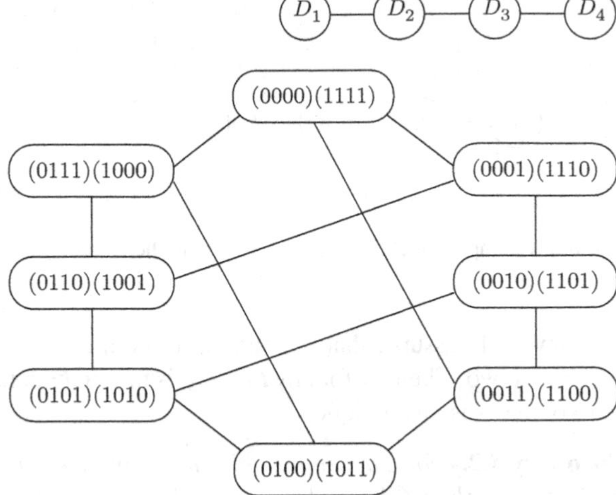

Fig. 8.4 $\mathscr{B}_{\boldsymbol{t}}^{\mathscr{T}}$ for $\boldsymbol{t}$ such that $\bar{\Delta}_{\boldsymbol{t}} = \{1,2,3,4\}$

$$\begin{aligned} \boldsymbol{x}_1 &= (000\,\boldsymbol{i}_{\{1,2,3\}^C})(111\,\boldsymbol{i}_{\{1,2,3\}^C}), \ \boldsymbol{x}_2 = (001\,\boldsymbol{i}_{\{1,2,3\}^C})(110\,\boldsymbol{i}_{\{1,2,3\}^C}), \\ \boldsymbol{x}_3 &= (010\,\boldsymbol{i}_{\{1,2,3\}^C})(101\,\boldsymbol{i}_{\{1,2,3\}^C}), \ \boldsymbol{x}_4 = (011\,\boldsymbol{i}_{\{1,2,3\}^C})(100\,\boldsymbol{i}_{\{1,2,3\}^C}). \end{aligned} \tag{8.9}$$

A minimal Markov basis connects these four elements by three moves. Let $\mathscr{T} = (\mathscr{D},\mathscr{E})$ be any clique tree for $\mathscr{G}$ and $\mathscr{T}' = (\mathscr{D}',\mathscr{E}')$ be the smallest subtree of $\mathscr{T}$ satisfying $D_l \in \mathscr{D}'$ for $l = 1,2$, and 3. Then we can assume that $\mathscr{T}'$ satisfies either of the following two conditions,

(i) D_2 is an interior point and D_1 and D_3 are endpoints on the path.
(ii) All of D_1, D_2, and D_3 are endpoints of $\mathscr{T}'$.

In both cases there exists $e \in \mathscr{E}$ such that $D_1, D_2 \subset V_e$ and $D_3 \subset V_e'$. Then $\mathscr{B}^{\mathscr{T}}(V_e,V_e')$ includes the following two moves,

$$\boldsymbol{z}_1 = \boldsymbol{x}_1 - \boldsymbol{x}_2, \quad \boldsymbol{z}_2 = \boldsymbol{x}_3 - \boldsymbol{x}_4.$$

On the other hand there also exists $e' \in \mathscr{E}$ such that $D_1 \subset V_{e'}$ and $D_2, D_3 \subset V_{e'}'$. In this case $\mathscr{B}^{\mathscr{T}}(V_{e'},V_{e'}')$ includes the following two moves,

$$\boldsymbol{z}_3 = \boldsymbol{x}_1 - \boldsymbol{x}_4, \quad \boldsymbol{z}_4 = \boldsymbol{x}_2 - \boldsymbol{x}_3.$$

Thus $\mathscr{B}^{\mathscr{T}}$ includes at least four moves for the fiber $\mathscr{F}_{\boldsymbol{t}}$, which implies that $\mathscr{B}^{\mathscr{T}}$ is not minimal for the model which does not have the unique minimal Markov basis. □

Example 8.2 (The complete independence model of $2\times2\times2\times2$ *contingency tables).* Consider the model $\mathscr{D} = \{\{1\},\{2\},\{3\},\{4\}\}$ and $D_l = \{l\}$ for $l = 1,\dots,4$. Let $\mathscr{F}_{\boldsymbol{t}}$ be the fiber with $\bar{\Delta}_{\boldsymbol{t}} = \{1,2,3,4\}$; that is, $c(\boldsymbol{t}) = 4$ and $|\mathscr{F}_{\boldsymbol{t}}| = 8$. Consider $\mathscr{B}^{\mathscr{T}}$ for $\mathscr{T}$ in Fig. 8.3. Denote the set of moves for $\mathscr{F}_{\boldsymbol{t}}$ belonging to $\mathscr{B}^{\mathscr{T}}$ by $\mathscr{B}_{\boldsymbol{t}}^{\mathscr{T}}$. Figure 8.4 shows $\mathscr{B}_{\boldsymbol{t}}^{\mathscr{T}}$. As seen from Fig. 8.4, $\mathscr{B}_{\boldsymbol{t}}^{\mathscr{T}}$ includes 12 moves. Because $|\mathscr{F}_{\boldsymbol{t}}| = 8$, 7 moves are sufficient to connect $\mathscr{F}_{\boldsymbol{t}}$.

8.5 Minimal Invariant Markov Bases

In this section we discuss Markov bases for decomposable models from the viewpoint of invariance under the action of the direct product of symmetric groups $G = G_{I_1,\dots,I_m} = S_{I_1} \times \cdots \times S_{I_m}$ on the levels of the variables and provide a minimal G-invariant Markov basis.

Here we denote $c = c(\boldsymbol{t})$ for simplicity. Let Γ_l, $l = 1,\dots,c$, be connected components of $\mathscr{G}(\bar{\Delta}_{\boldsymbol{t}})$ and let $\Gamma_{c+1} = \Delta \setminus \bar{\Delta}_{\boldsymbol{t}}$. For a subset of vertices $V \subset \Delta$, denote

$$\mathbf{0}_V := \overbrace{0\cdots0}^{|V|}, \quad \mathbf{1}_V := \overbrace{1\cdots1}^{|V|}.$$

As a representative fiber $\mathscr{F}_{\boldsymbol{t}}^0$, we can consider $\boldsymbol{t}$ such that the levels of all degenerate variables are determined as 0:

$$\mathscr{F}_{\boldsymbol{t}}^0 \ni \boldsymbol{x}_0^{\boldsymbol{t}} \equiv (\mathbf{0}_\Delta)(\mathbf{1}_{\bar{\Delta}_{\boldsymbol{t}}}\,\mathbf{0}_{\Gamma_{c+1}}).$$

Then any $\boldsymbol{x} \in \mathscr{F}_{\boldsymbol{t}}^0$ is expressed as follows,

$$\boldsymbol{x} = (\mathbf{0}_{\Gamma_1}\,\boldsymbol{i}_{\Gamma_2}\cdots\boldsymbol{i}_{\Gamma_c}\,\mathbf{0}_{\Gamma_{c+1}})(\mathbf{1}_{\Gamma_1}\,\boldsymbol{i}_{\Gamma_2}^*\cdots\boldsymbol{i}_{\Gamma_c}^*\,\mathbf{0}_{\Gamma_{c+1}}),$$
$$\boldsymbol{i}_{\Gamma_l} = \mathbf{0}_{\Gamma_l} \quad \text{or} \quad \boldsymbol{i}_{\Gamma_l} = \mathbf{1}_{\Gamma_l}, \quad l = 2,\dots,c.$$

Let G^{Γ_l}, $l = 2,\dots,c$, be the diagonal subgroup of $S_2^{|\Gamma_l|}$ defined by

$$G^{\Gamma_l} = \{\bar{g} = (g,\dots,g) \mid g \in S_2\} \subset S_2^{|\Gamma_l|}.$$

Define $G_{\boldsymbol{t}} = G^{\Gamma_2} \times \cdots \times G^{\Gamma_c}$ and let $g \in G_{\boldsymbol{t}}$ act on $\boldsymbol{x} \in \mathscr{F}_{\boldsymbol{t}}^0$ by

$$g(\boldsymbol{x}) = (\mathbf{0}_{\Gamma_1}\,\bar{g}_2(\boldsymbol{i}_{\Gamma_2})\cdots\bar{g}_c(\boldsymbol{i}_{\Gamma_c})\,\mathbf{0}_{\Gamma_{c+1}})(\mathbf{1}_{\Gamma_1}\,\bar{g}_2(\boldsymbol{i}_{\Gamma_2}^*)\cdots\bar{g}_c(\boldsymbol{i}_{\Gamma_c}^*)\,\mathbf{0}_{\Gamma_{c+1}}).$$

Clearly $g(\boldsymbol{x}) \in \mathscr{F}_{\boldsymbol{t}}^0$ for $\boldsymbol{x} \in \mathscr{F}_{\boldsymbol{t}}^0$ and furthermore for any $\boldsymbol{x} \in \mathscr{F}_{\boldsymbol{t}}^0$ there exists $g \in G_{\boldsymbol{t}}$ such that $\boldsymbol{x} = g(\boldsymbol{x}_0^{\boldsymbol{t}})$. This shows that $G_{\boldsymbol{t}} \subset G_{I_1,\dots,I_m}$ is the setwise stabilizer of $\mathscr{F}_{\boldsymbol{t}}^0$ acting transitively on $\mathscr{F}_{\boldsymbol{t}}^0$. Then $G_{\boldsymbol{t}} \subset G_{I_1,\dots,I_m}$ is isomorphic to a c-fold direct product of S_2s:

$$S_2^c = S_2 \times \cdots \times S_2.$$

Therefore the structure of $\mathscr{F}_{\boldsymbol{t}}$ is equivalent to the structure of the fiber $\mathscr{F}_{\boldsymbol{t}'}$ with $\Delta = \bar{\Delta}_{\boldsymbol{t}'} = \{1,\dots,c\}$.

Let $\mathscr{B}_{G_{\boldsymbol{t}}}$ be a minimal $G_{\boldsymbol{t}}$-invariant set of moves that connects $\mathscr{F}_{\boldsymbol{t}}^0$. Let $\kappa(\boldsymbol{t})$ be the number of $G_{\boldsymbol{t}}$-orbits included in $\mathscr{B}_{G_{\boldsymbol{t}}}$. As representative moves of $G_{\boldsymbol{t}}$-orbits in $\mathscr{B}_{G_{\boldsymbol{t}}}$ we can consider

$$\boldsymbol{z}_k^{\boldsymbol{t}} = \boldsymbol{x}_0^{\boldsymbol{t}} - \boldsymbol{x}_k^{\boldsymbol{t}} \in \mathscr{B}_{\boldsymbol{t}}, \quad \boldsymbol{x}_k^{\boldsymbol{t}} \in \mathscr{F}_{\boldsymbol{t}}^0, \quad k = 1,\dots,\kappa(\boldsymbol{t}).$$

This is because we can always send $\boldsymbol{x}$ in $\boldsymbol{z}=\boldsymbol{x}-\boldsymbol{x}'$ to $\boldsymbol{x}_0^{\boldsymbol{t}}$ by the transitivity of $G_{\boldsymbol{t}}$. Denote $\mathscr{B}_{G_{\boldsymbol{t}}}^0=\{\boldsymbol{z}_1^{\boldsymbol{t}},\dots,\boldsymbol{z}_{\kappa(\boldsymbol{t})}^{\boldsymbol{t}}\}$. Define the set of $\boldsymbol{t}$ that induces representative fibers by

$$\mathscr{B}_{\mathrm{nd}}^0=\{\boldsymbol{t}\mid \boldsymbol{x}_0^{\boldsymbol{t}}\in\mathscr{F}_{\boldsymbol{t}}^0\}\subset\mathscr{B}_{\mathrm{nd}}. \tag{8.10}$$

By Theorem 7.1 a minimal $G_{I_1,\dots,I_m}$-invariant Markov basis can be expressed by

$$\mathscr{B}_G=\bigcup_{\boldsymbol{t}\in\mathscr{B}_{\mathrm{nd}}^0}\bigcup_{k=1}^{\kappa(\boldsymbol{t})}G_{I_1,\dots,I_m}(\boldsymbol{z}_k^{\boldsymbol{t}}), \tag{8.11}$$

where $G_{I_1,\dots,I_m}(\boldsymbol{z}_k^{\boldsymbol{t}})$ denotes the $G_{I_1,\dots,I_m}$-orbit through $\boldsymbol{z}_k^{\boldsymbol{t}}$. Hence in order to clarify the structure of $\mathscr{B}_G$, it suffices to consider $2\times\cdots\times 2$ tables and investigate $\kappa(\boldsymbol{t})$ and $\mathscr{B}_{G_{\boldsymbol{t}}}^0$ for each $\mathscr{F}_{\boldsymbol{t}}^0$.

As mentioned above, the structure of $\mathscr{F}_{\boldsymbol{t}}^0$ is equivalent to the one of the fiber with $\bar{\Delta}_{\boldsymbol{t}}=\Delta=\{1,\dots,c\}$ and $\mathscr{G}(\bar{\Delta}_{\boldsymbol{t}})$ is totally disconnected. We first consider the structure of such a fiber. $\mathscr{F}_{\boldsymbol{t}}^0$ satisfies

$$\mathscr{F}_{\boldsymbol{t}}^0=\{(0\,\boldsymbol{i}_2\cdots\boldsymbol{i}_c)(1\,\boldsymbol{i}_2^*\cdots\boldsymbol{i}_c^*)\mid(\boldsymbol{i}_2\cdots\boldsymbol{i}_c)=\boldsymbol{i}_{\Delta\setminus\{1\}}\in\mathscr{I}_{\Delta\setminus\{1\}}\} \tag{8.12}$$

and $(0\cdots0)(1\cdots1)\in\mathscr{F}_{\boldsymbol{t}}^0$. Then we can identify $G_{\boldsymbol{t}}$ with S_2^{c-1}. For $g\in S_2^{c-1}$, we write $g=(g_1,\dots,g_c)$, where $g_l\in S_2$ for $l=1,\dots,c$. A representative move of an S_2^{c-1}-orbit is written by

$$\boldsymbol{z}^{\boldsymbol{t}}=(0\cdots0)(1\cdots1)-(0\,\boldsymbol{i}_{\Delta\setminus\{1\}})(1\,\boldsymbol{i}_{\Delta\setminus\{1\}}^*)$$

for some $\boldsymbol{i}_{\Delta\setminus\{1\}}\in\mathscr{I}_{\Delta\setminus\{1\}}$. We first consider deriving $\kappa(\boldsymbol{t})$ and $\mathscr{B}_{G_{\boldsymbol{t}}}$. Let $\mathscr{V}^{c-1}=\{0,1\}^{c-1}$ denote the $(c-1)$-dimensional vector space over the finite field GF(2), where the addition of two vectors is defined to be XOR of the elements. Let $\oplus$ denote the XOR operation. Let $\circ$ denote the group operation of S_2^{c-1}.

Lemma 8.3. *S_2^{c-1} is isomorphic to $\mathscr{V}^{c-1}$.*

Proof. Consider the map $\phi:S_2^{c-1}\to\mathscr{V}^{c-1}$ such that $\phi(g)=\boldsymbol{v}=(v_2,\dots,v_c)\in\mathscr{V}^{c-1}$, where

$$v_l=\begin{cases}0, & \text{if } g_l(\boldsymbol{i}_l)=\boldsymbol{i}_l,\\ 1, & \text{if } g_l(\boldsymbol{i}_l)=\boldsymbol{i}_l^*,\end{cases}$$

for $l=2,\dots,c$ and $\{\boldsymbol{i}_l,\boldsymbol{i}_l^*\}=\{0,1\}$. For $g'=(g_2',\dots,g_c')\in S_2^{c-1}$, $g_l'\in S_2$, and $\boldsymbol{v}'\in\mathscr{V}^{c-1}$, define $\phi(g')=\boldsymbol{v}'=(v_2',\dots,v_c')$. Then we have $\phi(g\circ g')=\tilde{\boldsymbol{v}}=(\tilde{v}_2,\dots,\tilde{v}_c)$, $\tilde{\boldsymbol{v}}\in\mathscr{V}^{c-1}$, where

$$\tilde{v}_l=\begin{cases}0, & \text{if } g_l\circ g_l'(\boldsymbol{i}_l)=\boldsymbol{i}_l,\\ 1, & \text{if } g_l\circ g_l'(\boldsymbol{i}_l)=\boldsymbol{i}_l^*\end{cases}$$

for $l=2,\dots,c$. Hence we have

$$\tilde{v}_l=v_l\oplus v_l',\quad l=2,\dots,c.$$

Therefore ϕ is a homomorphism. It is obvious that ϕ is a bijection. Therefore S_2^{c-1} is isomorphic to $\mathscr{V}^{c-1}$. □

Based on this lemma, we can show the equivalence between S_2^c-orbits in a minimal S_2^c-invariant set of moves that connects $\mathscr{F}_{\boldsymbol{t}}^0$ and a (vector space) basis of $\mathscr{V}^{c-1}$.

Theorem 8.5. *Let $\mathscr{V}^0 = \{\boldsymbol{v}_k = (v_{k2},\dots,v_{kc}),\ k=2,\dots,c\}$ be any basis of $\mathscr{V}^{c-1}$. Define $\boldsymbol{x}_0^{\boldsymbol{t}}, \boldsymbol{x}_{\boldsymbol{v}_k}^{\boldsymbol{t}} \in \mathscr{F}_{\boldsymbol{t}}^0$ by*

$$\boldsymbol{x}_0^{\boldsymbol{t}} = (00\cdots0)(11\cdots1), \quad \boldsymbol{x}_{\boldsymbol{v}_k}^{\boldsymbol{t}} = (0\,v_{k2}\cdots v_{kc})(1\,v_{k2}^*\cdots v_{kc}^*),$$

where $v_{kl}^* = 1 \oplus v_{kl}$. Let $\mathscr{B}_{G_{\boldsymbol{t}}}$ be an S_2^{c-1}-invariant set of moves in $\mathscr{F}_{\boldsymbol{t}}^0$. Then $\mathscr{B}_{G_{\boldsymbol{t}}}$ is a minimal S_2^{c-1}-invariant set of moves that connects $\mathscr{F}_{\boldsymbol{t}}^0$ if and only if the representative moves of the S_2^{c-1}-orbits in $\mathscr{B}_{G_{\boldsymbol{t}}}$ are expressed by $\boldsymbol{z}_{\boldsymbol{v}_k}^{\boldsymbol{t}} = \boldsymbol{x}_0^{\boldsymbol{t}} - \boldsymbol{x}_{\boldsymbol{v}_k}^{\boldsymbol{t}}$, $k=2,\dots,c$. Hence $\kappa(\boldsymbol{t}) = c-1$.

Proof. Suppose that $\mathscr{B}_{G_{\boldsymbol{t}}}$ is a minimal S_2^{c-1}-invariant set of moves that connects $\mathscr{F}_{\boldsymbol{t}}$ and that $\mathscr{B}_{G_{\boldsymbol{t}}}$ includes $\kappa(\boldsymbol{t})$ orbits $S_2^{c-1}(\boldsymbol{z}_1^{\boldsymbol{t}}),\dots,S_2^{c-1}(\boldsymbol{z}_{\kappa(\boldsymbol{t})}^{\boldsymbol{t}})$, where

$$\boldsymbol{z}_k^{\boldsymbol{t}} = \boldsymbol{x}_0^{\boldsymbol{t}} - \boldsymbol{x}_k^{\boldsymbol{t}}, \quad \boldsymbol{x}_k^{\boldsymbol{t}} = (0\,\boldsymbol{i}_{k2}\cdots\boldsymbol{i}_{kc})(1\,\boldsymbol{i}_{k2}^*\cdots\boldsymbol{i}_{kc}^*)$$

for $\boldsymbol{i}_{kl} \in \mathscr{I}_l$, $k=1,\dots,\kappa(\boldsymbol{t})$, $l=2,\dots,c$. Let $g^k \in S_2^{c-1}$ satisfy $g^k(\boldsymbol{x}_0^{\boldsymbol{t}}) = \boldsymbol{x}_k^{\boldsymbol{t}}$ for $k=1,\dots,\kappa(\boldsymbol{t})$. We write $g^k = (g_{k2},\dots,g_{kc})$, $g_{kl} \in S_2$ for $l=2,\dots,c$. Let $H_{\boldsymbol{t}} = \{g^1,\dots,g^{\kappa(\boldsymbol{t})}\} \subset S_2^{c-1}$ be a subset of S_2^{c-1}. As mentioned above, $\mathscr{F}_{\boldsymbol{t}}^0$ can be expressed as in (8.12). Hence for any $\boldsymbol{x} \in \mathscr{F}_{\boldsymbol{t}}^0$ there exists $g \in S_2^{c-1}$ satisfying $\boldsymbol{x} = g(\boldsymbol{x}_0^{\boldsymbol{t}})$. $\mathscr{B}_{G_{\boldsymbol{t}}}$ connects $\mathscr{F}_{\boldsymbol{t}}^0$ if and only if there exists $p \le \kappa(\boldsymbol{t})$ such that

$$\boldsymbol{x} = \boldsymbol{x}_0^{\boldsymbol{t}} - \boldsymbol{z}_{k_1}^{\boldsymbol{t}} - g^{k_1}(\boldsymbol{z}_{k_2}^{\boldsymbol{t}}) - \cdots - g^{k_{p-1}} \circ \cdots \circ g^{k_1}(\boldsymbol{z}_{k_p}^{\boldsymbol{t}})$$

and $g = g^{k_p} \circ \cdots \circ g^{k_1}$. Hence $\mathscr{B}_{G_{\boldsymbol{t}}}$ is a minimal S_2^{c-1}-invariant set of moves that connects $\mathscr{F}_{\boldsymbol{t}}$ if and only if $H_{\boldsymbol{t}}$ satisfies

$$\forall g \in S_2^{c-1},\ \exists p \le \kappa(\boldsymbol{t}),\ \exists g^{k_1} \in H_{\boldsymbol{t}},\dots,\exists g^{k_p} \in H_{\boldsymbol{t}} \quad \text{s.t. } g = g^{k_p} \circ \cdots \circ g^{k_1} \tag{8.13}$$

and no proper subset of $H_{\boldsymbol{t}}$ satisfies (8.13).

Denote $\mathscr{V}^0 = \phi(H_{\boldsymbol{t}}) \subset \mathscr{V}^{c-1}$. From Lemma 8.3, (8.13) is equivalent to

$$\forall \boldsymbol{v} \in \mathscr{V}, \quad \exists \boldsymbol{v}^1 \in \mathscr{V}^0,\dots,\exists \boldsymbol{v}^p \in \mathscr{V}^0 \quad \text{s.t.} \quad \boldsymbol{v} = \boldsymbol{v}^1 \oplus \cdots \oplus \boldsymbol{v}^p. \tag{8.14}$$

From the minimality of $\mathscr{B}_{G_{\boldsymbol{t}}}$ no proper subset of $\mathscr{V}^0$ satisfies (8.14). This implies that $\mathscr{V}^0$ is a basis of $\mathscr{V}^{c-1}$ and hence $\kappa(\boldsymbol{t}) = c-1$. If we define $g^k = \phi^{-1}(\boldsymbol{v}_{k+1})$ for $k=1,\dots,c-1$, we have $g_{kl}(0) = v_{k+1,l}$ and hence $g^k(\boldsymbol{x}_0^{\boldsymbol{t}}) = \boldsymbol{x}_k^{\boldsymbol{t}} = \boldsymbol{x}_{\boldsymbol{v}_{k+1}}^{\boldsymbol{t}}$. Therefore $\boldsymbol{z}_{\boldsymbol{v}_k}^{\boldsymbol{t}}$, $k=2,\dots,c$, are the representative moves of the S_2^{c-1}-orbits in $\mathscr{B}_{G_{\boldsymbol{t}}}$.

Conversely suppose that the representative moves of $\mathscr{B}_{G_t}$ are $z^t_{v_k}$, $k = 2,\dots,c$. $\mathscr{V}^0$ satisfies (8.14) and no proper subset of $\mathscr{V}^0$ satisfies (8.14). Hence if we define $g^k = \phi^{-1}(v_{k+1})$ and $H_t = \{g^1,\dots,g^{c-1}\}$, H_t satisfies (8.14) and no proper subset of H_t satisfies (8.14). Hence $\mathscr{B}_{G_t}$ is a minimal S_2^{c-1}-invariant set of moves that connects $\mathscr{F}_t$. □

For example, we can set $\mathscr{V}^0 = \{v_2,\dots,v_c\}$ as

$$v_2 = (11\cdots11), \quad v_3 = (01\cdots11), \quad \dots, \quad v_{c-1} = (00\cdots011), \quad v_c = (00\cdots01),$$

and then the representative moves in a minimal G-invariant Markov basis are

$$\begin{aligned} z^0_2 &= (00\cdots0)(11\cdots1) - (011\cdots11)(100\cdots00), \\ z^0_3 &= (00\cdots0)(11\cdots1) - (001\cdots11)(110\cdots00), \\ &\vdots \qquad\qquad \vdots \qquad\qquad\qquad \vdots \\ z^0_c &= (00\cdots0)(11\cdots1) - (000\cdots01)(111\cdots10). \end{aligned} \tag{8.15}$$

So far we have focused on $\mathscr{F}_t$ such that $\bar{\Delta}_t = \Delta = \{1,\dots,c\}$ and $\mathscr{G}(\bar{\Delta}_t)$ is totally disconnected. Now we consider a fiber for a general t of a general decomposable model. Define $\bar{g}_{kl} \in G^{\Gamma_l}$ by

$$\bar{g}_{kl}(\mathbf{0}_{\Gamma_l}) = \begin{cases} 0\cdots0 & \text{if} \quad v_{kl} = 0, \\ 1\cdots1 & \text{if} \quad v_{kl} = 1 \end{cases} \tag{8.16}$$

for $k = 2,\dots,c$ and $l = 2,\dots,c$ and define $g^k \in G_t$ by

$$g^k(x) = (\mathbf{0}_{\Gamma_1}\ \bar{g}_{k2}(i_{\Gamma_2})\cdots\bar{g}_{kc}(i_{\Gamma_c})\ \mathbf{0}_{\Gamma_{c+1}})(\mathbf{1}_{\Gamma_1}\ \bar{g}_{k2}(i^*_{\Gamma_2})\cdots\bar{g}_{kc}(i^*_{\Gamma_c})\ \mathbf{0}_{\Gamma_{c+1}}). \tag{8.17}$$

Denote $x^t_{v_k} = g^k(x^t_0)$ and $z^t_{v_k} = x^t_0 - x^t_{v_k}$. By following (8.11) and Theorem 8.5, we can easily obtain the following result.

Theorem 8.6. *$\mathscr{B}_{G_t}$ is a minimal S_2^{c-1}-invariant set of moves that connects $\mathscr{F}^0_t$ if and only if the representative moves of the S_2^{c-1}-orbits in $\mathscr{B}_{G_t}$ are expressed as $z^t_{v_k}$, $k = 2,\dots,c$. Hence $\kappa(t) = c-1$. Then*

$$\mathscr{B}_G = \bigcup_{t\in\mathscr{B}^0_{\mathrm{nd}}} \bigcup_{k=2}^{c} G_{I_1,\dots,I_m}(z^t_k)$$

is a minimal $G_{I_1,\dots,I_m}$-invariant Markov basis. Conversely every minimal $G_{I_1,\dots,I_m}$-invariant Markov basis can be written in this form.

Example 8.3 (The complete independence model of three-way contingency tables). Define $t_1,\dots,t_7$ as in Fig. 8.1 of Example 8.1. Then $\mathscr{B}^0_{\mathrm{nd}} = \{t_1,t_2,t_4,t_6\}$. Figure 8.5

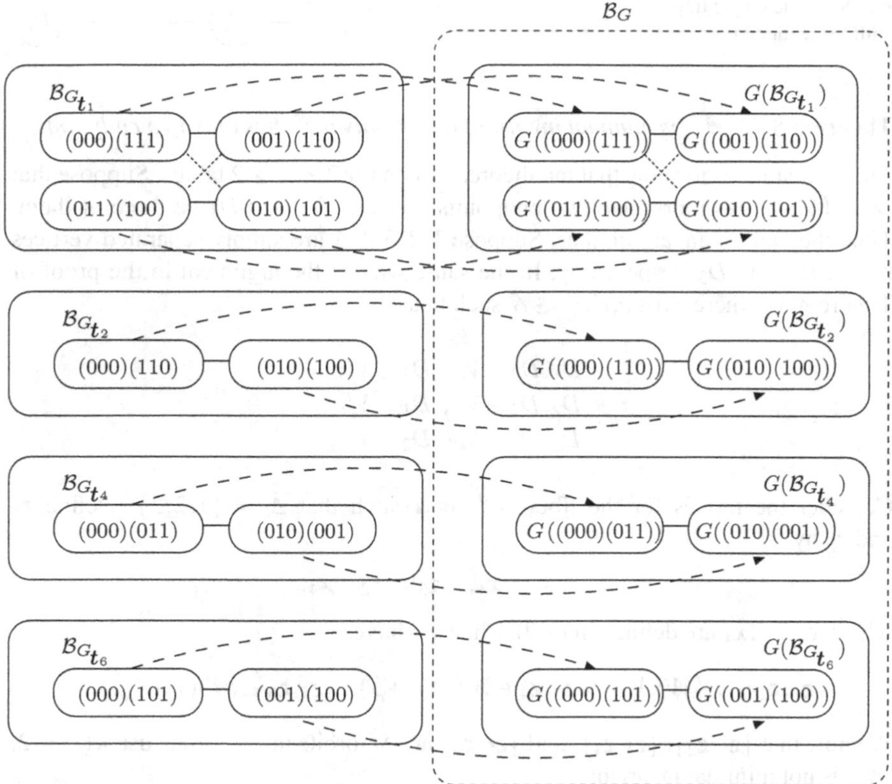

Fig. 8.5 The structure of minimal $G_{2,2,2}$-invariant Markov bases for the complete independence model of three-way contingency tables

shows a structure of $\mathscr{B}_G$ for the $I_1 \times I_2 \times I_3$ complete independence model of three-way contingency tables. The left half of the figure shows the structure of $\mathscr{B}_{G_{t_l}}$ for $2 \times 2 \times 2$ tables.

$c(\boldsymbol{t}_1) = 3$ and hence $\kappa(\boldsymbol{t}_1) = 2$. If we set $\boldsymbol{v}_1^{\boldsymbol{t}} = (10)$ and $\boldsymbol{v}_2^{\boldsymbol{t}} = (01)$, we have

$$\boldsymbol{z}_1^{\boldsymbol{t}_1} = (000)(111) - (010)(101), \quad \boldsymbol{z}_2^{\boldsymbol{t}_1} = (000)(111) - (001)(110).$$

The orbits $S_2^2(\boldsymbol{z}_1^{\boldsymbol{t}_1})$ and $S_2^2(\boldsymbol{z}_2^{\boldsymbol{t}_1})$ are expressed in dotted lines and solid lines, respectively, in the figure.

$c(\boldsymbol{t}_l) = 2$ and $\kappa(\boldsymbol{t}_l) = 1$ for $l = 2, 4, 6$. There exists one orbit in $\mathscr{B}_{G_{t_l}}$ for $l = 2, 4, 6$. Then from Theorem 8.6 a minimal $G_{2,2,2}$-invariant Markov basis is expressed by

$$\mathscr{B}_G = G(\boldsymbol{z}_1^{\boldsymbol{t}_1}) \cup G(\boldsymbol{z}_2^{\boldsymbol{t}_1}) \cup G(\boldsymbol{z}_1^{\boldsymbol{t}_2}) \cup G(\boldsymbol{z}_1^{\boldsymbol{t}_4}) \cup G(\boldsymbol{z}_1^{\boldsymbol{t}_6}).$$

Dobra's Markov basis $\mathscr{B}^{\mathscr{T}}$ is characterized from a viewpoint of invariance as follows. Because $\mathscr{B}^{\mathscr{T}}$ does not depend on the levels of the variables, $\mathscr{B}^{\mathscr{T}}$ is $G_{I_1,\dots,I_m}$-invariant. Based on the result of Theorem 8.5, we can show that $\mathscr{B}^{\mathscr{T}}$ is not always a minimal invariant Markov basis.

Fig. 8.6 The clique tree with two endpoints

Theorem 8.7. *$\mathscr{B}^{\mathscr{T}}$ is minimal invariant if and only if $\mathscr{T}$ has only two endpoints.*

Proof. It suffices to show that the theorem holds for $2 \times \cdots \times 2$ tables. Suppose that $\mathscr{T} = (\mathscr{D}, \mathscr{E})$ has more than two endpoints. Let D_1, D_2, and D_3 be three of them. Then they are boundary cliques. Suppose $1,2,3 \in \Delta$ are simply separated vertices in D_1, D_2, and D_3, respectively. In the same way as the argument in the proof of Theorem 8.4, there exist $e, e', e'' \in \mathscr{E}$ such that

$$\begin{aligned} D_1, D_2 \in V_e, \quad D_3 \in V'_e, \\ D_2, D_3 \in V_{e'}, \quad D_1 \in V'_{e'}, \\ D_3, D_1 \in V_{e''}, \quad D_2 \in V'_{e''}. \end{aligned}$$

Consider the moves for the fiber $\mathscr{F}_{\boldsymbol{t}}^0$ for $\boldsymbol{t}$ such that $\bar{\Delta}_{\boldsymbol{t}} = \{1,2,3\}$. Define $\boldsymbol{z}_5$ and $\boldsymbol{z}_6$ by

$$\boldsymbol{z}_5 = \boldsymbol{x}_1 - \boldsymbol{x}_3, \quad \boldsymbol{z}_6 = \boldsymbol{x}_2 - \boldsymbol{x}_4,$$

where $\boldsymbol{x}_1, \ldots, \boldsymbol{x}_4$ are defined in (8.9). Then we have

$$\boldsymbol{z}_1, \boldsymbol{z}_2 \in \mathscr{B}^T(V_e, V'_e), \quad \boldsymbol{z}_3, \boldsymbol{z}_4 \in \mathscr{B}^T(V_{e'}, V'_{e'}), \quad \boldsymbol{z}_5, \boldsymbol{z}_6 \in \mathscr{B}^T(V_{e''}, V'_{e''}).$$

We note that $\{\boldsymbol{z}_1, \boldsymbol{z}_2\}$, $\{\boldsymbol{z}_3, \boldsymbol{z}_4\}$, and $\{\boldsymbol{z}_5, \boldsymbol{z}_6\}$ are S_2^2-orbits in $\mathscr{B}_{\boldsymbol{t}}^{\mathscr{T}}$. Because $\kappa(\boldsymbol{t}) = 2$, $\mathscr{B}^{\mathscr{T}}$ is not minimal invariant.

Suppose that $\mathscr{T}$ has only two endpoints. Then $\mathscr{T}$ is expressed as in Fig. 8.6. Let $\Gamma_1, \ldots, \Gamma_c$ be the c connected components of $\mathscr{G}(\bar{\Delta}_{\boldsymbol{t}})$. Suppose that $\delta_l \in \Gamma_l$. The structure of $\mathscr{F}_{\boldsymbol{t}}^0$ is equivalent to the one of $\mathscr{F}_{\boldsymbol{t}'}^0$ such that $\bar{\Delta}_{\boldsymbol{t}'} = \{\delta_1, \ldots, \delta_{c-1}\}$ and $\mathscr{G}(\bar{\Delta}_{\boldsymbol{t}'})$ is totally disconnected. So we restrict our consideration to such a fiber. Denote by $\mathscr{F}_{\boldsymbol{t}'}^0$ the representative fiber for $\boldsymbol{t}'$. Let

$$\mathscr{B}_{\boldsymbol{t}'} = \{\boldsymbol{x} - \boldsymbol{x}' \mid \boldsymbol{x}, \boldsymbol{x}' \in \mathscr{F}_{\boldsymbol{t}'}^0, \ \boldsymbol{x} \neq \boldsymbol{x}'\}$$

denote the set of all moves in $\mathscr{F}_{\boldsymbol{t}'}^0$. Without loss of generality we can assume that $\delta_l \in D_{\pi(l)}$, where $\pi(1) < \cdots < \pi(c(\boldsymbol{t}'))$. Define $e_l = (D_{l-1}, D_l) \in \mathscr{E}$, $S_l = D_{l-1} \cap D_l$, $V_l = V_{e_l} \setminus S_l$ and $V'_l = V'_{e_l} \setminus S_l$ for $l = 2, \ldots, c(\boldsymbol{t}')$. Then the moves in $\mathscr{B}^{\mathscr{T}}(V_l, V'_l)$ are expressed as

$$\begin{aligned} \boldsymbol{z} = (\boldsymbol{i}_{V_l} \boldsymbol{i}_{V'_l} \boldsymbol{i}_{S_l})(\boldsymbol{j}_{V_l} \boldsymbol{j}_{V'_l}, \boldsymbol{i}_{S_l}) - (\boldsymbol{i}_{V_l} \boldsymbol{j}_{V'_l} \boldsymbol{i}_{S_l})(\boldsymbol{j}_{V_l} \boldsymbol{i}_{V'_l}, \boldsymbol{i}_{S_l}), \\ \boldsymbol{i}_{V_l}, \boldsymbol{j}_{V_l} \in \mathscr{I}_{V_l}, \quad \boldsymbol{i}_{V'_l}, \boldsymbol{j}_{V'_l} \in \mathscr{I}_{V'_l}, \quad \boldsymbol{i}_{S_l} \in \mathscr{I}_{S_l}. \end{aligned} \tag{8.18}$$

If $V_{e_l} \cap \bar{\Delta}_{\boldsymbol{t}'} = \emptyset$ or $V'_{e_l} \cap \bar{\Delta}_{\boldsymbol{t}'} = \emptyset$, then we have $\mathscr{B}^{\mathscr{T}}(V_{e_l}, V'_{e_l}) \cap \mathscr{B}_{\boldsymbol{t}'} = \emptyset$. If $V_{e_l} \cap \bar{\Delta}_{\boldsymbol{t}'} \neq \emptyset$ and $V'_{e_l} \cap \bar{\Delta}_{\boldsymbol{t}'} \neq \emptyset$, then there exists $2 \leq k(e_l) \leq c(\boldsymbol{t}')$ satisfying $\delta_k \in V_l$ for all $k < k(e_l)$ and $\delta_k \in V'_l$ for all $k \geq k(e_l)$. Then

Fig. 8.7 Clique trees for the four-way complete independence model

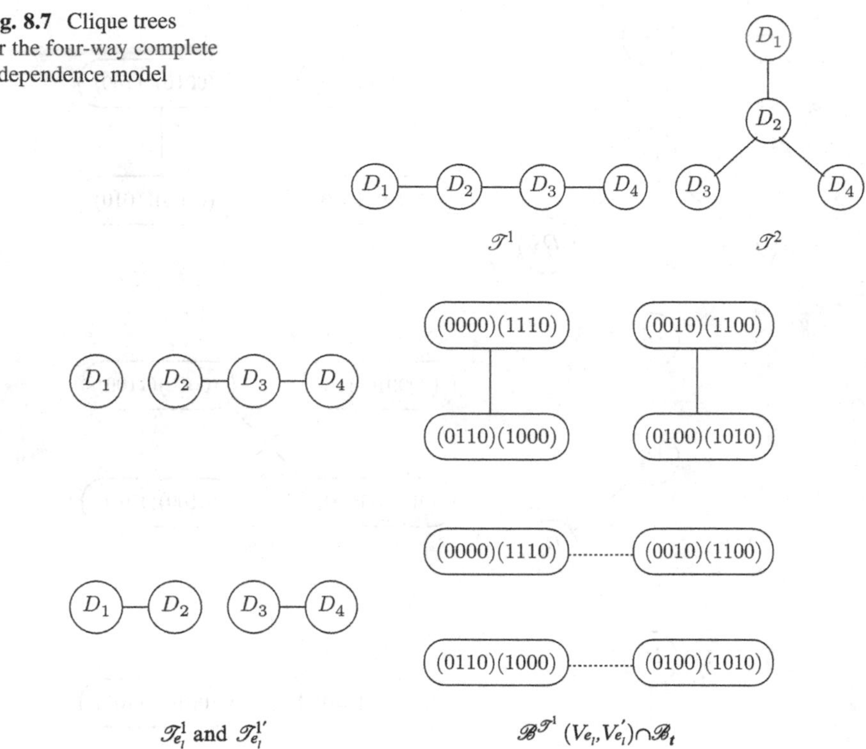

Fig. 8.8 The structure of $\mathscr{B}_t^{\mathscr{T}^1}$

$$\mathscr{B}^{\mathscr{T}}(V_{e_l}, V'_{e_l}) \cap \mathscr{B}_{t'} = S_2^{c-1}(z^0_{k(e_l)}),$$

where $z^0_{k(e_l)}$ is defined as in (8.15). Hence we have

$$\mathscr{B}^{\mathscr{T}}_{t'} = \bigcup_{e_l \in \mathscr{E}} \mathscr{B}^{\mathscr{T}}(V_{e_l}, V'_{e_l}) \cap \mathscr{B}_{t'} = \bigcup_{k=2}^{c(t')} S_2^{c-1}(z^0_k),$$

which contains $c(t') - 1$ orbits for all $t' \in \mathscr{B}^0_{\text{nd}}$. Hence $\mathscr{B}^{\mathscr{T}}$ is minimal $G_{I_1,\ldots,I_m}$-invariant. □

Example 8.4 (The complete independence model of four-way contingency tables). As an example we consider the $2 \times 2 \times 2 \times 2$ complete independence model $\mathscr{D} = \{D_l = \{i\}, i = 1, \ldots, 4\}$. Both $\mathscr{T}^1$ and $\mathscr{T}^2$ in Fig. 8.7 are clique trees for $\mathscr{D}$. From Theorem 8.7, $\mathscr{B}^{\mathscr{T}^1}$ is a minimal S_2^3-invariant Markov basis. Consider the representative fiber $\mathscr{F}^0_t$ such that $\bar{\Delta}_t = \{1,2,3\}$. For $j = 1,2$, denote the two induced subtrees of $\mathscr{T}^j$ obtained by removing the edge e_l by $\mathscr{T}^j_{e_l}$ and $\mathscr{T}^{j'}_{e_l}$. Figure 8.8 shows

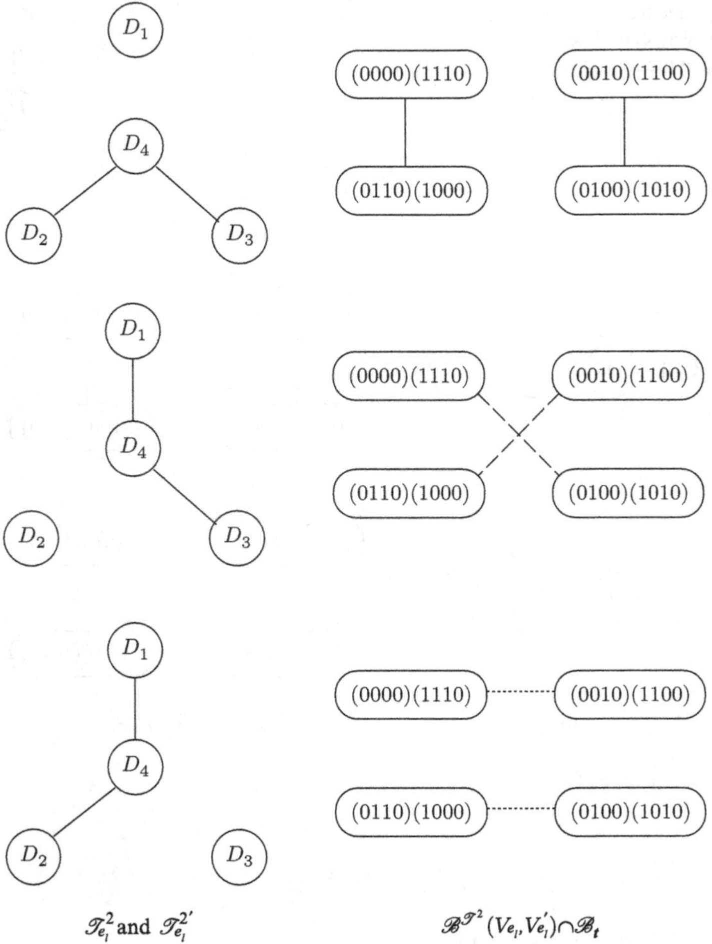

Fig. 8.9 The structure of $\mathscr{B}_t^{\mathscr{T}^2}$

$\mathscr{T}^1_{e_l}$, $\mathscr{T}^{1'}_{e_l}$ and $\mathscr{B}^{\mathscr{T}^1}(V_{e_l}, V'_{e_l}) \cap \mathscr{B}_t$. If we remove e_3 from $\mathscr{T}^1$, 1, 2, and 3 are still connected and hence $\mathscr{B}^{\mathscr{T}^1}(V_{e_3}, V'_{e_3}) \cap \mathscr{B}_t = \emptyset$. Therefore $\mathscr{B}_t^{\mathscr{T}^1}$ contains $\kappa(t) = 2$ orbits.

On the other hand because $\mathscr{T}^2$ has three endpoints, $\mathscr{B}^{\mathscr{T}^2}$ is not a minimal S_2^3-invariant Markov basis. Figure 8.9 shows $\mathscr{T}^2_{e_l}$, $\mathscr{T}^{2'}_{e_l}$, and $\mathscr{B}^{\mathscr{T}^2}(V_{e_l}, V'_{e_l}) \cap \mathscr{B}_t$. We can see that $\mathscr{B}_t^{\mathscr{T}^2}$ contains three orbits. As seen from this example, in general the minimality of $\mathscr{M}^{\mathscr{T}}$ depends on clique trees $\mathscr{T}$.

Example 8.5. We consider the model defined by the chordal graph in Fig. 8.10. The clique tree of this graph is uniquely determined by $\mathscr{T}^2$ in Fig. 8.7. As seen from this example, there exist decomposable models such that $\mathscr{B}^{\mathscr{T}}$ for every clique tree $\mathscr{T}$ is not minimal $G_{I_1,\ldots,I_m}$-invariant.

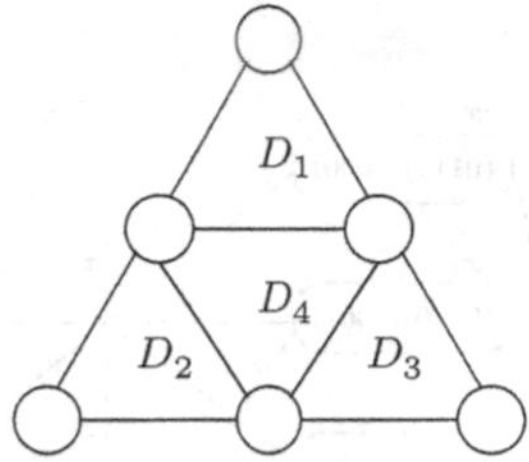

Fig. 8.10 A chordal graph whose clique tree is uniquely determined

8.6 The Relation Between Minimal and Minimal Invariant Markov Bases

From a practical point of view a $G_{I_1,\dots,I_m}$-invariant Markov basis is useful because its representative moves give the most concise expression of a Markov basis. On the other hand a minimal Markov basis is also important because the number of moves contained in it is minimum among Markov bases. Here we consider the relation between a minimal and a minimal $G_{I_1,\dots,I_m}$-invariant Markov basis and give an algorithm to obtain a minimal Markov basis from representative moves of a minimal $G_{I_1,\dots,I_m}$-invariant Markov basis.

As mentioned in the previous section, the set of $G_{\boldsymbol{t}}$-orbits in a minimal $G_{\boldsymbol{t}}$-invariant set $\mathscr{B}_{G_{\boldsymbol{t}}}$ of moves that connects $\mathscr{F}_{\boldsymbol{t}}^0$ has a one-to-one correspondence to a basis $\mathscr{V}^0$ of $\mathscr{V}^{c-1}$. Define $\bar{g}_{kl} \in G_{\Gamma_l}$ and $g^k \in G_{\boldsymbol{t}}$ as in (8.16) and (8.17). Let $H_{\boldsymbol{t}} = \{g^1,\dots,g^{c-1}\} \subset G_{\boldsymbol{t}}$. Now we consider generating a set of moves $\mathscr{B}_{\boldsymbol{t}}^*$ in $\mathscr{F}_{\boldsymbol{t}}$ by the following algorithm.

Algorithm 8.2
Input: $\mathscr{F}_{\boldsymbol{t}}$, $H_{\boldsymbol{t}} = \{g^1,\dots,g^{c-1}\}$
Output: $\mathscr{B}_{\boldsymbol{t}}^*$

```
begin
  𝓑*_t ← ∅;
  Choose any element x_1 in 𝓕_t;
  for k = 2 to c do
  begin
    for l = 1 to 2^{k-2} do
    begin
      x_{l+2^{k-2}} := g^{k-1}(x_l);
      z_{l+2^{k-2}} := x_l − x_{l+2^{k-2}};
      𝓑*_t ← 𝓑*_t ∪ {z_{l+2^{k-2}}};
    end
  end
  return 𝓑*_t;
end
```

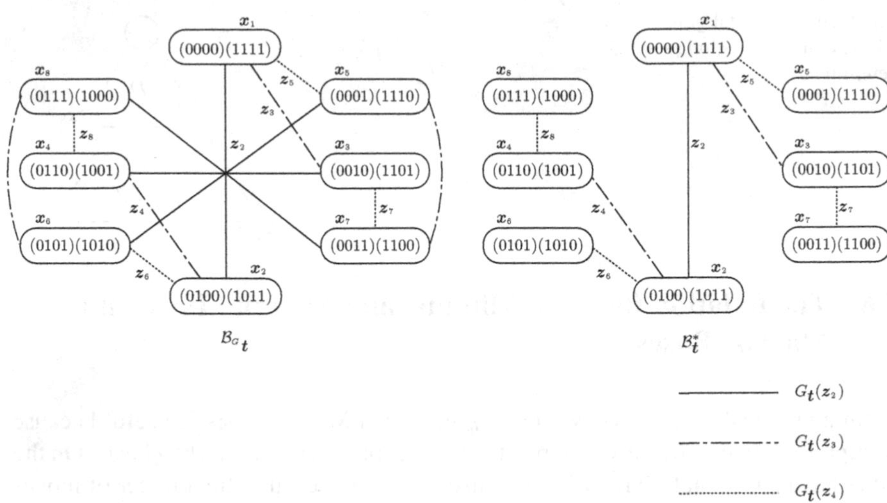

Fig. 8.11 $\mathscr{B}_{G_t}$ and $\mathscr{B}_{\boldsymbol{t}}^*$ generated by Algorithm 8.2

Theorem 8.8. *$\mathscr{B}_{\boldsymbol{t}}^*$ generated by Algorithm 8.2 is a minimal set of moves that connects $\mathscr{F}_{\boldsymbol{t}}$.*

Proof. Inasmuch as $|\mathscr{B}_{\boldsymbol{t}}^*| = 2^0 + 2^1 + \cdots + 2^{c-1} = 2^c - 1$, it suffices to show that $\boldsymbol{x}_l \neq \boldsymbol{x}_{l'}$ for $l \neq l'$. Suppose that there exist l and l' such that $l \neq l'$ and $\boldsymbol{x}_l = \boldsymbol{x}_{l'}$ and that $\boldsymbol{x}_l$ and $\boldsymbol{x}_{l'}$ are expressed as

$$\boldsymbol{x}_l = g^{k_p} \circ \cdots \circ g^{k_1}(\boldsymbol{x}_1), \quad \boldsymbol{x}_{l'} = g^{k'_{p'}} \circ \cdots \circ g^{k'_1}(\boldsymbol{x}_1),$$

where $k_1 < k_2 < \cdots < k_p \leq c-1$ and $k'_1 < k'_2 < \cdots < k'_{p'} \leq c-1$. Without loss of generality we can assume $p \leq p'$. Then we have

$$g^{k_p} \circ \cdots \circ g^{k_1} = g^{k'_{p'}} \circ \cdots \circ g^{k'_1} \tag{8.19}$$

and there exists $l \leq p$ such that $k_l \neq k'_l$. From Lemma 8.3 (8.19) is equivalent to

$$\boldsymbol{v}_{k_1} \oplus \cdots \oplus \boldsymbol{v}_{k_p} = \boldsymbol{v}_{k'_1} \oplus \cdots \oplus \boldsymbol{v}_{k'_{p'}},$$

which contradicts that $\mathscr{V}^0$ is a basis of $\mathscr{V}^{c-1}$. Hence we have $\boldsymbol{x}_l \neq \boldsymbol{x}_{l'}$ for $l \neq l'$. □

From (8.6) we obtain the following result.

Corollary 8.3. *$\mathscr{B}^* = \bigcup_{\boldsymbol{t} \in \mathscr{B}_{\mathrm{nd}}} \mathscr{B}_{\boldsymbol{t}}^*$ is a minimal Markov basis.*

Example 8.6 (The complete independence model of a four-way contingency table). We consider the same fiber as in Example 8.2. Define $\mathscr{V}^0 = \{\boldsymbol{v}_2, \boldsymbol{v}_3, \boldsymbol{v}_4\}$ by $\boldsymbol{v}_2 = (100)$, $\boldsymbol{v}_3 = (010)$, and $\boldsymbol{v}_4 = (001)$. Figure 8.11 shows $\mathscr{B}_{G_t}$ and $\mathscr{B}_{\boldsymbol{t}}^*$ generated by Algorithm 8.2 with $\boldsymbol{x}_1 = (0000)(1111)$.

Chapter 9
Markov Basis for No-Three-Factor Interaction Models and Some Other Hierarchical Models

9.1 No-Three-Factor Interaction Models for $3 \times 3 \times K$ Contingency Tables

The no-three-factor interaction model for three-way contingency tables is one of the simplest nondecomposable hierarchical models. In this chapter, we write $I \times J \times K$ contingency tables as $\boldsymbol{x} = \{x_{ijk} \mid \boldsymbol{i} = (ijk) \in \mathscr{I}\}$ where $\mathscr{I} = \{1,\ldots,I\} \times \{1,\ldots,J\} \times \{1,\ldots,K\}$. The generating class of no-three-factor interaction models is $\mathscr{D} = \{\{1,2\},\{1,3\},\{2,3\}\}$. Therefore the cell probability for $\boldsymbol{i} = (ijk)$ is written as

$$\log p_{ijk} = \mu_{\{1,2\}}(ij) + \mu_{\{1,3\}}(ik) + \mu_{\{2,3\}}(jk).$$

With lexicographic ordering of indices, the configuration A for this model is written as

$$A = \begin{pmatrix} E_I \otimes E_J \otimes \mathbf{1}'_K \\ E_I \otimes \mathbf{1}'_J \otimes E_K \\ \mathbf{1}'_I \otimes E_J \otimes E_K \end{pmatrix}.$$

As we see below, the structure of Markov bases for this model is very complicated. In fact, the closed-form expression of Markov bases for this model of general $I \times J \times K$ tables is not yet obtained at present. Instead, we show the structure of minimal Markov basis for $I = J = 3$ cases (i.e., $3 \times 3 \times K$ contingency tables) given in [10]. The arguments in [10] are based on the distance-reducing proofs in Chap. 6: first we give a candidate set of moves $\mathscr{B}^*$, and in order to show that $\mathscr{F}_{\boldsymbol{t}}$ constitutes one $\mathscr{B}^*$-equivalence class for any $\boldsymbol{t}$, we suppose $\mathscr{F}_1$ and $\mathscr{F}_2$ are different $\mathscr{B}^*$-equivalence classes of $\mathscr{F}_{\boldsymbol{t}}$ for some $\boldsymbol{t}$, then choose $\boldsymbol{x} \in \mathscr{F}_1$ and $\boldsymbol{y} \in \mathscr{F}_2$ such that

$$|\boldsymbol{y} - \boldsymbol{x}| = \sum_{i,j,k} |y_{ijk} - x_{ijk}| > 0$$

is minimized, and finally derive a contradiction.

S. Aoki et al., *Markov Bases in Algebraic Statistics*, Springer Series in Statistics 199, DOI 10.1007/978-1-4614-3719-2_9,

For an $I \times J \times K$ table $\boldsymbol{x} = \{x_{ijk}\}$, i-slice (or $i = i_0$ slice) of $\boldsymbol{x}$ is the two-dimensional slice $\{x_{i_0jk}\}_{1\le j\le J, 1\le k\le K}$, where $i = i_0$ is fixed. We similarly define j-slice and k-slice. In this chapter, to display $I \times J \times K$ contingency tables, we write I i-slices of size $J \times K$ as follows.

$$\begin{array}{|ccc|}\hline x_{111} & \cdots & x_{11K} \\ \vdots & & \vdots \\ x_{1J1} & \cdots & x_{1JK} \\ \hline\end{array} \begin{array}{|ccc|}\hline x_{211} & \cdots & x_{21K} \\ \vdots & & \vdots \\ x_{2J1} & \cdots & x_{2JK} \\ \hline\end{array} \cdots \begin{array}{|ccc|}\hline x_{I11} & \cdots & x_{I1K} \\ \vdots & & \vdots \\ x_{IJ1} & \cdots & x_{IJK} \\ \hline\end{array}.$$

We also use the concise expression of moves in Chap. 7 by the locations of nonzero cells. For example, a move of $3 \times 3 \times 3$ table displayed as

$$\begin{array}{|ccc|}\hline +1 & -1 & 0 \\ -1 & +1 & 0 \\ 0 & 0 & 0 \\ \hline\end{array} \begin{array}{|ccc|}\hline -1 & +1 & 0 \\ 0 & -1 & +1 \\ +1 & 0 & -1 \\ \hline\end{array} \begin{array}{|ccc|}\hline 0 & 0 & 0 \\ +1 & 0 & -1 \\ -1 & 0 & +1 \\ \hline\end{array}$$

is also written as

$$(111)(122)(212)(223)(231)(321)(333) - (112)(121)(211)(222)(233)(323)(331).$$

In this chapter, it is always assumed that the indices are integers such that

$$\begin{aligned} &1 \le i_1, i_2, \ldots, i_I \le I, && i_1, i_2, \ldots, i_I \text{ all distinct,} \\ &1 \le j_1, j_2, \ldots, j_J \le J, && j_1, j_2, \ldots, j_J \text{ all distinct,} \\ &1 \le k_1, k_2, \ldots, k_K \le K, && k_1, k_2, \ldots, k_K \text{ all distinct.} \end{aligned}$$

9.2 Unique Minimal Markov Basis for $3 \times 3 \times 3$ Tables

First we define the most elementary eight-entry move.

Definition 9.1. A move of degree 4 is a move $\boldsymbol{m}_4(i_1i_2, j_1j_2, k_1k_2) \in \ker_{\mathbb{Z}} A$ written as

$$(i_1j_1k_1)(i_1j_2k_2)(i_2j_1k_2)(i_2j_2k_1) - (i_1j_1k_2)(i_1j_2k_1)(i_2j_1k_1)(i_2j_2k_2).$$

We call this move a *basic move* for the no-three-factor interaction model. Figure 9.1 gives a three-dimensional view of the basic move. From the definition, the relation

$$\boldsymbol{m}_4(i_1i_2, j_1j_2, k_1k_2) = \boldsymbol{m}_4(i_1i_2, j_2j_1, k_2k_1) = \boldsymbol{m}_4(i_2i_1, j_1j_2, k_2k_1) = -\boldsymbol{m}_4(i_2i_1, j_1j_2, k_1k_2)$$

holds. These moves of degree 4 are the most elementary moves in the sense that all the other moves of higher degrees in $\ker_{\mathbb{Z}} A$ are written as linear combinations

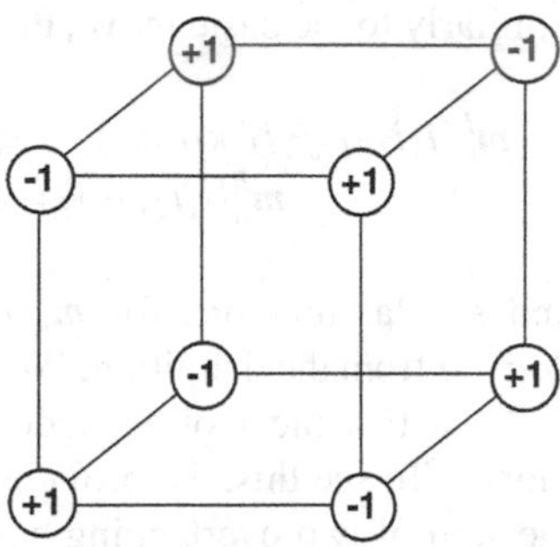

Fig. 9.1 $2 \times 2 \times 2$ move of degree 4 (basic move)

of these degree 4 moves with integral coefficients. Namely, the set of basic moves contains a lattice basis of $\ker_{\mathbb{Z}} A$. It is seen that the basic moves are indispensable (see Definition 5.1) because

$$\{(i_1 j_1 k_1)(i_1 j_2 k_2)(i_2 j_1 k_2)(i_2 j_2 k_1),\ (i_1 j_1 k_2)(i_1 j_2 k_1)(i_2 j_1 k_1)(i_2 j_2 k_2)\}$$

constitutes a two-element fiber.

For $I \times J \times K$ tables with fixed two-dimensional marginals, the set of basic moves is not a Markov basis when at least two of I, J, K are larger than 2. To see this, consider $3 \times 3 \times 3$ contingency tables having two-dimensional marginals as $x_{ij\cdot} = x_{i\cdot k} = x_{\cdot jk} = 2$ for all $1 \le i, j, k \le 3$. There are 132 elements in this fiber, but elements such as

$$\begin{array}{|ccc|}\hline 2&0&0\\ 0&2&0\\ 0&0&2\\ \hline\end{array}\quad \begin{array}{|ccc|}\hline 0&2&0\\ 0&0&2\\ 2&0&0\\ \hline\end{array}\quad \begin{array}{|ccc|}\hline 0&0&2\\ 2&0&0\\ 0&2&0\\ \hline\end{array}$$

are not connected to any other element in the fiber by the basic moves. This simple example suggests that the following moves of degree 6 are needed for the Markov basis.

Definition 9.2. Moves of degree 6 are a move $\boldsymbol{m}_6^I(i_1 i_2, j_1 j_2 j_3, k_1 k_2 k_3) \in \ker_{\mathbb{Z}} A$ written as

$$\begin{aligned}&(i_1 j_1 k_1)(i_1 j_2 k_2)(i_1 j_3 k_3)(i_2 j_1 k_2)(i_2 j_2 k_3)(i_2 j_3 k_1)\\ &\qquad -(i_1 j_1 k_2)(i_1 j_2 k_3)(i_1 j_3 k_1)(i_2 j_1 k_1)(i_2 j_2 k_2)(i_2 j_3 k_3),\end{aligned}$$

a move $\boldsymbol{m}_6^J(i_1 i_2 i_3, j_1 j_2, k_1 k_2 k_3) \in \ker_{\mathbb{Z}} A$ written as

$$\begin{aligned}&(i_1 j_1 k_1)(i_1 j_2 k_2)(i_2 j_1 k_2)(i_2 j_2 k_3)(i_3 j_1 k_3)(i_3 j_2 k_1)\\ &\qquad -(i_1 j_1 k_2)(i_1 j_2 k_1)(i_2 j_1 k_3)(i_2 j_2 k_2)(i_3 j_1 k_1)(i_3 j_2 k_3)\end{aligned}$$

and a move $\boldsymbol{m}_6^K(i_1 i_2 i_3, j_1 j_2 j_3, k_1 k_2) \in \ker_{\mathbb{Z}} A$ written as

$$\begin{aligned}&(i_1 j_1 k_1)(i_1 j_2 k_2)(i_2 j_2 k_1)(i_2 j_3 k_2)(i_3 j_1 k_2)(i_3 j_3 k_1)\\ &\qquad -(i_1 j_1 k_2)(i_1 j_2 k_1)(i_2 j_2 k_2)(i_2 j_3 k_1)(i_3 j_1 k_1)(i_3 j_3 k_2).\end{aligned}$$

Similarly to the basic move, the relations

$$\boldsymbol{m}_6^I(i_1i_2, j_1j_2j_3, k_1k_2k_3) = \boldsymbol{m}_6^I(i_1i_2, j_2j_3j_1, k_2k_3k_1) = \boldsymbol{m}_6^I(i_2i_1, j_1j_3j_2, k_2k_1k_3),$$
$$\boldsymbol{m}_6^I(i_1i_2, j_1j_2j_3, k_1k_2k_3) = -\boldsymbol{m}_6^I(i_2i_1, j_1j_2j_3, k_1k_2k_3)$$

and similar relations for $\boldsymbol{m}_6^J(i_1i_2i_3, j_1j_2, k_1k_2k_3)$ and $\boldsymbol{m}_6^K(i_1i_2i_3, j_1j_2j_3, k_1k_2)$ are derived from the definition. We see that all the moves of degree 6 are indispensable.

Note that the moves of degree 6 are obtained as combinations of two basic moves. To see this, we provide a complete list of the patterns that are obtained by the sum of two overlapping basic moves. For basic moves $\boldsymbol{m}_4(i_1i_2, j_1j_2, k_1k_2)$ and $\boldsymbol{m}_4(i'_1i'_2, j'_1j'_2, k'_1k'_2)$, define

$$\begin{aligned}\Delta_I &= \delta_{i_1i'_1} + \delta_{i_1i'_2} + \delta_{i_2i'_1} + \delta_{i_2i'_2},\\ \Delta_J &= \delta_{j_1j'_1} + \delta_{j_1j'_2} + \delta_{j_2j'_1} + \delta_{j_2j'_2},\\ \Delta_K &= \delta_{k_1k'_1} + \delta_{k_1k'_2} + \delta_{k_2k'_1} + \delta_{k_2k'_2}\end{aligned}$$

and

$$\Delta = \Delta_I + \Delta_J + \Delta_K,$$

where $\delta_{ij} = 1$ if $i = j$, and $= 0$ otherwise. Because two moves are overlapping, $\Delta_I, \Delta_J, \Delta_K \geq 1$. Furthermore, $\Delta_I \leq 2$, because $i_1 \neq i_2$ and $i'_1 \neq i'_2$. Similarly, $\Delta_J, \Delta_K \leq 2$, therefore $\Delta \in \{3,4,5,6\}$. Corresponding to the values of Δ, all the patterns are classified as follows.

- $\Delta = 3$: $\boldsymbol{m}_4(i_1i_2, j_1j_2, k_1k_2)$ and $\boldsymbol{m}_4(i'_1i'_2, j'_1j'_2, k'_1k'_2)$ overlap at one nonzero entry. In this chapter, we call this case a *combination of type 1* or a *type*-1 *combination*. If the signs of this overlapping cell are opposite, a move of degree 7 is obtained. Figure 9.2 gives a three-dimensional view of this type of move.
- $\Delta = 4$: $\boldsymbol{m}_4(i_1i_2, j_1j_2, k_1k_2)$ and $\boldsymbol{m}_4(i'_1i'_2, j'_1j'_2, k'_1k'_2)$ overlap at two nonzero entries. We call this case a *combination of type* 2 or a *type*-2 *combination*. If the pairs of signs of these two cells are opposite, an indispensable move of degree 6 in Definition 9.2 is obtained. Figure 9.3 gives a three-dimensional view of this type of move.
- $\Delta = 5$: $\boldsymbol{m}_4(i_1i_2, j_1j_2, k_1k_2)$ and $\boldsymbol{m}_4(i'_1i'_2, j'_1j'_2, k'_1k'_2)$ overlap at four nonzero entries along a two-dimensional rectangle. If all the pairs of signs are canceled, a basic move is again obtained as

$$\begin{aligned}\boldsymbol{m}_4(i_1i_2, j_1j_2, k_1k_2) &= \boldsymbol{m}_4(i_1i_2, j_1j_2, k_1k_3) + \boldsymbol{m}_4(i_1i_2, j_1j_2, k_3k_2)\\ &= \boldsymbol{m}_4(i_1i_2, j_1j_3, k_1k_2) + \boldsymbol{m}_4(i_1i_2, j_3j_2, k_1k_2)\\ &= \boldsymbol{m}_4(i_1i_3, j_1j_2, k_1k_2) + \boldsymbol{m}_4(i_3i_2, j_1j_2, k_1k_2).\end{aligned}$$

- $\Delta = 6$: $\boldsymbol{m}_4(i_1i_2, j_1j_2, k_1k_2)$ and $\boldsymbol{m}_4(i'_1i'_2, j'_1j'_2, k'_1k'_2)$ overlap completely.

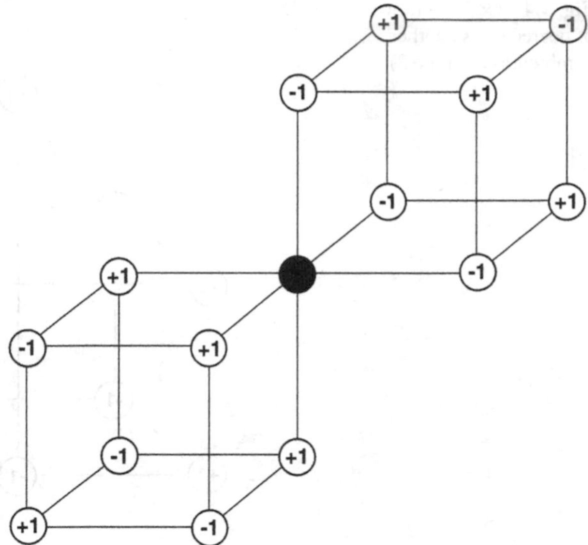

Fig. 9.2 $3 \times 3 \times 3$ move of degree 7

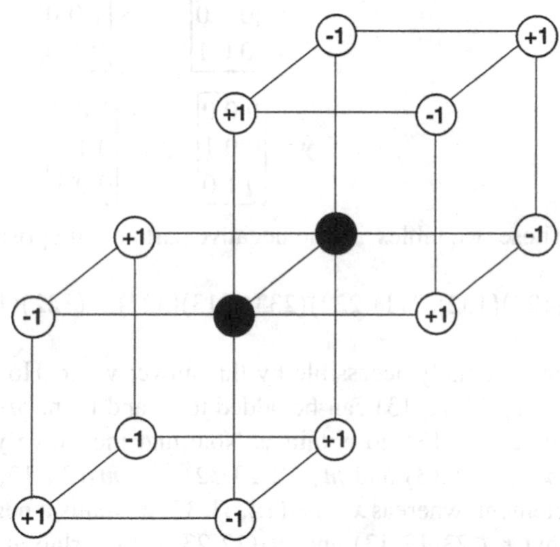

Fig. 9.3 $2 \times 3 \times 3$ move of degree 6

In the above list, the cases of $\Delta = 3$ and $\Delta = 4$ yield so called "two-step moves;" that is, two basic moves are needed to construct these moves of degree 6 and degree 7. As we see in Theorem 9.1 below, the moves of degree 7 in Fig. 9.2 are not needed for a minimal Markov basis. To demonstrate this point, consider the following two $3 \times 3 \times 3$ tables.

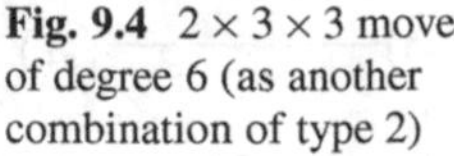

Fig. 9.4 $2 \times 3 \times 3$ move of degree 6 (as another combination of type 2)

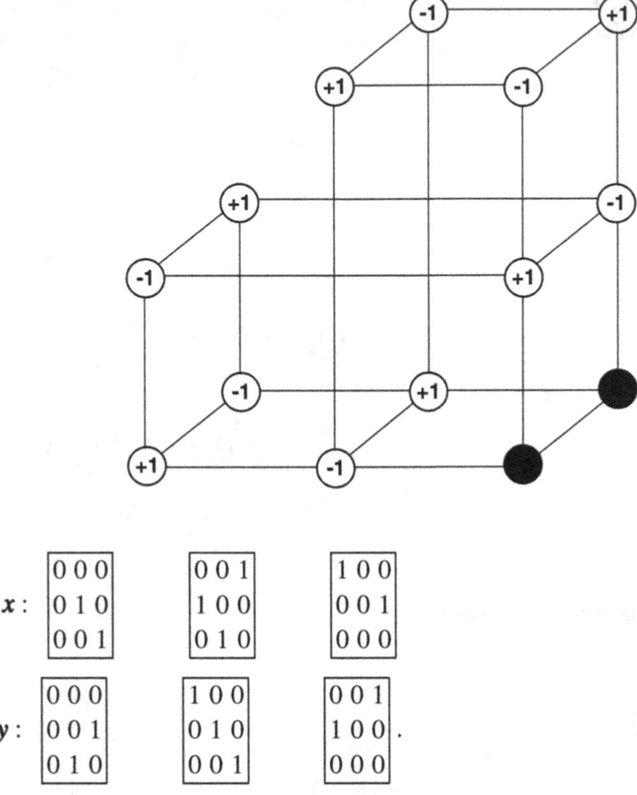

$$
\boldsymbol{x}: \quad \begin{array}{|ccc|}\hline 0&0&0\\ 0&1&0\\ 0&0&1\\ \hline\end{array} \quad \begin{array}{|ccc|}\hline 0&0&1\\ 1&0&0\\ 0&1&0\\ \hline\end{array} \quad \begin{array}{|ccc|}\hline 1&0&0\\ 0&0&1\\ 0&0&0\\ \hline\end{array}
$$

$$
\boldsymbol{y}: \quad \begin{array}{|ccc|}\hline 0&0&0\\ 0&0&1\\ 0&1&0\\ \hline\end{array} \quad \begin{array}{|ccc|}\hline 1&0&0\\ 0&1&0\\ 0&0&1\\ \hline\end{array} \quad \begin{array}{|ccc|}\hline 0&0&1\\ 1&0&0\\ 0&0&0\\ \hline\end{array} .
$$

These two tables are the negative part and the positive part of the move of degree 7,

$$
(123)(132)(211)(222)(233)(313)(321) - (122)(133)(213)(221)(232)(311)(323),
$$

and mutually accessible by this move: $\boldsymbol{y} - \boldsymbol{x}$. However, instead of adding $\boldsymbol{y} - \boldsymbol{x}$ to $\boldsymbol{x}$, $\boldsymbol{m}_4(23,12,13)$ can be added to $\boldsymbol{x}$, and then, $\boldsymbol{m}_4(12,23,32)$ can be added to $\boldsymbol{x} + \boldsymbol{m}_4(23,12,13)$, to obtain $\boldsymbol{y}$. Note that the move $\boldsymbol{y} - \boldsymbol{x}$ is a type-1 combination of $\boldsymbol{m}_4(23,12,13)$ and $\boldsymbol{m}_4(12,23,32)$. $\boldsymbol{x} + \boldsymbol{m}_4(23,12,13)$ does not contain a negative element, whereas $\boldsymbol{x} + \boldsymbol{m}_4(12,23,32)$ contains a negative element $(2,2,3)$. Note also that $\boldsymbol{m}_4(23,12,13)$ and $\boldsymbol{m}_4(12,23,32)$ overlap at this cell. Because the two basic moves are canceling at this cell, it is obvious that at least one of these basic moves (that has $+1$ at this cell) can be added without causing negative elements. On the other hand, because the type-2 combination has two overlapping cells, it cannot be avoided that one of these two elements becomes negative in adding basic moves one by one. For this reason, the type-2 combination is essential.

We also note that the expression of the move of degree 6 as a type-2 combination of two basic moves is not unique. Figure 9.4 illustrates the same move of degree 6 shown in Fig. 9.3, but the overlapping cells of the two basic moves are different.

Now we give a unique minimal Markov basis for $3 \times 3 \times 3$ case.

Theorem 9.1. *The set of basic moves* $\boldsymbol{m}_4(i_1i_2, j_1j_2, k_1k_2)$ *and moves of degree* 6, $\boldsymbol{m}_6^I(i_1i_2, j_1j_2j_3, k_1k_2k_3), \boldsymbol{m}_6^J(i_1i_2i_3, j_1j_2, k_1k_2k_3), \boldsymbol{m}_6^K(i_1i_2i_3, j_1j_2j_3, k_1k_2)$ *constitutes a unique minimal Markov basis for* $3 \times 3 \times 3$ *tables.*

Note that the minimality and the uniqueness directly hold if this set of moves constitutes a Markov basis, because it is composed of indispensable moves only. See Corollary 5.2 in Chap. 5.

Following the distance-reducing proofs in Chap. 6, we consider the pattern of $\boldsymbol{z} = \boldsymbol{y} - \boldsymbol{x}$ where $\boldsymbol{x}$ and $\boldsymbol{y}$ have the same two-dimensional marginals. Before we give a proof of Theorem 9.1, we show a useful lemma concerning the patterns of two-dimensional slices of $\boldsymbol{y} - \boldsymbol{x}$ for general $3 \times 3 \times K$ cases.

Definition 9.3. Let C be a two-dimensional matrix with elements c_{ij}. Then a *rectangle* is a set of four entries $(c_{i_1j_1}, c_{i_2j_1}, c_{i_2j_2}, c_{i_1j_2})$ with alternating signs. Similarly, a *6-cycle* is a set of six entries $(c_{i_1j_1}, c_{i_2j_1}, c_{i_2j_2}, c_{i_3j_2}, c_{i_3j_3}, c_{i_1j_3})$ with alternating signs.

Using the fact that all the marginal totals of $\boldsymbol{z} = \boldsymbol{y} - \boldsymbol{x}$ are zeros, it can be easily shown that any nonzero entry of $\boldsymbol{z}$ has to be a member of either a rectangle or a 6-cycle in all of the i-, j-, and k-slices when $\boldsymbol{x}$ and $\boldsymbol{y}$ are $3 \times 3 \times K$ contingency tables.

Lemma 9.1. *Let* $\boldsymbol{x}$ *and* $\boldsymbol{y}$ *be* $3 \times 3 \times K$ *contingency tables and let* $\boldsymbol{z} = \boldsymbol{y} - \boldsymbol{x}$*. Consider* $\boldsymbol{z}$ *after minimizing* $|\boldsymbol{z}|$ *by applying the basic moves and the moves of degree* 6 *without causing negative entries on the way. Then*

(a) No k-slice of $\boldsymbol{z}$ *contains* 6*-cycles.*
(b) There is at least one rectangle in either an i-slice or a j-slice unless $\boldsymbol{z} = 0$.

Proof. In the proof of this lemma, we display k-slices of $\boldsymbol{z}$ instead of our usual display of i-slices.

To prove (a), suppose that, without loss of generality, $k = 1$ slice of $\boldsymbol{z}$ contains the following 6-cycle

$$\begin{array}{c|ccc|} i\backslash j & 1 & 2 & 3 \\ \hline 1 & + & - & * \\ 2 & - & * & + \\ 3 & * & + & - \\ \hline \end{array}\,.$$

Because $z_{11\cdot} = 0$, there exists at least one negative element in $z_{112}, z_{113}, \ldots, z_{11K}$. Let $z_{112} < 0$ without loss of generality. As is shown above, z_{112} has to be an element of either a rectangle or a 6-cycle in the $k = 2$ slice. These two cases are considered, respectively, as follows.

Case 1. z_{112} is an element of a 6-cycle.
It is seen that the negative entries in the 6-cycle in the $k = 2$ slice, which includes z_{112}, can be either (i) $(z_{112}, z_{222}, z_{332})$ or (ii) $(z_{112}, z_{232}, z_{322})$. In case (i), $\boldsymbol{m}_4(12, 12, 12)$ can be added to $\boldsymbol{x}$ without causing negative entries to make $|\boldsymbol{z}|$ smaller

because $x_{121}, x_{211}, x_{112}, x_{222} > 0$. On the other hand, in case (ii), $\boldsymbol{m}_6^K(132, 123, 12)$ can be added to $\boldsymbol{x}$ without causing negative entries to make $|\boldsymbol{z}|$ smaller because $x_{121}, x_{211}, x_{331}, x_{112}, x_{232}, x_{322} > 0$. These imply that Case 1 is a contradiction.

Case 2. z_{112} is an element of a rectangle.
It is seen that the negative entries in the rectangle, which includes z_{112}, can be either (i) (z_{112}, z_{222}), (ii) (z_{112}, z_{232}), (iii) (z_{112}, z_{322}), or (iv) (z_{112}, z_{332}). In case (i), $\boldsymbol{m}_4(12, 12, 12)$ can be added to $\boldsymbol{x}$ without causing negative entries and $|\boldsymbol{z}|$ can be made smaller as in (*i*) of Case 1. In case (ii), it follows that $z_{132}, z_{212} > 0$ and $\boldsymbol{m}_4(12, 13, 21)$ can be added to $\boldsymbol{y}$ without causing negative entries and make $|\boldsymbol{z}|$ smaller because $y_{111}, y_{231}, y_{132}, y_{212} > 0$. Case (iii) is the symmetric case of (ii).

In case (iv), the two k-slices, $\{z_{ij1}\}$ and $\{z_{ij2}\}$, are represented as

$$\{z_{ij1}\}: \begin{array}{c|ccc|} i\backslash j & 1 & 2 & 3 \\ \hline 1 & + & - & * \\ 2 & - & * & + \\ 3 & * & + & - \\ \hline \end{array} \qquad \{z_{ij2}\}: \begin{array}{c|ccc|} i\backslash j & 1 & 2 & 3 \\ \hline 1 & - & * & + \\ 2 & * & * & * \\ 3 & + & * & - \\ \hline \end{array}.$$

In this case, because $z_{331}, z_{332} < 0$, at least one of $z_{333}, \ldots, z_{33K}$ has to be positive. Let $z_{333} > 0$ without loss of generality. Here, z_{333} is again an element of either a rectangle or a 6-cycle. But as already seen in Case 1, there cannot be another 6-cycle in the $k \neq 1$ slice. Thus z_{333} has to be a member of a rectangle. Moreover, for the same reason as (i)–(iii) of Case 2, the $k = 3$ slice has to be a mirror image of the $k = 2$ slice:

$$\{z_{ij1}\}: \begin{array}{c|ccc|} i\backslash j & 1 & 2 & 3 \\ \hline 1 & + & - & * \\ 2 & - & * & + \\ 3 & * & + & - \\ \hline \end{array} \qquad \{z_{ij2}\}: \begin{array}{c|ccc|} i\backslash j & 1 & 2 & 3 \\ \hline 1 & - & * & + \\ 2 & * & * & * \\ 3 & + & * & - \\ \hline \end{array} \qquad \{z_{ij3}\}: \begin{array}{c|ccc|} i\backslash j & 1 & 2 & 3 \\ \hline 1 & + & * & - \\ 2 & * & * & * \\ 3 & - & * & + \\ \hline \end{array}.$$

However, $\boldsymbol{m}_4(13, 13, 23)$ can be added to $\boldsymbol{y}$ or $\boldsymbol{m}_4(13, 13, 32)$ can be added to $\boldsymbol{x}$ without causing negative entries and $|\boldsymbol{z}|$ can be made smaller, which contradicts the assumption. These imply that Case 2 also is a contradiction.

These considerations indicate that the 6-cycle cannot be included in any 3×3 slices and the proof of (a) is completed.

Next (b) is proved. Suppose $\boldsymbol{z}$ has nonzero entries and let $z_{111} > 0$ without loss of generality. It is known that z_{111} is a member of a rectangle in the $k = 1$ slice from (a). Then let the $k = 1$ slice be represented as

$$\begin{array}{c|ccc|} i\backslash j & 1 & 2 & 3 \\ \hline 1 & + & - & * \\ 2 & - & + & * \\ 3 & * & * & * \\ \hline \end{array}$$

without loss of generality. We are assuming that there exists no rectangle in the $3 \times K$ i-slices or j-slices of $\boldsymbol{z}$. Write $z_{112} < 0$ without loss of generality because $z_{11\cdot} = 0$.

$$\{z_{ij1}\}: \begin{array}{c|ccc|} i\backslash j & 1 & 2 & 3 \\ \hline 1 & + & - & * \\ 2 & - & + & * \\ 3 & * & * & * \\ \hline \end{array} \qquad \{z_{ij2}\}: \begin{array}{c|ccc|} i\backslash j & 1 & 2 & 3 \\ \hline 1 & - & * & * \\ 2 & * & * & * \\ 3 & * & * & * \\ \hline \end{array}.$$

From the assumption, it follows that $z_{122}, z_{212} \leq 0$ because otherwise either $i = 1$ slice or $j = 1$ slice has a rectangle. We also have $z_{222} \geq 0$ because otherwise we can add $\boldsymbol{m}_4(12,12,12)$ to $\boldsymbol{x}$ without causing negative entries and make $|\boldsymbol{z}|$ smaller. Hereafter we display nonnegative elements by $0+$ and nonpositive elements by $0-$.

$$\{z_{ij1}\}: \begin{array}{c|ccc|} i\backslash j & 1 & 2 & 3 \\ \hline 1 & + & - & * \\ 2 & - & + & * \\ 3 & * & * & * \\ \hline \end{array} \qquad \{z_{ij2}\}: \begin{array}{c|ccc|} i\backslash j & 1 & 2 & 3 \\ \hline 1 & - & 0- & * \\ 2 & 0- & 0+ & * \\ 3 & * & * & * \\ \hline \end{array}.$$

Inasmuch as z_{112} has to be an element of a rectangle in a $k = 2$ slice, $z_{132} > 0, z_{312} > 0$ and $z_{332} < 0$ are derived.

$$\{z_{ij1}\}: \begin{array}{c|ccc|} i\backslash j & 1 & 2 & 3 \\ \hline 1 & + & - & * \\ 2 & - & + & * \\ 3 & * & * & * \\ \hline \end{array} \qquad \{z_{ij2}\}: \begin{array}{c|ccc|} i\backslash j & 1 & 2 & 3 \\ \hline 1 & - & 0- & + \\ 2 & 0- & 0+ & * \\ 3 & + & * & - \\ \hline \end{array}.$$

It is seen that if $z_{131} < 0$, there appears a rectangle in the $i = 1$ slice; and if $z_{311} < 0$, there appears a rectangle in the $j = 1$ slice. These contradict the assumption.

Then it follows that $z_{131}, z_{311} \geq 0$. Here we write $z_{123} > 0$ without loss of generality, because $z_{12\cdot} = 0$.

$$\{z_{ij1}\}: \begin{array}{c|ccc|} i\backslash j & 1 & 2 & 3 \\ \hline 1 & + & - & 0+ \\ 2 & - & + & * \\ 3 & 0+ & * & * \\ \hline \end{array} \qquad \{z_{ij2}\}: \begin{array}{c|ccc|} i\backslash j & 1 & 2 & 3 \\ \hline 1 & - & 0- & + \\ 2 & 0- & 0+ & * \\ 3 & + & * & - \\ \hline \end{array} \qquad \{z_{ij3}\}: \begin{array}{c|ccc|} i\backslash j & 1 & 2 & 3 \\ \hline 1 & * & + & * \\ 2 & * & * & * \\ 3 & * & * & * \\ \hline \end{array}.$$

It is seen that if $z_{113} < 0$, there appears a rectangle in the $i = 1$ slice; and if $z_{223} < 0$, there appears a rectangle in the $j = 2$ slice. These contradict the assumption. Then it follows that $z_{113}, z_{223} \geq 0$.

$$\{z_{ij1}\}: \begin{array}{c|ccc|} i\backslash j & 1 & 2 & 3 \\ \hline 1 & + & - & 0+ \\ 2 & - & + & * \\ 3 & 0+ & * & * \\ \hline \end{array} \qquad \{z_{ij2}\}: \begin{array}{c|ccc|} i\backslash j & 1 & 2 & 3 \\ \hline 1 & - & 0- & + \\ 2 & 0- & 0+ & * \\ 3 & + & * & - \\ \hline \end{array} \qquad \{z_{ij3}\}: \begin{array}{c|ccc|} i\backslash j & 1 & 2 & 3 \\ \hline 1 & 0+ & + & * \\ 2 & * & 0+ & * \\ 3 & * & * & * \\ \hline \end{array}.$$

z_{123} has to be an element of a rectangle in the $k=3$ slice, therfore $z_{133}, z_{323} < 0$ and $z_{333} > 0$ are derived.

$$
\{z_{ij1}\}: \begin{array}{c|ccc|} i\backslash j & 1 & 2 & 3 \\ \cline{2-4} 1 & + & - & 0+ \\ 2 & - & + & * \\ 3 & 0+ & * & * \\ \cline{2-4} \end{array} \quad \{z_{ij2}\}: \begin{array}{c|ccc|} i\backslash j & 1 & 2 & 3 \\ \cline{2-4} 1 & - & 0- & + \\ 2 & 0- & 0+ & * \\ 3 & + & * & - \\ \cline{2-4} \end{array} \quad \{z_{ij3}\}: \begin{array}{c|ccc|} i\backslash j & 1 & 2 & 3 \\ \cline{2-4} 1 & 0+ & + & - \\ 2 & * & 0+ & * \\ 3 & * & - & + \\ \cline{2-4} \end{array}.
$$

But then a rectangle $(z_{132}, z_{133}, z_{333}, z_{332})$ appears in the $j=3$ slice, which contradicts the assumption and the proof of (b) is completed. □

Now we carry out a proof of Theorem 9.1 using Lemma 9.1.

Proof (Theorem 9.1). As we have stated, we only need to show that the elements of $\boldsymbol{z} = \boldsymbol{y} - \boldsymbol{x}$ have to be all zero after minimizing $|\boldsymbol{z}|$ by applying the basic moves or the moves of degree 6 without causing negative entries on the way.

Suppose $\boldsymbol{z}$ has nonzero entries. Let $z_{111} > 0$ without loss of generality. From Lemma 9.1(a), z_{111} has to be an element of rectangles, in each of the $i=1, j=1$, and $k=1$ slices. We can take one of these rectangles in the $i=1$ slice as $(z_{111}, z_{112}, z_{122}, z_{121})$ without loss of generality.

$$
\begin{array}{|ccc|} \hline + & - & * \\ - & + & * \\ * & * & * \\ \hline \end{array} \quad \begin{array}{|ccc|} \hline * & * & * \\ * & * & * \\ * & * & * \\ \hline \end{array} \quad \begin{array}{|ccc|} \hline * & * & * \\ * & * & * \\ * & * & * \\ \hline \end{array}.
$$

Next consider the $j=1$ slice. We claim that z_{111} and z_{112} are elements of the same rectangle in $j=1$ slice. To prove this, consider the sign of z_{113}. If $z_{113} \geq 0$, the rectangle containing z_{111} in the $j=1$ slice contains z_{112}, and if $z_{113} < 0$, the rectangle containing z_{112} in the $j=1$ slice contains z_{111}. Therefore, z_{111} and z_{112} are elements of the same rectangle in the $j=1$ slice and the rectangle can be taken as $(z_{111}, z_{112}, z_{212}, z_{211})$ without loss of generality.

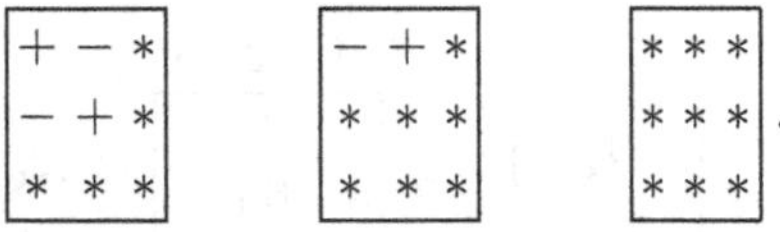

Now consider the rectangle in the $k=1$ slice containing z_{111}. For a similar reason as above, this rectangle also contains z_{121}. In addition, if $z_{221} > 0$, $\boldsymbol{m}_4(12,12,21)$ can be added to $\boldsymbol{y}$ without causing negative entries and $|\boldsymbol{z}|$ can be made smaller, which contradicts the assumption. Hence, the rectangle in the $k=1$ slice including z_{111} has to be $(z_{111}, z_{121}, z_{321}, z_{311})$.

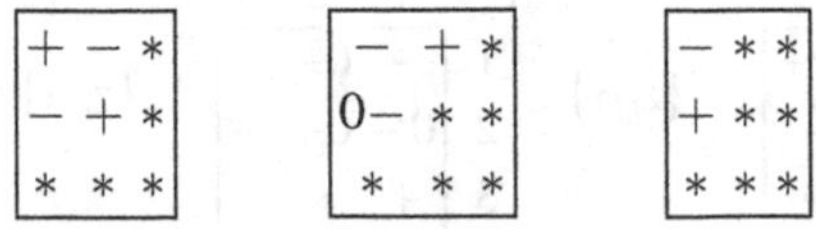

Next consider the rectangle in the $j = 2$ slice including z_{121}. For a similar reason as above, this rectangle also contains z_{122}. Hence, the rectangle in the $j = 2$ slice including z_{121} has to be $(z_{121}, z_{122}, z_{322}, z_{321})$.

$$\begin{array}{|ccc|}\hline + & - & * \\ - & + & * \\ * & * & * \\ \hline\end{array} \quad \begin{array}{|ccc|}\hline - & + & * \\ 0- & * & * \\ * & * & * \\ \hline\end{array} \quad \begin{array}{|ccc|}\hline - & * & * \\ + & - & * \\ * & * & * \\ \hline\end{array}.$$

However, $\boldsymbol{m}_4(13,12,12)$ can be added to $\boldsymbol{x}$ without causing negative entries and $|\boldsymbol{z}|$ can be made smaller, which contradicts the assumption. From these considerations, a set of the basic moves and the moves of degree 6 is shown to be a Markov basis for the $3 \times 3 \times 3$ case. All these moves are indispensable, therefore the minimality and the uniqueness also follow. This completes the proof of Theorem 9.1. □

9.3 Unique Minimal Markov Basis for $3 \times 3 \times 4$ Tables

The next indispensable move is constructed as a three-step move. For the case of a general $I \times J \times K$ contingency table, there are several types of such a move. One is a $2 \times 4 \times 4$ move of degree 8 and another is a $3 \times 4 \times 4$ move of degree 9. We consider these moves in Sect. 9.5. For the $3 \times 3 \times K$ case, the following type of move is needed.

Definition 9.4. A move of degree 8 is a move $\boldsymbol{m}_8(i_1i_2i_3, j_1j_2j_3, k_1k_2k_3k_4) \in \ker_{\mathbb{Z}} A$ written as

$$\begin{aligned}&(i_1j_1k_1)(i_1j_2k_2)(i_2j_1k_3)(i_2j_2k_1)(i_2j_3k_4)(i_3j_1k_2)(i_3j_2k_4)(i_3j_3k_3)\\ &\qquad -(i_1j_1k_2)(i_1j_2k_1)(i_2j_1k_1)(i_2j_2k_4)(i_2j_3k_3)(i_3j_1k_3)(i_3j_2k_2)(i_3j_3k_4).\end{aligned}$$

Figure 9.5 gives a three-dimensional view of this type of move. From the definition, the relation

$$\begin{aligned}\boldsymbol{m}_8(i_1i_2i_3, j_1j_2j_3, k_1k_2k_3k_4) &= -\boldsymbol{m}_8(i_1i_3i_2, j_2j_1j_3, k_2k_1k_4k_3)\\ &= \boldsymbol{m}_8(i_1i_3i_2, j_1j_2j_3, k_2k_1k_3k_4)\end{aligned}$$

is derived.

Now we state a theorem for the $3 \times 3 \times 4$ case.

Theorem 9.2. *The set of basic moves $\boldsymbol{m}_4(i_1i_2, j_1j_2, k_1k_2)$, moves of degree 6, $\boldsymbol{m}_6^I(i_1i_2, j_1j_2j_3, k_1k_2k_3), \boldsymbol{m}_6^J(i_1i_2i_3, j_1j_2, k_1k_2k_3), \boldsymbol{m}_6^K(i_1i_2i_3, j_1j_2j_3, k_1k_2)$, and moves of degree 8, $\boldsymbol{m}_8(i_1i_2i_3, j_1j_2j_3, k_1k_2k_3k_4)$ constitutes a unique minimal Markov basis for $3 \times 3 \times 4$ tables.*

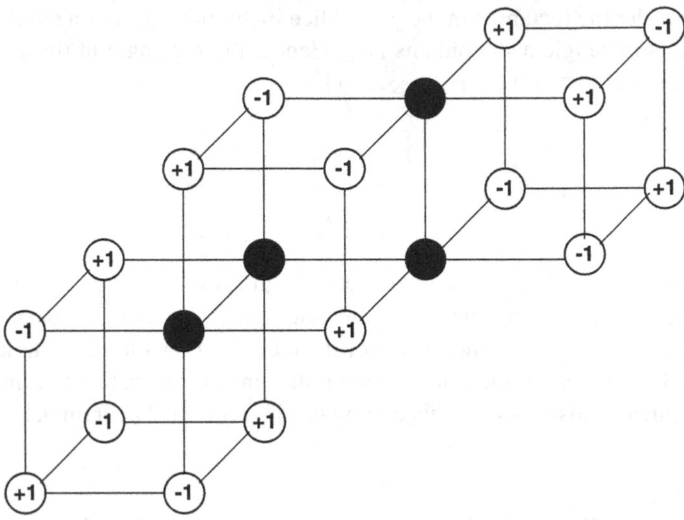

Fig. 9.5 $3 \times 3 \times 4$ move of degree 8

Proof. Similarly to the proof of Theorem 9.1, we only need to show that the set of the moves of degree 4, 6, and 8 above constitutes a Markov basis; that is, the pattern of $\boldsymbol{z} = \boldsymbol{y} - \boldsymbol{x}$ has to be of all zero entries after minimizing $|\boldsymbol{z}|$ by adding the basic moves, the moves of degree 6 or degree 8, without causing negative entries on the way.

Suppose $\boldsymbol{z}$ has nonzero entries. Let $z_{111} > 0$ without loss of generality. From Lemma 9.1(b), we can also assume that there is a rectangle including z_{111} in either an $i = 1$ slice or a $j = 1$ slice. We can take one of these rectangles in the $i = 1$ slice as $(z_{111}, z_{112}, z_{121}, z_{122})$ without loss of generality. Moreover, $z_{211} < 0, z_{221} > 0$ without loss of generality because it is known from Lemma 9.1(a) that z_{111} is an element of a rectangle in the $k = 1$ slice.

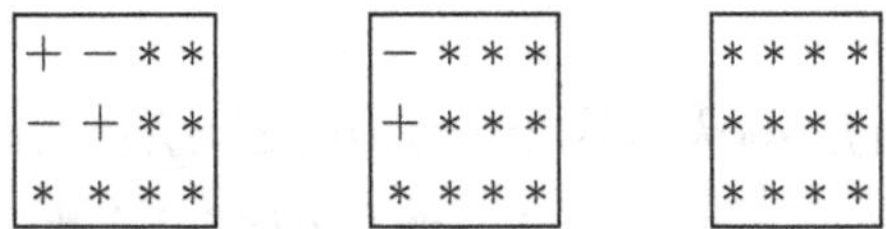

+	−	∗	∗
−	+	∗	∗
∗	∗	∗	∗

−	∗	∗	∗
+	∗	∗	∗
∗	∗	∗	∗

∗	∗	∗	∗
∗	∗	∗	∗
∗	∗	∗	∗

.

As in the proof of Theorem 9.1, by considering the sign of z_{132}, we see that z_{112} and z_{122} are members of the same rectangle in the $k = 2$ slice. Then (z_{212}, z_{222}) and/or (z_{312}, z_{322}) has to be $(+, -)$. But if $z_{212} > 0$, $\boldsymbol{m}_4(12, 12, 21)$ can be added to $\boldsymbol{y}$ without causing negative entries; and if $z_{222} < 0$, $\boldsymbol{m}_4(12, 12, 12)$ can be added to $\boldsymbol{x}$ without causing negative entries; and $|\boldsymbol{z}|$ can be made smaller. These imply that $z_{312} > 0, z_{322} < 0$ and $z_{212} \leq 0, z_{222} \geq 0$. Similarly, if $z_{311} < 0$, $\boldsymbol{m}_4(13, 12, 12)$ can be added to $\boldsymbol{x}$ without causing negative entries; and if $z_{321} > 0$, $\boldsymbol{m}_4(13, 12, 21)$ can be added to $\boldsymbol{y}$ without causing negative entries; and $|\boldsymbol{z}|$ can be made smaller, which forces $z_{311} \geq 0$ and $z_{321} \leq 0$.

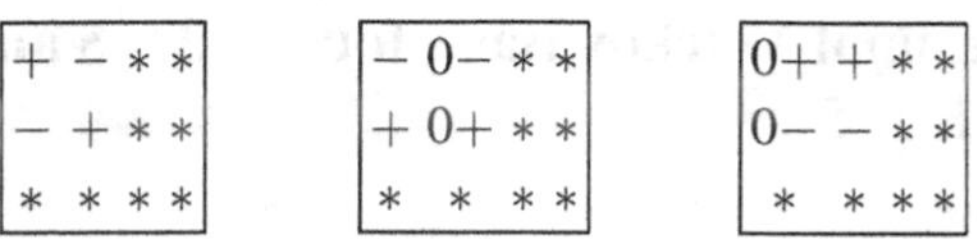

Inasmuch as $z_{21\cdot} = 0$, let $z_{213} > 0$ without loss of generality, which forces $z_{123} \le 0$, otherwise, $\boldsymbol{m}_4(12,12,31)$ can be added to $\boldsymbol{y}$ without causing negative entries and $|\boldsymbol{z}|$ can be made smaller. The fact that $z_{213} > 0$ also forces $z_{323} \le 0$, otherwise, $\boldsymbol{m}_6^J(132,21,123)$ can be added to $\boldsymbol{y}$ without causing negative entries and $|\boldsymbol{z}|$ can be made smaller.

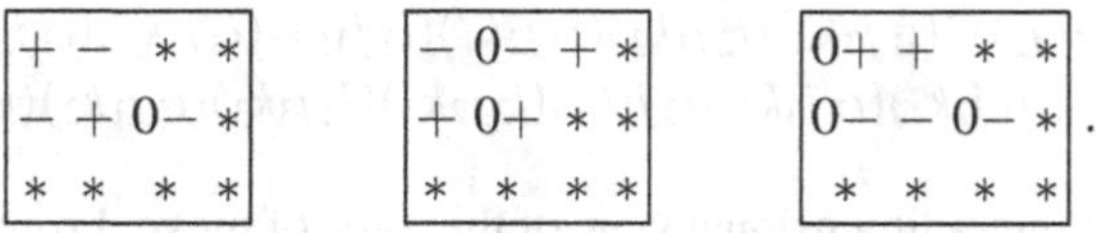

Because $z_{\cdot 23} = 0$, it follows that $z_{223} \ge 0$. This implies $z_{224}, z_{233} < 0$ because $z_{22\cdot} = z_{2\cdot 3} = 0$.

$$\begin{array}{|cccc|}\hline + & - & * & * \\ - & + & 0- & * \\ * & * & * & * \\ \hline\end{array}\quad \begin{array}{|cccc|}\hline - & 0- & + & * \\ + & 0+ & 0+ & - \\ * & * & - & * \\ \hline\end{array}\quad \begin{array}{|cccc|}\hline 0+ & + & * & * \\ 0- & - & 0- & * \\ * & * & * & * \\ \hline\end{array}\,.$$

From symmetry (in interchanging roles of $+$ and $-$), $z_{114}, z_{314} \ge 0$, otherwise, $\boldsymbol{m}_4(12,12,14)$ can be added to $\boldsymbol{x}$ without causing negative entries or $\boldsymbol{m}_6^J(132,12,124)$ can be added to $\boldsymbol{x}$ without causing negative entries and $|\boldsymbol{z}|$ can be made smaller. These also imply $z_{214} \le 0, z_{234} > 0$ because $z_{\cdot 14} = z_{2\cdot 4} = 0$.

$$\begin{array}{|cccc|}\hline + & - & * & 0+ \\ - & + & 0- & * \\ * & * & * & * \\ \hline\end{array}\quad \begin{array}{|cccc|}\hline - & 0- & + & 0- \\ + & 0+ & 0+ & - \\ * & * & - & + \\ \hline\end{array}\quad \begin{array}{|cccc|}\hline 0+ & + & * & 0+ \\ 0- & - & 0- & * \\ * & * & * & * \\ \hline\end{array}\,. \tag{9.1}$$

Because $z_{31\cdot} = z_{32\cdot} = z_{3\cdot 3} = z_{3\cdot 4} = 0$, it follows that $z_{313} < 0, z_{324} > 0, z_{333} > 0$, $z_{334} < 0$.

$$\begin{array}{|cccc|}\hline + & - & * & 0+ \\ - & + & 0- & * \\ * & * & * & * \\ \hline\end{array}\quad \begin{array}{|cccc|}\hline - & 0- & + & 0- \\ + & 0+ & 0+ & - \\ * & * & - & + \\ \hline\end{array}\quad \begin{array}{|cccc|}\hline 0+ & + & - & 0+ \\ 0- & - & 0- & + \\ * & * & + & - \\ \hline\end{array}\,.$$

But $\boldsymbol{m}_8(132,123,2134)$ can be added to $\boldsymbol{y}$ (or $\boldsymbol{m}_8(123,123,1234)$ can be added to $\boldsymbol{x}$) without causing negative entries and $|\boldsymbol{z}|$ can be made smaller.

From these considerations, a set of the basic moves, the moves of degree 6 and degree 8, is shown to be a Markov basis for the $3 \times 3 \times 4$ case. This is a set of indispensable moves, thus this is a unique minimal Markov basis and Theorem 9.2 is proved. □

9.4 Unique Minimal Markov Basis for $3 \times 3 \times 5$ and $3 \times 3 \times K$ Tables for $K > 5$

Continuing the above discussion, next we consider a *four-step move*. For the case of a $3 \times 3 \times K$ contingency table, only a move of the following type needs to be considered.

Definition 9.5. A move of degree 10 is a move $\boldsymbol{m}_{10}(i_1i_2i_3, j_1j_2j_3, k_1k_2k_3k_4k_5) \in \ker_{\mathbb{Z}} A$ written as

$$(i_1j_1k_1)(i_1j_2k_2)(i_1j_2k_5)(i_1j_3k_4)(i_2j_1k_3)(i_2j_2k_1)(i_2j_3k_5)(i_3j_1k_2)(i_3j_2k_4)(i_3j_3k_3)$$
$$-(i_1j_1k_2)(i_1j_2k_1)(i_1j_2k_4)(i_1j_3k_5)(i_2j_1k_1)(i_2j_2k_5)(i_2j_3k_3)(i_3j_1k_3)(i_3j_2k_2)(i_3j_3k_4).$$

Figure 9.6 gives a three-dimensional view of this type of move. From the definition, the relation

$$\begin{aligned}\boldsymbol{m}_{10}(i_1i_2i_3, j_1j_2j_3, k_1k_2k_3k_4k_5) &= \boldsymbol{m}_{10}(i_1i_3i_2, j_3j_2j_1, k_4k_5k_3k_1k_2)\\ &= -\boldsymbol{m}_{10}(i_1i_2i_3, j_3j_2j_1, k_5k_4k_3k_2k_1)\end{aligned}$$

is derived.

As for a connected Markov chain, the next theorem holds for the $3 \times 3 \times 5$ case.

Theorem 9.3. *The set of basic moves* $\boldsymbol{m}_4(i_1i_2, j_1j_2, k_1k_2)$, *moves of degree* 6, $\boldsymbol{m}_6^I(i_1i_2, j_1j_2j_3, k_1k_2k_3)$, $\boldsymbol{m}_6^J(i_1i_2i_3, j_1j_2, k_1k_2k_3)$, $\boldsymbol{m}_6^K(i_1i_2i_3, j_1j_2j_3, k_1k_2)$, *moves of degree* 8, $\boldsymbol{m}_8(i_1i_2i_3, j_1j_2j_3, k_1k_2k_3k_4)$, *and moves of degree* 10, $\boldsymbol{m}_{10}(i_1i_2i_3, j_1j_2j_3, k_1k_2k_3k_4k_5)$, *constitutes a unique minimal Markov basis for the* $3 \times 3 \times 5$ *tables.*

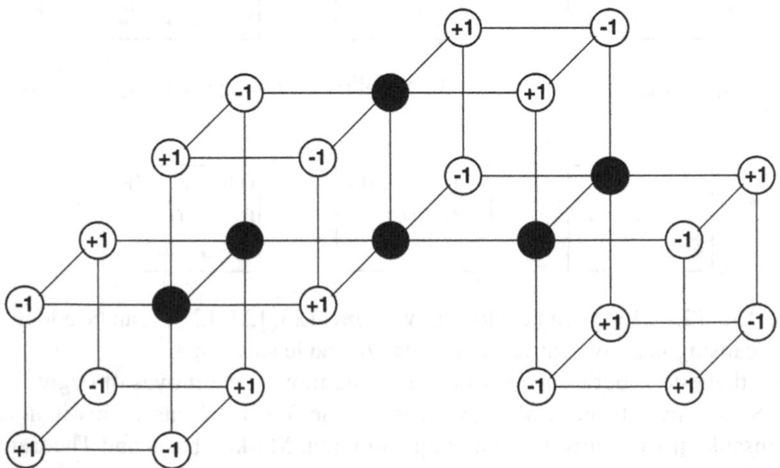

Fig. 9.6 $3 \times 3 \times 5$ move of degree 10

Interestingly, this set of moves is shown to be a unique minimal Markov basis for the general $3 \times 3 \times K (K \geq 5)$ case. We give a main result of this section.

Theorem 9.4. *The set of basic moves* $\boldsymbol{m}_4(i_1i_2, j_1j_2, k_1k_2)$, *moves of degree* 6, $\boldsymbol{m}_6^I(i_1i_2, j_1j_2j_3, k_1k_2k_3)$, $\boldsymbol{m}_6^J(i_1i_2i_3, j_1j_2, k_1k_2k_3)$, $\boldsymbol{m}_6^K(i_1i_2i_3, j_1j_2j_3, k_1k_2)$, *moves of degree* 8, $\boldsymbol{m}_8(i_1i_2i_3, j_1j_2j_3, k_1k_2k_3k_4)$, *and moves of degree* 10, $\boldsymbol{m}_{10}(i_1i_2i_3, j_1j_2j_3, k_1k_2k_3k_4k_5)$, *constitutes a unique minimal Markov basis for* $3 \times 3 \times K (K \geq 5)$ *tables.*

Proof (Theorem 9.3). Again all we have to show is that the pattern of $\boldsymbol{z} = \boldsymbol{y} - \boldsymbol{x}$ must be of all zero entries after minimizing $|\boldsymbol{z}|$ by adding the basic moves, the moves of degree 6, degree 8, or degree 10, without causing negative entries on the way.

Suppose $\boldsymbol{z}$ has nonzero entries. For a similar reason leading to (9.1) in the proof of Theorem 9.2, the patterns can be restricted to

$$\boxed{\begin{array}{ccccc} + & - & \ast & 0+ & \ast \\ - & + & 0- & \ast & \ast \\ \ast & \ast & \ast & \ast & \ast \end{array}} \qquad \boxed{\begin{array}{ccccc} - & 0- & + & 0- & \ast \\ + & 0+ & 0+ & - & \ast \\ \ast & \ast & - & + & \ast \end{array}} \qquad \boxed{\begin{array}{ccccc} 0+ & + & \ast & 0+ & \ast \\ 0- & - & 0- & \ast & \ast \\ \ast & \ast & \ast & \ast & \ast \end{array}}$$

without loss of generality. Because $z_{31\cdot} = z_{32\cdot} = 0$, at least one of z_{313} and z_{315} has to be negative and at least one of z_{324} and z_{325} has to be positive. But we have already seen that $(z_{313}, z_{324}) = (-, +)$ contradicts the assumption. In addition, if $(z_{315}, z_{325}) = (-, +)$, it follows that $z_{115} \leq 0$, $z_{125} \geq 0$ (otherwise $\boldsymbol{m}_4(13, 12, 25)$ can be added to $\boldsymbol{y}$ without causing negative entries and $\boldsymbol{m}_4(13, 12, 52)$ can be added to $\boldsymbol{x}$ without causing negative entries and $|\boldsymbol{z}|$ can be made smaller) and $(z_{215}, z_{225}) = (+, -)$ because $z_{\cdot 15} = z_{\cdot 25} = 0$. But $\boldsymbol{m}_6^J(132, 21, 125)$ can be added to $\boldsymbol{y}$ without causing negative entries and $\boldsymbol{m}_6^J(132, 12, 125)$ can be added to $\boldsymbol{x}$ without causing negative entries and $|\boldsymbol{z}|$ can be made smaller. All of these contradict the assumption.

The remaining patterns are $(z_{313}, z_{325}) = (-, +)$ or $(z_{315}, z_{324}) = (-, +)$. Considering the symmetry, we write $(z_{313}, z_{325}) = (-, +)$ without loss of generality. Then the patterns are, without loss of generality, summarized as

$$\boxed{\begin{array}{ccccc} + & - & \ast & 0+ & \ast \\ - & + & 0- & \ast & \ast \\ \ast & \ast & \ast & \ast & \ast \end{array}} \qquad \boxed{\begin{array}{ccccc} - & 0- & + & 0- & \ast \\ + & 0+ & 0+ & - & \ast \\ \ast & \ast & - & + & \ast \end{array}} \qquad \boxed{\begin{array}{ccccc} 0+ & + & - & 0+ & 0+ \\ 0- & - & 0- & 0- & + \\ \ast & \ast & \ast & \ast & \ast \end{array}}.$$

$z_{\cdot 24} = z_{1\cdot 4} = z_{3\cdot 3} = z_{3\cdot 5} = 0$, thus it follows that $z_{124} > 0, z_{134} < 0, z_{333} > 0, z_{335} < 0$.

$$\boxed{\begin{array}{ccccc} + & - & \ast & 0+ & \ast \\ - & + & 0- & + & \ast \\ \ast & \ast & \ast & - & \ast \end{array}} \qquad \boxed{\begin{array}{ccccc} - & 0- & + & 0- & \ast \\ + & 0+ & 0+ & - & \ast \\ \ast & \ast & - & + & \ast \end{array}} \qquad \boxed{\begin{array}{ccccc} 0+ & + & - & 0+ & 0+ \\ 0- & - & 0- & 0- & + \\ \ast & \ast & + & \ast & - \end{array}}.$$

If $z_{225} < 0$, $\boldsymbol{m}_8(123,123,1235)$ can be added to $\boldsymbol{x}$ without causing negative entries and $|\boldsymbol{z}|$ can be made smaller, which contradicts the assumption. Similarly, if $z_{235} > 0$, $\boldsymbol{m}_8(123,213,1253)$ can be added to $\boldsymbol{y}$ without causing negative entries and $|\boldsymbol{z}|$ can be made smaller, which contradicts the assumption. These imply $z_{225} \geq 0, z_{234} \leq 0$, which also imply $z_{125} < 0, z_{135} > 0$ inasmuch as $z_{\cdot 25} = z_{\cdot 35} = 0$.

$$\boxed{\begin{array}{ccccc} + & - & * & 0+ & * \\ - & + & 0- & + & - \\ * & * & * & - & + \end{array}} \quad \boxed{\begin{array}{ccccc} - & 0- & + & 0- & * \\ + & 0+ & 0+ & - & 0+ \\ * & * & - & + & 0- \end{array}} \quad \boxed{\begin{array}{ccccc} 0+ & + & - & 0+ & 0+ \\ 0- & - & 0- & 0- & + \\ * & * & + & * & - \end{array}}\,.$$

But $\boldsymbol{m}_{10}(123,321,45321)$ can be added to $\boldsymbol{y}$ (or $\boldsymbol{m}_{10}(123,123,12354)$ can be added to $\boldsymbol{x}$) without causing negative entries and $|\boldsymbol{z}|$ can be made smaller, which contradicts the assumption.

From these considerations, the set of the basic moves, the moves of degree 6, degree 8, and degree 10 is shown to be a Markov basis for the $3 \times 3 \times 5$ case. Because this is a set of indispensable moves, this is a unique minimal Markov basis and Theorem 9.3 is proved. □

Proof (Theorem 9.4). Again we can begin with the following pattern.

$$\boxed{\begin{array}{cccccc} + & - & * & 0+ & * & * \\ - & + & 0- & * & * & * \\ * & * & * & * & * & * \end{array}} \quad \boxed{\begin{array}{cccccc} - & 0- & + & 0- & * & * \\ + & 0+ & 0+ & - & * & * \\ * & * & - & + & * & * \end{array}} \quad \boxed{\begin{array}{cccccc} 0+ & + & * & 0+ & * & * \\ 0- & - & 0- & * & * & * \\ * & * & * & * & * & * \end{array}}\,.$$

As we have seen in the proof of Theorem 9.3, z_{313} has to be nonnegative and z_{324} has to be nonpositive, because either one of $(z_{313}, z_{326}) = (-,+)$ and $(z_{316}, z_{324}) = (-,+)$ also contradicts the assumption. The case of $(z_{316}, z_{326}) = (-,+)$ also contradicts the assumption for a similar reason to that of $(z_{315}, z_{325}) = (-,+)$. Hence the remaining patterns are $(z_{315}, z_{326}) = (-,+)$ and $(z_{316}, z_{325}) = (-,+)$. We write $(z_{315}, z_{326}) = (-,+)$ without loss of generality.

$$\boxed{\begin{array}{cccccc} + & - & * & 0+ & * & * \\ - & + & 0- & * & * & * \\ * & * & * & * & * & * \end{array}} \quad \boxed{\begin{array}{cccccc} - & 0- & + & 0- & * & * \\ + & 0+ & 0+ & - & * & * \\ * & * & - & + & * & * \end{array}} \quad \boxed{\begin{array}{cccccc} 0+ & + & 0+ & 0+ & - & * \\ 0- & - & 0- & 0- & * & + \\ * & * & * & * & * & * \end{array}}\,.$$

According to the symmetry in interchanging the roles of $\{+,-\}$, the roles of $\{z_{2jk}, z_{3jk}\}$, and the roles of $\{(z_{ij3}, z_{ij4}), (z_{ij5}, z_{ij6})\}$, the patterns can be restricted to

$$\boxed{\begin{array}{cccccc} + & - & * & 0+ & * & 0- \\ - & + & 0- & * & 0+ & * \\ * & * & * & * & * & * \end{array}} \quad \boxed{\begin{array}{cccccc} - & 0- & + & 0- & 0- & 0- \\ + & 0+ & 0+ & - & 0+ & 0+ \\ * & * & - & + & * & * \end{array}} \quad \boxed{\begin{array}{cccccc} 0+ & + & 0+ & 0+ & - & 0+ \\ 0- & - & 0- & 0- & 0- & + \\ * & * & * & * & + & - \end{array}}$$

for a similar reason to the proof of Theorem 9.2. Because $z_{\cdot 13} = z_{\cdot 15} = z_{\cdot 24} = z_{\cdot 26} = 0$, it follows that $z_{113} < 0, z_{115} > 0, z_{124} > 0$ and $z_{126} < 0$. $z_{1\cdot 3} = z_{1\cdot 4} = z_{1\cdot 5} = z_{1\cdot 6} = 0$ also forces $z_{133} > 0, z_{134} < 0, z_{135} < 0$ and $z_{136} > 0$.

$$\begin{array}{|cccccc|}\hline + & - & - & 0+ & + & 0- \\ - & + & 0- & + & 0+ & - \\ * & * & + & - & - & + \\ \hline \end{array}\quad \begin{array}{|cccccc|}\hline - & 0- & + & 0- & 0- & 0- \\ + & 0+ & 0+ & - & 0+ & 0+ \\ * & * & - & + & * & * \\ \hline \end{array}\quad \begin{array}{|cccccc|}\hline 0+ & + & 0+ & 0+ & - & 0+ \\ 0- & - & 0- & 0- & 0- & + \\ * & * & * & * & + & - \\ \hline \end{array}\ .$$

But this pattern includes moves of degree 6. We can add $\boldsymbol{m}_6^I(21,132,134)$ to $\boldsymbol{y}$, $\boldsymbol{m}_6^I(12,132,134)$ to $\boldsymbol{x}$, $\boldsymbol{m}_6^I(13,132,256)$ to $\boldsymbol{y}$, or $\boldsymbol{m}_6^I(31,132,256)$ to $\boldsymbol{x}$ without causing negative entries and make $|\boldsymbol{z}|$ smaller, which contradicts the assumption.

From these considerations, it is shown that the set of the basic moves, the moves of degree 6, degree 8, and degree 10 is also a Markov basis for the $3 \times 3 \times K$ $(K \geq 5)$ case. The minimality and the uniqueness directly hold again. Note that although we have displayed $3 \times 3 \times 6$ tables, the above argument does not involve k slices for $k \geq 7$. Therefore we obtain the same contradiction for the $3 \times 3 \times K$ $(K \geq 7)$ tables. This completes the proof. □

A result corresponding to Theorem 9.4 for Gröbner bases was given in [27].

9.5 Indispensable Moves for Larger Tables

The fact that the structure of a minimal Markov basis for $3 \times 3 \times K$ tables is essentially explained by $3 \times 3 \times 5$ tables is very attractive. Such a theoretical result seems important because even if we can obtain the reduced Gröbner basis for the $3 \times 3 \times 6$ table by an algebraic algorithm, we have to carry out new calculations to obtain results for $3 \times 3 \times 7$ or $3 \times 3 \times 8$ problems. In fact, the following result is shown in [131] as a special case of the Graver complexity of the higher Lawrence lifting (see Sect. 9.8 below).

Proposition 9.1 (Corollary 2 of [131]). *For any positive integers I, J, there exists a positive integer m such that every element of a minimal Markov basis for the $I \times J \times K$ tables with fixed two-dimensional marginal frequencies is included in $I \times J \times m$.*

The above m is called a *Markov complexity* for the configuration A. The values of m for some cases are computed in [131]. For the example of the $3 \times 3 \times K$ table with fixed two-dimensional marginal frequencies, an upper bound of the Markov complexity is given by the Graver complexity, which is computed to be 9 by [131]. Therefore no new type of conformally primitive move appears for $K \geq 10$.

The results of the previous section are not derived by algebraic algorithms but by "thoroughly checking symmetry by inspection" ([10]) based on the distance-reducing proofs in Sect. 6. Of course, if we carry out a similar method for problems of larger sizes, the number of the cases we have to consider becomes huge. In [9], a similar approach of distance-reducing proofs is carried out for the $3 \times 4 \times K$ cases. The result of [9] is that "for $3 \times 4 \times K$ problems, it is sufficient to consider up to $K = 8$, and the set of the 20 kinds of moves up to 16 degree forms a unique minimal Markov basis." The truth of this proposition has not yet been confirmed by algebraic algorithms. On the other hand, as for the development of the algorithms for calculating Gröbner bases, calculating Markov bases for the $4 \times 4 \times 4$ tables has been used as a benchmark problem since about 2002. This problem is first solved completely by [83] using 4ti2 ([1]). In [83], it is reported that "148,654 elements in 15 kinds of moves form a minimal Markov basis for the $4 \times 4 \times 4$ problem." See [15] for an overview of the history of calculating Markov bases for $4 \times 4 \times 4$ problems.

We have already pointed out that the type-2 combination is essential in Sect. 9.2. In fact, from the three-dimensional views of the indispensable moves in the unique minimal Markov basis for $3 \times 3 \times K$ cases in Figs. 9.3, 9.5, and 9.6, we see that they are constructed as the type-2 combination of several basic moves. However, structure of the indispensable moves for general $I \times J \times K$ cases can be more complicated. To see this point, we show the unique minimal Markov basis for the $3 \times 4 \times 4$ case. It is composed of basic moves, moves of degree 6 ($2 \times 3 \times 3, 3 \times 2 \times 3, 3 \times 3 \times 2$), and moves of degree 8 ($3 \times 3 \times 4, 3 \times 4 \times 3$), and moves of degree 8 ($2 \times 4 \times 4$) like

$$\begin{array}{|cccc|}\hline +1 & -1 & 0 & 0 \\ 0 & +1 & -1 & 0 \\ 0 & 0 & +1 & -1 \\ -1 & 0 & 0 & +1 \\ \hline \end{array} \quad \begin{array}{|cccc|}\hline -1 & +1 & 0 & 0 \\ 0 & -1 & +1 & 0 \\ 0 & 0 & -1 & +1 \\ +1 & 0 & 0 & -1 \\ \hline \end{array} \quad \begin{array}{|cccc|}\hline 0 & 0 & 0 & 0 \\ 0 & 0 & 0 & 0 \\ 0 & 0 & 0 & 0 \\ 0 & 0 & 0 & 0 \\ \hline \end{array}, \tag{9.2}$$

moves of degree 9 ($3 \times 4 \times 4$, Fig. 9.7) like

$$\begin{array}{|cccc|}\hline +1 & -1 & 0 & 0 \\ -1 & 0 & +1 & 0 \\ 0 & +1 & -1 & 0 \\ 0 & 0 & 0 & 0 \\ \hline \end{array} \quad \begin{array}{|cccc|}\hline -1 & +1 & 0 & 0 \\ +1 & 0 & 0 & -1 \\ 0 & 0 & 0 & 0 \\ 0 & -1 & 0 & +1 \\ \hline \end{array} \quad \begin{array}{|cccc|}\hline 0 & 0 & 0 & 0 \\ 0 & 0 & -1 & +1 \\ 0 & -1 & +1 & 0 \\ 0 & +1 & 0 & -1 \\ \hline \end{array},$$

and moves of degree 10 ($3 \times 4 \times 4$, Fig. 9.8) like

$$\begin{array}{|cccc|}\hline +1 & -1 & 0 & 0 \\ -1 & +1 & 0 & 0 \\ 0 & 0 & +1 & -1 \\ 0 & 0 & -1 & +1 \\ \hline \end{array} \quad \begin{array}{|cccc|}\hline -1 & +1 & 0 & 0 \\ 0 & 0 & 0 & 0 \\ +1 & 0 & -1 & 0 \\ 0 & -1 & +1 & 0 \\ \hline \end{array} \quad \begin{array}{|cccc|}\hline 0 & 0 & 0 & 0 \\ +1 & -1 & 0 & 0 \\ -1 & 0 & 0 & +1 \\ 0 & +1 & 0 & -1 \\ \hline \end{array}.$$

Among the newly obtained moves, the $3 \times 4 \times 4$ move of degree 10 is interpreted as a type-2 combination of a basic move and a move of degree 8, which is similar to the $3 \times 3 \times 5$ move of degree 10 shown in Sect. 9.4. However, the $3 \times 4 \times 4$ move of degree 9 is new in the sense that this is a type-2 combination of a basic move and a move of degree 7. Recall that the move of degree 7 itself is not needed to construct

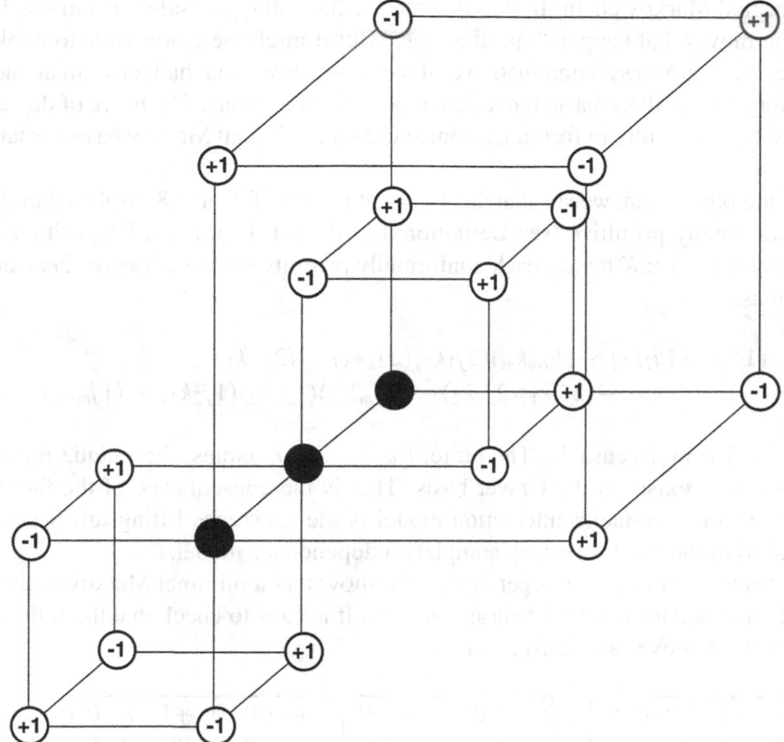

Fig. 9.7 $3 \times 4 \times 4$ move of degree 9

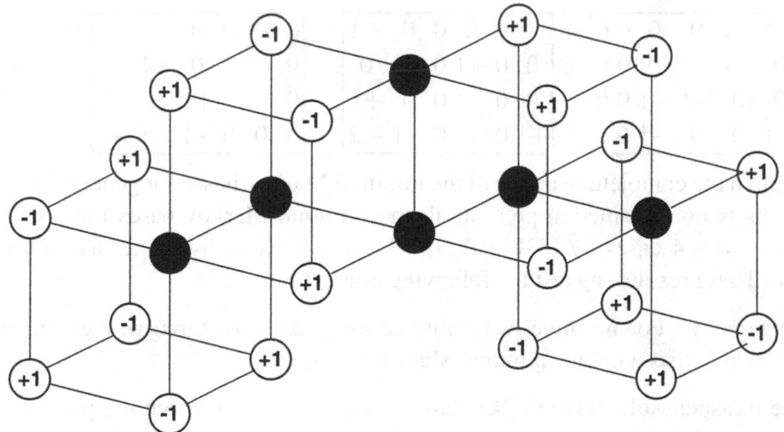

Fig. 9.8 $3 \times 4 \times 4$ move of degree 10

a connected Markov chain. In this chapter, we have only considered combinations of basic moves that happen "one at a time." But it might be worthwhile to think of this degree 9 move as a combination of three basic moves that happens "all at once," and every two of these basic moves are type-1 combinations. The move of degree 9 suggests the difficulty in forming a conjecture on a minimal Markov basis for larger tables.

On the other hand, we see that the $2 \times 4 \times 4$ moves of degree 8 displayed in (9.2) are conformally primitive (see Definition 4.2 of Sect. 4.6). From Proposition 4.2, for general $2 \times J \times K$ tables, each conformally primitive move of degree $2m$ can be written as

$$\begin{aligned}&(1 j_1 k_1)(1 j_2 k_2) \cdots (1 j_m k_m)(2 j_1 k_2)(2 j_2 k_3) \cdots (2 j_m k_1)\\ &\qquad - (2 j_1 k_1)(2 j_2 k_2) \cdots (2 j_m k_m)(1 j_1 k_2)(1 j_2 k_3) \cdots (1 j_m k_1)\end{aligned}$$

and it is also indispensable. Therefore, for $2 \times J \times K$ tables, the unique minimal Markov basis exists as the Graver basis. This is the consequence of the fact that $2 \times J \times K$ no-three-factor interaction model is the Lawrence lifting (cf. Sects. 4.6 and 5.4.3) of the $J \times K$ two-way complete independence model.

As another difficulty in larger tables, we show that a minimal Markov basis can include non-square-free indispensable moves. It is easy to check that the following two $3 \times 4 \times 6$ moves are indispensable.

$$\begin{array}{|cccccc|}\hline +1 & -1 & 0 & 0 & 0 & 0 \\ 0 & +1 & -1 & 0 & 0 & 0 \\ 0 & 0 & +1 & 0 & 0 & -1 \\ -1 & 0 & 0 & 0 & 0 & +1 \\ \hline\end{array} \quad \begin{array}{|cccccc|}\hline 0 & +1 & 0 & -1 & 0 & 0 \\ 0 & -1 & 0 & 0 & +1 & 0 \\ 0 & 0 & 0 & +1 & 0 & -1 \\ 0 & 0 & 0 & 0 & -1 & +1 \\ \hline\end{array} \quad \begin{array}{|cccccc|}\hline -1 & 0 & 0 & +1 & 0 & 0 \\ 0 & 0 & +1 & 0 & -1 & 0 \\ 0 & 0 & -1 & -1 & 0 & +2 \\ +1 & 0 & 0 & 0 & +1 & -2 \\ \hline\end{array}\,, \tag{9.3}$$

$$\begin{array}{|cccccc|}\hline +1 & -1 & 0 & 0 & 0 & 0 \\ 0 & +1 & -1 & 0 & 0 & 0 \\ 0 & 0 & +1 & -1 & 0 & 0 \\ -1 & 0 & 0 & +1 & 0 & 0 \\ \hline\end{array} \quad \begin{array}{|cccccc|}\hline -1 & 0 & 0 & 0 & 0 & +1 \\ 0 & 0 & +1 & 0 & -1 & 0 \\ 0 & 0 & -1 & 0 & 0 & +1 \\ +1 & 0 & 0 & 0 & +1 & -2 \\ \hline\end{array} \quad \begin{array}{|cccccc|}\hline 0 & +1 & 0 & 0 & 0 & -1 \\ 0 & -1 & 0 & 0 & +1 & 0 \\ 0 & 0 & 0 & +1 & 0 & -1 \\ 0 & 0 & 0 & -1 & -1 & +2 \\ \hline\end{array}\,.$$

Though the complete structure of the minimal Markov bases for general $I \times J \times K$ problems is not obtained at present, all the minimal Markov bases for $3 \times 3 \times K$ cases, $4 \times 4 \times 4$ cases by [83], and also $3 \times 4 \times K$ cases by [9] turned out to be unique. These results suggest the following conjecture.

Conjecture 9.1. For no-three-factor interaction models of three-way contingency tables, there exists a unique minimal Markov basis.

The indispensable move in (9.3) has $+2$ and -2; that is, both the positive part and the negative part are non-square-free. It is known (cf. Lemma 6.1 of [111]) that existence of an indispensable move with both parts non-square-free implies that the semigroup associated with the configuration is not normal. The example of a hole for the $3 \times 4 \times 6$ contingency table in Sect. 10 of [148] corresponds to the indispensable move (9.3). Results on normality of semigroups for larger tables are summarized in [113]. Normality for the $3 \times 5 \times 5$ case was recently established by [29].

9.6 Reducible Models

Let $[m] = \{1,\dots,m\}$ be the set of variables and let $\mathscr{K}$ denote a simplicial complex on $[m]$. Denote by $\mathscr{D}$ the set of maximal elements (i.e., facets) of $\mathscr{K}$. Then, as seen in Sects. 1.5 and 8.1, the hierarchical model associated with $\mathscr{K}$ is defined as

$$\log p(\boldsymbol{i}) = \sum_{D\in\mathscr{D}} \mu_D(\boldsymbol{i}_D). \tag{9.4}$$

A sufficient statistic $\boldsymbol{t}$ for this model is the set of marginal frequencies for each facet

$$\boldsymbol{t} = \{x_D(\boldsymbol{i}_D), \boldsymbol{i}_D \in \mathscr{I}_D, D \in \mathscr{D}\}.$$

In the following we identify $\mathscr{K}$ with the hierarchical model (9.4).

We note that $\mathscr{D}$ is considered as a hypergraph such that each facet in $\mathscr{D}$ is a hyperedge of $\mathscr{D}$. Here we introduce some notions on hypergraphs according to Badsberg and Malvestuto [20] and Malvestuto and Moscarini [101]. A subset D of a hyperedge of $\mathscr{D}$ is called a *partial edge*. We note that the submodel induced by a partial edge is saturated. A partial edge S is a separator of $\mathscr{D}$ if the subhypergraph of $\mathscr{D}$ induced by $[m]\setminus S$ is disconnected. A partial edge separator S of $\mathscr{D}$ is called a *divider* if there exist two vertices $u,v \in [m]$ that are separated by S but by no proper subset of S. When $\mathscr{D}$ is graphical, a partial edge separator and a divider are the clique separator and clique minimal vertex separator [97], respectively (e.g., Hara and Takemura [78], Leimer [99]).

If two vertices $u,v \in [m]$ are not separated by any partial edge, u and v are called *tightly connected*. A subset $C \subset [m]$ is called a *compact component* if every pair of variables in C is tightly connected. Denote the set of maximal compact components of $\mathscr{D}$ by $\mathscr{C}$. Then there exists a sequence of maximal compact components $C_1,\dots,C_{|\mathscr{C}|}$ such that

$$(C_1 \cup \dots \cup C_{k-1}) \cap C_k = S_k \tag{9.5}$$

and S_k, $k = 2,\dots,|\mathscr{C}|$, are dividers of $\mathscr{D}$. We denote $\mathscr{S} = \{S_2,\dots,S_{|\mathscr{C}|}\}$. $\mathscr{S}$ is a multiset in general. The property (9.5) is called the running intersection property. Then cell probability $p(\boldsymbol{i})$ is expressed as a rational form of marginal probabilities,

$$p(\boldsymbol{i}) = \frac{\prod_{C\in\mathscr{C}} p(\boldsymbol{i}_C)}{\prod_{S\in\mathscr{S}} p(\boldsymbol{i}_S)}.$$

MLE is expressed as a rational form of the MLE of marginal probabilities,

$$\hat{p}(\boldsymbol{i}) = \frac{\prod_{C\in\mathscr{C}} \hat{p}(\boldsymbol{i}_C)}{\prod_{S\in\mathscr{S}} \hat{p}(\boldsymbol{i}_S)} = \frac{\prod_{C\in\mathscr{C}} \hat{p}(\boldsymbol{i}_C)}{\prod_{S\in\mathscr{S}} (x(\boldsymbol{i}_S)/n)}. \tag{9.6}$$

If $\mathscr{D}$ does not have a divider (i.e., if $|\mathscr{C}| = 1$), the corresponding hierarchical model (9.4) is called *prime*. On the other hand, if $\mathscr{D}$ has a divider, the corresponding hierarchical model is called *reducible* (e.g., Develin and Sullivant [49], Hoşten and Sullivant [88]).

9.7 Markov Basis for Reducible Models

From (9.6), in order to compute the MLE of a reducible model, it suffices to compute the MLEs of marginal models $p(\boldsymbol{i}_C)$ for each maximal compact component. In the same way, a Markov basis for a reducible hierarchical model is also constructed from Markov bases of marginal models for maximal compact components. In this section we discuss the divide-and-conquer approach to the computation of a Markov basis for reducible models by Hoşten and Sullivant [88] and Dobra and Sullivant [54].

For a subset of variables $D \subset [m]$, denote by $\mathscr{K}(D)$ the submodel induced by D. Let (A_1, A_2, S) be a decomposition of $\mathscr{D}$ and define $V_1 := A_1 \cup S$ and $V_2 := A_2 \cup S$. Denote by $A_{V_1} = \{\boldsymbol{a}_{V_1}(\boldsymbol{i}_{V_1})\}_{\boldsymbol{i}_{V_1} \in \mathscr{I}_{V_1}}$ and $A_{V_2} = \{\boldsymbol{a}_{V_2}(\boldsymbol{i}_{V_2})\}_{\boldsymbol{i}_{V_2} \in \mathscr{I}_{V_2}}$ the configurations for the marginal models $\mathscr{K}(V_1)$ and $\mathscr{K}(V_2)$, where $\boldsymbol{a}_{V_1}(\boldsymbol{i}_{V_1})$ and $\boldsymbol{a}_{V_2}(\boldsymbol{i}_{V_2})$ denote column vectors of A_{V_1} and A_{V_2}, respectively. Noting that $\boldsymbol{i}_{V_1} = (\boldsymbol{i}_{A_1}\boldsymbol{i}_S)$ and $\boldsymbol{i}_{V_2} = (\boldsymbol{i}_S\boldsymbol{i}_{A_2})$, the configuration A for a reducible model is written by

$$A = A_{V_1} \oplus_S A_{V_2} = \{\boldsymbol{a}_{V_1}(\boldsymbol{i}_{A_1}\boldsymbol{i}_S) \oplus \boldsymbol{a}_{V_2}(\boldsymbol{i}_S\boldsymbol{i}_{A_2})\}_{\boldsymbol{i}_{A_1} \in \mathscr{I}_{A_1}, \boldsymbol{i}_S \in \mathscr{I}_S, \boldsymbol{i}_{A_2} \in \mathscr{I}_{A_2}},$$

where

$$\boldsymbol{a}_{V_1}(\boldsymbol{i}_{A_1}\boldsymbol{i}_S) \oplus \boldsymbol{a}_{V_2}(\boldsymbol{i}_S\boldsymbol{i}_{A_2}) = \begin{pmatrix} \boldsymbol{a}_{V_1}(\boldsymbol{i}_{A_1}\boldsymbol{i}_S) \\ \boldsymbol{a}_{V_2}(\boldsymbol{i}_S\boldsymbol{i}_{A_2}) \end{pmatrix}$$

denotes the stacked vector (1.21).

As in previous sections we denote a move $\boldsymbol{z}$ with degree d by

$$\boldsymbol{z} = \boldsymbol{i}_1 \cdots \boldsymbol{i}_d - \boldsymbol{i}'_1 \cdots \boldsymbol{i}'_d,$$

where $\boldsymbol{i}_1, \ldots, \boldsymbol{i}_d \in \mathscr{I}$ are cells of positive elements of $\boldsymbol{z}$ and $\boldsymbol{i}'_1, \ldots, \boldsymbol{i}'_d \in \mathscr{I}$ are cells of negative elements of $\boldsymbol{z}$. $\boldsymbol{i}_k$ appears $z(\boldsymbol{i}_k)$ times in $\{\boldsymbol{i}_1, \ldots, \boldsymbol{i}_d\}$ and in the same way $\boldsymbol{i}'_k$ appears $|z(\boldsymbol{i}'_k)|$ times in $\{\boldsymbol{i}'_1, \ldots, \boldsymbol{i}'_d\}$.

Assume that $\mathscr{B}(V_1)$ and $\mathscr{B}(V_2)$ are Markov bases for $\mathscr{K}(V_1)$ and $\mathscr{K}(V_2)$, respectively. Let $\boldsymbol{z}_1 = \{z_1(\boldsymbol{i}_{V_1})\}_{\boldsymbol{i}_{V_1} \in \mathscr{I}_{V_1}}$ and $\boldsymbol{z}_2 = \{z_2(\boldsymbol{i}_{V_2})\}_{\boldsymbol{i}_{V_2} \in \mathscr{I}_{V_2}}$ be degree d moves in $\mathscr{B}(V_1)$ and $\mathscr{B}(V_2)$, respectively. Because S is a partial edge separator, $\mathscr{K}(S)$ is saturated and we have

$$\sum_{\boldsymbol{i}_{V_1 \setminus S} \in \mathscr{I}_{V_1 \setminus S}} z_1(\boldsymbol{i}_{V_1}) = 0, \qquad \sum_{\boldsymbol{i}_{V_2 \setminus S} \in \mathscr{I}_{V_2 \setminus S}} z_2(\boldsymbol{i}_{V_2}) = 0.$$

Hence z_1 and z_2 can be written as

$$\begin{aligned} z_1 &= (\boldsymbol{i}_1\boldsymbol{j}_1)\cdots(\boldsymbol{i}_d\boldsymbol{j}_d) - (\boldsymbol{i}'_1\boldsymbol{j}_1)\cdots(\boldsymbol{i}'_d\boldsymbol{j}_d), \\ z_2 &= (\boldsymbol{j}_1\boldsymbol{k}_1)\cdots(\boldsymbol{j}_d\boldsymbol{k}_d) - (\boldsymbol{j}_1\boldsymbol{k}'_1)\cdots(\boldsymbol{j}_d\boldsymbol{k}'_d), \end{aligned} \tag{9.7}$$

respectively, where $\boldsymbol{i}_l, \boldsymbol{i}'_l \in \mathscr{I}_{A_1}$, $\boldsymbol{j}_l \in \mathscr{I}_S$ and $\boldsymbol{k}_l, \boldsymbol{k}'_l \in \mathscr{I}_{A_2}$ for $l = 1, \ldots, d$.

Definition 9.6 (Dobra and Sullivant [54]). Define $z_1 \in \mathscr{B}(V_1)$ as in (9.7). Let $\boldsymbol{k} := \{\boldsymbol{k}_1, \ldots, \boldsymbol{k}_d\} \in \mathscr{I}_{A_2} \times \cdots \times \mathscr{I}_{A_2}$. Define $z_1^{\boldsymbol{k}}$ by

$$z_1^{\boldsymbol{k}} := (\boldsymbol{i}_1\boldsymbol{j}_1\boldsymbol{k}_1)\cdots(\boldsymbol{i}_d\boldsymbol{j}_d\boldsymbol{k}_d) - (\boldsymbol{i}'_1\boldsymbol{j}_1\boldsymbol{k}_1)\cdots(\boldsymbol{i}'_d\boldsymbol{j}_d\boldsymbol{k}_d).$$

Then we define $\mathrm{Ext}(\mathscr{B}(V_1) \to \mathscr{K})$ by

$$\mathrm{Ext}(\mathscr{B}(V_1) \to \mathscr{K}) := \{z_1^{\boldsymbol{k}} \mid \boldsymbol{k} \in \mathscr{I}_{A_2} \times \cdots \times \mathscr{I}_{A_2}\}.$$

Lemma 9.2. *Suppose that $z_1 \in \mathscr{B}(V_1)$ as in (9.7). Then $\mathrm{Ext}(\mathscr{B}(V_1) \to \mathscr{K})$ is the set of moves for $\mathscr{K}$.*

Proof. Let $z \in \mathrm{Ext}(\mathscr{B}(V_1) \to \mathscr{K})$. Then we have

$$Az = \begin{pmatrix} \sum_{\boldsymbol{i}_{V_1} \in \mathscr{I}_{V_1}} a_{V_1}(\boldsymbol{i}_{V_1}) z_{V_1}(\boldsymbol{i}_{V_1}) \\ \sum_{\boldsymbol{i}_{V_2} \in \mathscr{I}_{V_2}} a_{V_2}(\boldsymbol{i}_{V_2}) z_{V_2}(\boldsymbol{i}_{V_2}) \end{pmatrix},$$

where

$$z_{V_1}(\boldsymbol{i}_{V_1}) = \sum_{\boldsymbol{i}_{V_1^C} \in \mathscr{I}_{V_1^C}} z(\boldsymbol{i}), \quad z_{V_2}(\boldsymbol{i}_{V_2}) = \sum_{\boldsymbol{i}_{V_2^C} \in \mathscr{I}_{V_2^C}} z(\boldsymbol{i}).$$

Because $z_{V_1}(\boldsymbol{i}_{V_1}) = z_1(\boldsymbol{i}_{V_1})$ and $z_1 \in \mathscr{B}(V_1)$, $\sum_{\boldsymbol{i}_{V_1} \in \mathscr{I}_{V_1}} a_{V_1}(\boldsymbol{i}_{V_1}) z_{V_1}(\boldsymbol{i}_{V_1}) = 0$. From Definition 9.6, $z_{V_2}(\boldsymbol{i}_{V_2}) = 0$ for all $\boldsymbol{i}_{V_2} \in \mathscr{I}_{V_2}$. Hence $Az = 0$.

The following theorem by Dobra and Sullivant [54] shows that a Markov basis for $\mathscr{K}$ is computed recursively from Markov basis for the induced submodel $\mathscr{K}(C)$, $C \in \mathscr{C}$.

Theorem 9.5 (Dobra and Sullivant [54]). *Let $\mathscr{B}(V_1)$ and $\mathscr{B}(V_2)$ be Markov bases for $\mathscr{K}(V_1)$ and $\mathscr{K}(V_2)$, respectively. Let $\mathscr{B}_{V_1,V_2}$ be a Markov basis for the decomposable model with two cliques V_1 and V_2. Then*

$$\mathscr{B} := \mathrm{Ext}(\mathscr{B}(V_1) \to \mathscr{K}) \cup \mathrm{Ext}(\mathscr{B}(V_2) \to \mathscr{K}) \cup \mathscr{B}_{V_1,V_2} \tag{9.8}$$

is a Markov basis for $\mathscr{K}$.

Proof. Let $\boldsymbol{x}, \boldsymbol{x}' \in \mathscr{F}_{\boldsymbol{t}}$ be two tables in the same fiber $\mathscr{F}_{\boldsymbol{t}}$. As in the previous sections we denote $\boldsymbol{x}$ and $\boldsymbol{x}'$ as

$$x_1 = (\boldsymbol{i}_1\boldsymbol{j}_1\boldsymbol{k}_1)\cdots(\boldsymbol{i}_n\boldsymbol{j}_n\boldsymbol{k}_n), \quad x_2 = (\boldsymbol{i}'_1\boldsymbol{j}'_1\boldsymbol{k}'_1)\cdots(\boldsymbol{i}'_n\boldsymbol{j}'_n\boldsymbol{k}'_n),$$

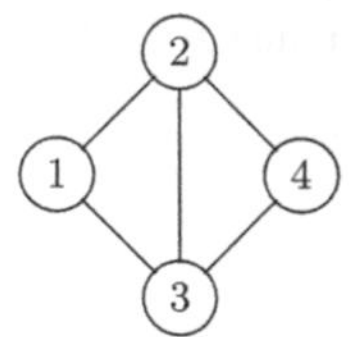

Fig. 9.9 The independence graph for $\mathscr{D}$

where $\boldsymbol{i}_k, \boldsymbol{i}'_k \in \mathscr{I}_{A_1}$, $\boldsymbol{j}_k, \boldsymbol{j}'_k \in \mathscr{I}_S$, $\boldsymbol{k}_k, \boldsymbol{k}'_k \in \mathscr{I}_{A_2}$. Let $\boldsymbol{x}_{V_1}$ and $\boldsymbol{x}'_{V_1}$ be V_1-marginal sum of $\boldsymbol{x}$ and $\boldsymbol{x}'$, respectively. Because $\boldsymbol{x}_{V_1}$ and $\boldsymbol{x}'_{V_1}$ belong to the same fiber for $\mathscr{K}(V_1)$, there exists a sequence of moves $\boldsymbol{z}^1_{V_1}, \ldots, \boldsymbol{z}^t_{V_1}$ connecting $\boldsymbol{x}_{V_1}$ and $\boldsymbol{x}'_{V_1}$. We note that $\boldsymbol{z}^1_{V_1}$ is written as

$$\boldsymbol{z}^1_{V_1} = (\boldsymbol{i}''_1 \boldsymbol{j}_1) \cdots (\boldsymbol{i}''_m \boldsymbol{j}_m) - (\boldsymbol{i}_1 \boldsymbol{j}_1) \cdots (\boldsymbol{i}_m \boldsymbol{j}_m),$$

where $0 < m < n$. Define $\boldsymbol{z}^1$ by

$$\boldsymbol{z}^1 = (\boldsymbol{i}''_1 \boldsymbol{j}_1 \boldsymbol{k}_1) \cdots (\boldsymbol{i}''_m \boldsymbol{j}_m \boldsymbol{k}_m) - (\boldsymbol{i}_1 \boldsymbol{j}_1 \boldsymbol{k}_1) \cdots (\boldsymbol{i}_m \boldsymbol{j}_m \boldsymbol{k}_m). \tag{9.9}$$

Then $\boldsymbol{z}^1 \in \mathrm{Ext}(\mathscr{B}(V_1) \to \mathscr{K})$ and $\boldsymbol{x} + \boldsymbol{z}^1 \geq \boldsymbol{0}$. By iterating this procedure with $\boldsymbol{z}^2_{V_1}, \ldots, \boldsymbol{z}^t_{V_1}$, we can define $\boldsymbol{z}^2, \ldots, \boldsymbol{z}^t \in \mathrm{Ext}(\mathscr{B}(V_1) \to \mathscr{K})$ in the same way as (9.9) such that $\boldsymbol{x}_1 + \sum_{l=1}^{t'} \boldsymbol{z}^l \geq \boldsymbol{0}$ for all $t' \leq t$ and V_1- and V_2-marginal sums of $\boldsymbol{y} := \boldsymbol{x} + \sum_{l=1}^{t} \boldsymbol{z}_s$ are $\boldsymbol{y}_{V_1} = \boldsymbol{x}'_{V_1}$ and $\boldsymbol{y}_{V_2} = \boldsymbol{x}_{V_2}$, respectively.

On the other hand $\boldsymbol{x}'_{V_2}$ and $\boldsymbol{y}_{V_2}$ also belong to the same fiber for $\mathscr{K}(V_2)$ and hence $\boldsymbol{x}'_{V_2}$ and $\boldsymbol{y}_{V_2}$ are connected by moves in $\mathscr{B}(V_2)$. By using the same argument as above, there exist moves $\boldsymbol{w}^1, \ldots, \boldsymbol{w}^s \in \mathrm{Ext}(\mathscr{B}(V_2) \to \mathscr{K})$ such that $\boldsymbol{x}_1 + \sum_{l=1}^{s'} \boldsymbol{w}^l \geq 0$ for all $s' \leq s$ and V_1- and V_2-marginal sums of $\boldsymbol{y}' := \boldsymbol{x} + \sum_{l=1}^{s} \boldsymbol{w}^l$ are $\boldsymbol{y}_{V_1} = \boldsymbol{x}'_{V_1}$ and $\boldsymbol{y}_{V_2} = \boldsymbol{x}_{V_2}$, respectively. $\square$

In general it is not easy to obtain an explicit list of a Markov basis for a hierarchical model. This theorem shows that when we study the structure of Markov bases for hierarchical models, we only need to focus on Markov bases for prime models. Hara et al. [73] extend this result to a more general class of log affine models which is called the hierarchical subspace model.

Example 9.1. Consider the model

$$\mathscr{D} = \{\{1,2\},\{1,3\},\{2,3\},\{2,4\},\{3,4\}\}.$$

Let $\mathscr{K}$ be the corresponding simplicial complex. Assume that the number of levels for all variables are two; that is, $I_k = 2$ for $k = 1,2,3,4$. The independence graph for $\mathscr{D}$ is described as in Fig. 9.9. In this model $S := \{2,3\}$ is a divider and hence $\mathscr{K}$ is reducible. Then $\mathscr{D}$ is decomposed by S into two no-three-factor interaction models $\mathscr{D}_1 := \{\{1,2\},\{1,3\},\{2,3\}\}$ and $\mathscr{D}_2 := \{\{2,3\},\{2,4\},\{3,4\}\}$. Let $V_1 := \{1,2,3\}$ and $V_2 := \{2,3,4\}$.

A Markov basis for $\mathscr{D}_1$ consists of one move

$$\begin{array}{c|cc} i_1\backslash i_2 & 1 & 2 \\ \hline 1 & 1 & -1 \\ 2 & -1 & 1 \end{array}\quad \begin{array}{c|cc} i_1\backslash i_2 & 1 & 2 \\ \hline 1 & -1 & 1 \\ 2 & 1 & -1 \end{array}$$
$$\qquad i_3 = 1 \qquad\qquad i_3 = 2$$

(Develin and Sullivant [49]). This move are described as

$$(111)(212)(221)(122) - (211)(112)(121)(222).$$

By Definition 9.6, moves in $\mathrm{Ext}(\mathscr{B}(V_1) \to \mathscr{K})$ are described as

$$(111k_1)(212k_2)(221k_3)(122k_4) - (211k_1)(112k_2)(121k_3)(222k_4),$$

where $k_1, k_2, k_3, k_4 \in \mathscr{I}_4 = \{1, 2\}$. For example, when $(k_1, k_2, k_3, k_4) = (1, 1, 2, 2)$, the corresponding move in $\mathrm{Ext}(\mathscr{B}(V_1) \to \mathscr{K})$ is

$$\underbrace{\begin{array}{c|cc} i_1\backslash i_2 & 1 & 2 \\ \hline 1 & 1 & 0 \\ 2 & -1 & 0 \\ \multicolumn{3}{c}{i_3=1} \end{array}\quad \begin{array}{c|cc} i_1\backslash i_2 & 1 & 2 \\ \hline 1 & -1 & 0 \\ 2 & 1 & 0 \\ \multicolumn{3}{c}{i_3=2} \end{array}}_{i_4=1}\qquad \underbrace{\begin{array}{c|cc} i_1\backslash i_2 & 1 & 2 \\ \hline 1 & 0 & -1 \\ 2 & 0 & 1 \\ \multicolumn{3}{c}{i_3=1} \end{array}\quad \begin{array}{c|cc} i_1\backslash i_2 & 1 & 2 \\ \hline 1 & 0 & 1 \\ 2 & 0 & -1 \\ \multicolumn{3}{c}{i_3=2} \end{array}}_{i_4=2}.$$

Moves in $\mathrm{Ext}(\mathscr{B}(V_2) \to \mathscr{K})$ are obtained in the same way.

$\mathscr{B}_{V_1,V_2}$ is a Markov basis for the decomposable model associated with the graph in Fig. 9.9. $\mathscr{B}_{V_1,V_2}$ is obtained by following the argument in Chap. 8. Then

$$\mathscr{B} := \mathrm{Ext}(\mathscr{B}(V_1) \to \mathscr{K}) \cup \mathrm{Ext}(\mathscr{B}(V_2) \to \mathscr{K}) \cup \mathscr{B}_{V_1,V_2}$$

forms a Markov basis for $\mathscr{D}$.

9.8 Markov Complexity and Graver Complexity

In this section we give a brief review of the recent progress on the evaluation of complexity of Markov bases and the Graver basis for hierarchical models.

In Sect. 4.6 we discussed the Lawrence lifting of a configuration $A : \nu \times \eta$. Santos and Sturmfels [131] introduced the rth Lawrence lifting (or the rth Lawrence configuration) $\Lambda^{(r)}(A)$ as the following configuration.

$$
\Lambda^{(r)}(A) = \begin{pmatrix} A & 0 & 0 & \cdots & 0 \\ 0 & A & 0 & \cdots & 0 \\ \vdots & \vdots & \vdots & \ddots & \vdots \\ 0 & 0 & 0 & \cdots & A \\ E_\eta & E_\eta & E_\eta & \cdots & E_\eta \end{pmatrix} : (r\nu + \eta) \times (r\eta). \tag{9.10}
$$

As in the ordinary Lawrence lifting, we can omit one block of rows from $\Lambda^{(r)}(A)$ and write the the rth Lawrence lifting also as

$$
\tilde{\Lambda}^{(r)}(A) = \begin{pmatrix} \overbrace{A \;\; 0 \;\; \cdots \;\; 0}^{r-1} & 0 \\ 0 \;\; A \;\; 0 & \cdots \;\; 0 \\ \vdots \;\; \ddots \;\; \ddots & \ddots \;\; \vdots \\ 0 \;\; \cdots \;\; 0 & A \;\; 0 \\ E_\eta \;\; E_\eta \;\; \cdots & E_\eta \;\; E_\eta \end{pmatrix} : ((r-1)\nu + \eta) \times (r\eta). \tag{9.11}
$$

It can be easily seen that the configuration of no-three-factor interaction model for $I \times J \times r$ tables is the rth Lawrence lifting of the configuration A for the $I \times J$ two-way independence model.

From a statistical viewpoint, as we saw in Sect. 4.6, the Lawrence lifting corresponds to a logistic regression. Similarly the rth Lawrence lifting corresponds to an unordered multinomial logistic regression, where the response variable $Y = Y_{\boldsymbol{i}}$ can take r different levels for each cell $\boldsymbol{i}$ and the probability $P(Y_{\boldsymbol{i}} = k)$, $k = 1, \dots, r$, is expressed as

$$
P(Y_{\boldsymbol{i}} = k) = p(\boldsymbol{i}, k) = \frac{\exp(\boldsymbol{\theta}'_k a(\boldsymbol{i}))}{\sum_{h=1}^r \exp(\boldsymbol{\theta}'_h a(\boldsymbol{i}))}, \tag{9.12}
$$

where $\boldsymbol{\theta}_k$ is a ν-dimensional parameter vector, $k = 1, \dots, r$. For each cell $\boldsymbol{i}$ we observe $n_{\boldsymbol{i}}$ independent trials, each trial taking one of the r levels. Let $x(\boldsymbol{i}, k)$, $k = 1, \dots, r$, denote the frequency of the level k among $n_{\boldsymbol{i}}$ trials. Then $(x(\boldsymbol{i}, 1), \dots, x(\boldsymbol{i}, r))$ has the multinomial distribution $\mathrm{Mult}(n_{\boldsymbol{i}}, (p(\boldsymbol{i}, 1), \dots, p(\boldsymbol{i}, r)))$. For each cell $\boldsymbol{i}$, the lowest block of rows $(E_\eta, \dots, E_\eta)$ of $\Lambda^{(r)}(A)$ corresponds to fixing the total number of frequencies

$$
n_{\boldsymbol{i}} = x(\boldsymbol{i}, 1) + \cdots + x(\boldsymbol{i}, r).
$$

As explained in Sect. 4.6, when $n_{\boldsymbol{i}}$s are allowed to vary over nonnegative integers, the rth Lawrence lifting is the configuration for the multinomial logistic regression model.

The result on the bound of complexity of Markov basis for $3 \times 3 \times K$ tables in Theorem 9.4 led to many investigations of its generalizations. Santos and Sturmfels [131] gave a first general result on the complexity of the Graver basis of $\Lambda^{(r)}(A)$, which was already mentioned in Proposition 9.1. Note that $\boldsymbol{z} \in \ker_{\mathbb{Z}} \Lambda^{(r)}(A)$ is an

ηr-dimensional integer column vector and we can express it as $z' = (z'_1, \dots, z'_r)$. z belongs to $\ker_{\mathbb{Z}} \Lambda^{(r)}(A)$ if and only if

$$Az_i = 0,\ i = 1, \dots, r, \ \text{and} \ \sum_{i=1}^{r} z_i = 0.$$

We call z_i the ith *slice* of z. Define the type of z by

$$\text{type}(z) = |\{i \mid z_i \neq 0\}|,$$

which is the number of nonzero slices of z. Let $\mathscr{B}_{\text{Gr}}(\Lambda^{(r)}(A))$ denote the Graver basis of $\Lambda^{(r)}(A)$. For a given configuration A define the Graver complexity $g(A)$ by

$$g(A) = \max_{r \geq 1} \{\text{type}(z) \mid z \in \mathscr{B}_{\text{Gr}}(\Lambda^{(r)}(A))\}.$$

As we see below in Proposition 9.2, there is an explicit characterization for $g(A)$ and $g(A)$ is indeed finite. Because the configuration of the no-three-factor interaction model is the higher Lawrence lifting of the two-way independence model and the Markov complexity mentioned in Proposition 9.1 is bounded from above by the Graver complexity $g(A)$, Proposition 9.2 also implies that the Markov complexity is bounded for the no-three-factor interaction model.

Let $\mathscr{B}_{\text{Gr}}(A)$ denote the Graver basis of A consisting of conformally primitive moves $z^{(1)}, \dots, z^{(\eta')}$ for A, where $\eta' = |\mathscr{B}_{\text{Gr}}(A)|$. Write each conformally primitive move $z^{(i)}$ as an η-dimensional integer column vector and let

$$\mathscr{G}(A) = (z^{(1)}, \dots, z^{(\eta')})$$

be an $\eta \times \eta'$ integral matrix. Furthermore let $\mathscr{B}_{\text{Gr}}(\mathscr{G}(A))$ denote the Graver basis of $\mathscr{G}(A)$.

Note that $\mathscr{G}(A)$ does not satisfy the homogeneity assumption in (3.7), because the sign of $z^{(i)}$ is arbitrary. However, conformally primitive moves and the Graver basis for $\mathscr{G}(A)$ can be defined without the homogeneity assumption.

Now we present a characterization of the Graver complexity, which shows that $g(A)$ is given by the maximum 1-norm of the elements of the Graver basis $\mathscr{B}_{\text{Gr}}(\mathscr{G}(A))$ of the Graver basis of A. Note that two Graver bases are nested in this result.

Proposition 9.2 (Theorem 3 of [131]). *$g(A)$ is given as*

$$g(A) = \max\{|\psi| \mid \psi \in \mathscr{B}_{\text{Gr}}(\mathscr{G}(A))\}, \tag{9.13}$$

where $|\cdot|$ denotes the 1-norm.

Proof. Let $z_1, \dots, z_r$ be slices of $z \in \ker_{\mathbb{Z}} \Lambda^{(r)}(A)$ and suppose that z is conformally primitive. Suppose that z_i has a nontrivial conformal decomposition, say,

$z_i = y_1 + \cdots + y_k$, where $y_1, \ldots, y_k$ are conformally primitive moves for A. In this case we can remove the ith slice from z and insert slices corresponding to $y_1, \ldots, y_k$. Then we obtain a conformally primitive move for $\Lambda^{(r+k-1)}$. This argument shows that in considering $g(A)$, we only need to look at conformally primitive moves $z \in \ker_{\mathbb{Z}} \Lambda^{(r)}(A)$ such that their slices are all conformally primitive moves for A. Also we can assume that no slice of z is a negative of another slice of z.

Write the Graver basis of A as $\mathscr{B}_{\mathrm{Gr}}(A) = \{z^{(1)}, \ldots, z^{(\eta')}\}$. For a given z whose slices are conformally primitive moves of A, let $\psi = (\psi_1 \ldots, \psi_{\eta'})$ be an integer vector such that ψ_i counts the number of times $z^{(i)}$ appears as a slice of z. Then $|\psi| = \mathrm{type}(z)$. Furthermore it is easily seen that z has a nontrivial conformal decomposition if and only if ψ has a nontrivial conformal decomposition. Therefore z is conformally primitive if and only if ψ belongs to $\mathscr{B}_{\mathrm{Gr}}(\mathscr{G}(A))$. This proves the proposition. □

Hoşten and Sullivant [89] introduced a generalized form of the higher Lawrence lifting. For two configurations A, B with the same number of columns, they considered a configuration of the following form,

$$\Lambda^{(r)}(A,B) = \begin{pmatrix} A & 0 & 0 & \cdots & 0 \\ 0 & A & 0 & \cdots & 0 \\ \vdots & \vdots & \vdots & \ddots & \vdots \\ 0 & 0 & 0 & \cdots & A \\ B & B & B & \cdots & B \end{pmatrix}, \tag{9.14}$$

and generalized Proposition 9.2 to this form. Furthermore they showed that the configuration for a hierarchical model for $I_1 \times \cdots \times I_m$ contingency tables can be written in this form by letting $r = I_1$ and considering the slices of the first axis. Their results imply that the number of slices appearing in the Graver basis for a hierarchical model is bounded if we increase the number of levels for a single axis.

The no-three-factor interaction model is often called the *three-way transportation problem* in integer programming. The importance of three-way transportation problems for the general integer programming problem is discussed in [47]. The Graver complexity of an integer matrix is studied in [45, 84] and [23] from the viewpoint of integer programming.

9.9 Markov Width for Some Hierarchical Models

The complexity of Markov bases is also evaluated by maximal degrees of minimal Markov bases, which is also called *Markov width*. In this section we summarize some important facts on maximal degrees of minimal Markov bases for hierarchical models.

A graphical model containing only two factor interaction effects corresponding to edges of an independence graph is called a *graph model*. Let $G = ([m], E)$ be a graph with the vertex set $[m]$ and the edge set E. We assume that the model is binary, that is, the number of levels for every variable is $I_v = 2$, $v \in [m]$. Denote by $\mu(G)$ the Markov width of the graph model corresponding to G.

Develin and Sullivant [49] discuss Markov width of some binary graph models.

Theorem 9.6 ([49]). *The binary graph model for the complete graph G with $m \geq 3$ has the Markov width $\mu(G) \geq 2m - 2$.*

Theorem 9.7 ([49]). *The binary graph model for the cycle graph G with $m \geq 3$ has a Markov basis consisting of moves with degrees two and four. Therefore $\mu(G) = 4$.*

Let $K_{m,n}$ denote the complete bipartite graph with partitions of sizes m and n.

Theorem 9.8 ([49]). *The binary graph model for the bipartite graph $K_{2,n}$ has a Markov basis consisting of moves with degrees two and four. Therefore $\mu(K_{2,n}) = 4$.*

Petrović and Stokes [117] characterize Markov width of some classes of hierarchical model in term of Betti numbers of Stanley–Reisner ideals for a simplicial complex Δ. For details, refer to Petrović and Stokes [117].

Chapter 10
Two-Way Tables with Structural Zeros and Fixed Subtable Sums

10.1 Markov Bases for Two-Way Tables with Structural Zeros

10.1.1 Quasi-Independence Model in Two-Way Incomplete Contingency Tables

Let $\boldsymbol{x} = \{x_{ij}\}$ be an $R \times C$ contingency table and denote by $S \subset \mathscr{I} = \{(i,j) \mid 1 \le i \le R, 1 \le j \le C\}$ the set of cells that are not structural zeros. In a structural zero cell, no frequency is observed by definition, such as the number of people with driver's licenses under the age of 10. We consider models for cell probabilities in incomplete contingency tables

$$\begin{cases} \log p_{ij} = \mu + \alpha_i + \beta_j + \gamma_{ij}, \ (i,j) \in S, \\ \quad\ p_{ij} = 0, \qquad\qquad\qquad\qquad \text{otherwise.} \end{cases} \tag{10.1}$$

As a null hypothesis for (10.1), we consider $H_0 : \gamma_{ij} = 0$ for $(i,j) \in S$; that is,

$$\begin{cases} \log p_{ij} = \mu + \alpha_i + \beta_j, \ (i,j) \in S, \\ \quad\ p_{ij} = 0, \qquad\qquad\quad \text{otherwise.} \end{cases} \tag{10.2}$$

The model (10.2) is called the *quasi-independence model* (Bishop et al. [26]). In this section we provide a full description of the unique minimal Markov basis for the quasi-independence model. Rapallo ([124,125]) discuss the quasi-independence model mainly from the viewpoint of Gröbner basis.

Denote by $\mathscr{B}(S)$ the set of moves for the quasi-independence model (10.2) on S. A sufficient statistic for the quasi-independence models is $\boldsymbol{t} = (x_{1+},\dots,x_{R+}, x_{+1},\dots,x_{+C})$. Therefore

$$\mathscr{B}(S) = \{\boldsymbol{z} = \{z_{ij}\} \mid z_{i+} = z_{+j} = 0,\ z_{ij} = 0 \ \text{ for } \ (i,j) \notin S\}.$$

We denote a structural zero by [0] to distinguish it from a sampling zero.

S. Aoki et al., *Markov Bases in Algebraic Statistics*, Springer Series in Statistics 199, DOI 10.1007/978-1-4614-3719-2_10,

When $S = \mathscr{I}$, (10.2) is equivalent to the two-way complete independence model. Then, as shown in Sec. 5.4 the set of all basic moves

$$\mathscr{B}_0 = \{z(i,i';j,j') \mid 1 \le i < i' \le R, 1 \le j < j' \le C\}$$

in (8.1) forms the unique minimal Markov basis for the two-way complete independence model. However when $S \ne \mathscr{I}$, $\mathscr{B}_0 \cap \mathscr{B}(S)$ is not always a Markov basis.

Example 10.1. Consider a fiber $\mathscr{F}_t$ of 3×3 contingency tables having structural zero cells as the diagonal elements; that is, $S = \{(i,j),\ i \ne j\}$ with $x_{i+} = x_{+j} = 1$ for all $1 \le i, j \le 3$. Then $\mathscr{F}_t$ contains only the following two elements,

$$\begin{array}{|ccc|}\hline [0] & 1 & 0 \\ 0 & [0] & 1 \\ 1 & 0 & [0] \\ \hline \end{array} \quad \text{and} \quad \begin{array}{|ccc|}\hline [0] & 0 & 1 \\ 1 & [0] & 0 \\ 0 & 1 & [0] \\ \hline \end{array},$$

which implies that the degree 3 move

$$\begin{array}{|ccc|}\hline [0] & -1 & +1 \\ +1 & [0] & -1 \\ -1 & +1 & [0] \\ \hline \end{array}$$

is indispensable.

In a two-way table, two cells (i,j) and (i',j') are *associated* if $(i,j),(i',j') \in S$ and either $i = i'$ or $j = j'$. $S' \subset S$ is *connected* if for every pair of cells (i,j) and (i',j') in S', there exists a chain of cells, any two consecutive members of which are associated. An incomplete two-way table is connected if its nonstructural zero cells form a connected set. An incomplete table that is not connected is said to be *separable* ([26, 102]). Separable two-way contingency tables can be rearranged to a block diagonal form with connected subtables after an appropriate interchange of rows and columns.

Example 10.2. Consider the following 4×8 contingency table,

$$\begin{array}{|cccccccc|}\hline x_{11} & [0] & [0] & x_{14} & x_{15} & [0] & [0] & x_{18} \\ [0] & x_{22} & x_{23} & [0] & [0] & x_{26} & x_{27} & [0] \\ [0] & x_{32} & x_{33} & [0] & [0] & x_{36} & x_{37} & [0] \\ x_{41} & [0] & [0] & x_{44} & x_{45} & [0] & [0] & x_{48} \\ \hline \end{array}.$$

By an appropriate interchange of rows and columns, we can obtain the following separable table with exactly two connected subtables

$$\begin{array}{|cccccccc|}\hline x_{11} & x_{14} & x_{15} & x_{18} & [0] & [0] & [0] & [0] \\ x_{41} & x_{44} & x_{45} & x_{48} & [0] & [0] & [0] & [0] \\ [0] & [0] & [0] & [0] & x_{22} & x_{23} & x_{26} & x_{27} \\ [0] & [0] & [0] & [0] & x_{32} & x_{33} & x_{36} & x_{37} \\ \hline \end{array}.$$

It is clear that the minimal Markov basis for this example consists of basic moves only. This is obvious from the fact that the two connected subtables do not contain structural zeros.

Example 10.3. The minimal Markov basis for the following separable 6×7 contingency table

$$\begin{array}{|ccccccc|}\hline x_{11} & x_{12} & x_{13} & [0] & [0] & [0] & [0] \\ [0] & [0] & x_{23} & x_{24} & [0] & [0] & [0] \\ x_{31} & x_{32} & [0] & x_{34} & [0] & [0] & [0] \\ [0] & [0] & [0] & [0] & x_{45} & x_{46} & [0] \\ [0] & [0] & [0] & [0] & [0] & x_{56} & x_{57} \\ [0] & [0] & [0] & [0] & x_{65} & [0] & x_{67} \\ \hline \end{array} \qquad (10.3)$$

is the union of the minimal Markov bases for two subtables,

$$\begin{array}{|cccc|}\hline x_{11} & x_{12} & x_{13} & [0] \\ [0] & [0] & x_{23} & x_{24} \\ x_{31} & x_{32} & [0] & x_{34} \\ \hline \end{array} \quad \text{and} \quad \begin{array}{|ccc|}\hline x_{45} & x_{46} & [0] \\ [0] & x_{56} & x_{57} \\ x_{65} & [0] & x_{67} \\ \hline \end{array}. \qquad (10.4)$$

Therefore we only need to consider the case where S is connected.

10.1.2 Unique Minimal Markov Basis for Two-Way Quasi-Independence Model

Assume that the level indices $i_1, i_2, \ldots$ and $j_1, j_2, \ldots$ are all distinct; that is,

$$i_m \neq i_n \quad \text{and} \quad j_m \neq j_n \quad \text{for all } m \neq n.$$

Denote $\boldsymbol{i}_{[r]} = (i_1, \ldots, i_r)$, $\boldsymbol{j}_{[r]} = (i_1, \ldots, i_r)$. The loop $\boldsymbol{z}_r(\boldsymbol{i}_{[r]}; \boldsymbol{j}_{[r]})$ was defined in Definition 4.3.

Definition 10.1. A loop of degree r in Definition 4.3 is a move on S if $\boldsymbol{z}_r(\boldsymbol{i}_{[r]}; \boldsymbol{j}_{[r]}) \in \mathscr{B}(S)$. The support of $\boldsymbol{z}_r(\boldsymbol{i}_{[r]}; \boldsymbol{j}_{[r]})$ is the set of its nonzero cells $\{(i_1, j_1), (i_1, j_2), \ldots, (i_r, j_1)\}$.

The following integer arrays are examples of loops of degree 2, 3, and 4 on some S. In fact they are df 1 loops as defined in Definition 10.2.

$$\begin{array}{|ccccc|}\hline +1 & -1 & 0 & 0 & 0 \\ -1 & +1 & 0 & 0 & 0 \\ 0 & 0 & 0 & 0 & 0 \\ 0 & 0 & 0 & 0 & 0 \\ \hline \end{array}, \quad \begin{array}{|ccccc|}\hline +1 & -1 & [0] & 0 & 0 \\ -1 & [0] & +1 & 0 & 0 \\ [0] & +1 & -1 & 0 & 0 \\ 0 & 0 & 0 & 0 & 0 \\ \hline \end{array}, \quad \begin{array}{|ccccc|}\hline +1 & -1 & [0] & [0] & 0 \\ -1 & [0] & +1 & [0] & 0 \\ [0] & +1 & [0] & -1 & 0 \\ [0] & [0] & -1 & +1 & 0 \\ \hline \end{array}. \qquad (10.5)$$

The following lemma was proved as Proposition 4.2.

Lemma 10.1. *Any $R \times C$ move $\boldsymbol{z} \in \mathscr{B}(S)$ is expressed as a finite sum of loops on S,*

$$\boldsymbol{z} = \sum_k a_k \boldsymbol{z}_{r(k)}(i_{1(k)}, \ldots, i_{r(k)}; j_{1(k)}, \ldots, j_{r(k)}),$$

where a_k is a positive integer, $r(k) \le \min\{R, C\}$, and there is no cancellation of signs in any cell.

Example 10.4. Let $\boldsymbol{z} \in \mathscr{B}(S)$ be 4×5 integer array expressed as follows,

$$\boldsymbol{z} = \begin{array}{|rrrrr|} \hline 3 & -2 & 0 & -2 & 1 \\ -2 & 3 & 0 & 0 & -1 \\ -1 & -1 & 2 & 0 & 0 \\ 0 & 0 & -2 & 2 & 0 \\ \hline \end{array}.$$

Then $\boldsymbol{z}$ has a decomposition

$$\boldsymbol{z} = \begin{array}{|rrrrr|} \hline 2 & -2 & 0 & 0 & 0 \\ -2 & 2 & 0 & 0 & 0 \\ 0 & 0 & 0 & 0 & 0 \\ 0 & 0 & 0 & 0 & 0 \\ \hline \end{array} + \begin{array}{|rrrrr|} \hline 1 & 0 & 0 & -1 & 0 \\ 0 & 0 & 0 & 0 & 0 \\ -1 & 0 & 1 & 0 & 0 \\ 0 & 0 & -1 & 1 & 0 \\ \hline \end{array} + \begin{array}{|rrrrr|} \hline 0 & 0 & 0 & -1 & 1 \\ 0 & 1 & 0 & 0 & -1 \\ 0 & -1 & 1 & 0 & 0 \\ 0 & 0 & -1 & 1 & 0 \\ \hline \end{array}$$

$$= 2z_2(1,2;1,2) + z_3(1,4,3;1,4,3) + z_4(1,4,3,2;5,4,3,2),$$

satisfying the condition of Lemma 10.1. We note that the decomposition is not unique in general. It is easy to check that

$$\boldsymbol{z} = z_2(1,2;1,2) + z_2(1,2;5,2) + z_3(1,4,3;1,4,3) + z_4(1,4,3,2;1,4,3,2)$$

is another decomposition of $\boldsymbol{z}$.

Suppose $\boldsymbol{x}, \boldsymbol{y} \in \mathscr{F}_{\boldsymbol{t}}$. Then the difference $\boldsymbol{z} = \boldsymbol{y} - \boldsymbol{x}$ is in $\mathscr{B}(S)$. Hence to move from $\boldsymbol{x}$ to $\boldsymbol{y}$, we can add a sequence of loops in Definition 10.1 to $\boldsymbol{x}$, without forcing negative entries on the way. In other words, the set of all the loops of degree $2, \ldots, \min\{I, J\}$ on S constitutes a trivial Markov basis.

Definition 10.2. A loop $z_r(\boldsymbol{i}_{[r]}; \boldsymbol{j}_{[r]})$ is called *df* 1 if $\mathscr{R}(\boldsymbol{i}_{[r]}; \boldsymbol{j}_{[r]})$ does not contain support of any loop on S of degree $2, \ldots, r-1$, where

$$\mathscr{R}(\boldsymbol{i}_{[r]}; \boldsymbol{j}_{[r]}) = \{(i,j) \mid i \in \{i_1, \ldots, i_r\}, j \in \{j_1, \ldots, j_r\}\}.$$

Lemma 10.2. *$z_r(\boldsymbol{i}_{[r]}; \boldsymbol{j}_{[r]})$ is df 1 if and only if $\mathscr{R}(\boldsymbol{i}_{[r]}; \boldsymbol{j}_{[r]})$ contains exactly two elements in S in every row and column.*

Proof. The case $r = 2$ is obvious. Suppose that $r \geq 3$.

First we show the necessity; that is, if $\boldsymbol{z}_r(\boldsymbol{i}_{[r]};\boldsymbol{j}_{[r]})$ is df 1 then $\mathscr{R}(\boldsymbol{i}_{[r]};\boldsymbol{j}_{[r]})$ contains exactly two elements in S in every row and column by showing its contraposition. Without loss of generality we can suppose that $\boldsymbol{z}_r([r];[r])$ is a degree r loop and that $(1,a) \in S$, $3 \leq \exists a \leq r$. Then this loop is decomposed into two loops on S as

$$\boldsymbol{z}_r([r];[r]) = \boldsymbol{z}_{r-a+2}([1,a:r];[1,a:r]) + \boldsymbol{z}_{a-1}([a-1];[a,2:a-1]), \qquad (10.6)$$

where we define $i:j = \{i, i+1, \ldots, j\}$ for $i < j$. An example for $r = 5$ and $a = 4$ is represented as follows.

$$\begin{array}{|ccccc|}\hline +1 & -1 & [0] & 0 & [0] \\ [0] & +1 & -1 & [0] & [0] \\ [0] & [0] & +1 & -1 & [0] \\ [0] & [0] & [0] & +1 & -1 \\ -1 & [0] & [0] & [0] & +1 \\ \hline \end{array} = \begin{array}{|ccccc|}\hline +1 & 0 & [0] & -1 & [0] \\ [0] & 0 & 0 & [0] & [0] \\ [0] & [0] & 0 & 0 & [0] \\ [0] & [0] & [0] & +1 & -1 \\ -1 & [0] & [0] & [0] & +1 \\ \hline \end{array} + \begin{array}{|ccccc|}\hline 0 & -1 & [0] & +1 & [0] \\ [0] & +1 & -1 & [0] & [0] \\ [0] & [0] & +1 & -1 & [0] \\ [0] & [0] & [0] & 0 & 0 \\ 0 & [0] & [0] & [0] & 0 \\ \hline \end{array}.$$

The nonzero cells of the two loops in the right-hand side of (10.6) overlap at $(1,a) \in S$ only. Hence $\mathscr{R}([r];[r])$ contains their supports, which contradicts the assumption.

Next we show the sufficiency. Suppose that $\boldsymbol{z}_r([r];[r])$ is a degree r loop such that $\mathscr{R}([r];[r])$ contains exactly two elements in S in every row and column. Then it is sufficient to show that $\mathscr{R}([r];[r])$ does not contain support of any loop of degree $2,\ldots,r-1$ on S. From the assumption, $(1,1)$ is the only cell in S in $\mathscr{R}([r-1];1)$, because $\boldsymbol{z}_r([r];[r])$ has exactly two nonzero elements there: $z_{11} = +1$ and $z_{r1} = -1$. Hence z_{11} is zero in any loop in $\mathscr{R}([r-1];[r])$. Moreover, by using the constraints $z_{1\cdot} = z_{\cdot 2} = z_{2\cdot} = \cdots = z_{r-1\cdot} = 0$, it is shown that the only element of $\mathscr{B}_0(S)$ that can be contained in $\mathscr{R}([r-1];[r])$ is the zero contingency table. □

The loops in (10.5) are examples of df 1 loops of degree 2, 3, and 4 on some S in 4×5 integer arrays. Denote the positive part and the negative part of a df 1 loop $\boldsymbol{z}_r(\boldsymbol{i}_{[r]};\boldsymbol{j}_{[r]})$ as $\boldsymbol{z}_r^+(\boldsymbol{i}_{[r]};\boldsymbol{j}_{[r]})$ and $\boldsymbol{z}_r^-(\boldsymbol{i}_{[r]};\boldsymbol{j}_{[r]})$, respectively. Then

$$\boldsymbol{z}_r(\boldsymbol{i}_{[r]};\boldsymbol{j}_{[r]}) = \boldsymbol{z}_r^+(\boldsymbol{i}_{[r]};\boldsymbol{j}_{[r]}) - \boldsymbol{z}_r^-(\boldsymbol{i}_{[r]};\boldsymbol{j}_{[r]}). \qquad (10.7)$$

Let $\mathscr{F}_{\boldsymbol{t}}$ be a fiber such that $\boldsymbol{z}_r^+(\boldsymbol{i}_{[r]};\boldsymbol{j}_{[r]}), \boldsymbol{z}_r^-(\boldsymbol{i}_{[r]};\boldsymbol{j}_{[r]}) \in \mathscr{F}_{\boldsymbol{t}}$. Then $\mathscr{F}_{\boldsymbol{t}}$ is a two-element fiber; that is, every df 1 move is an indispensable move for the quasi-independence model (10.2).

Theorem 10.1. *The set of df* 1 *loops of degree* $2,\ldots,\min\{R,C\}$ *constitutes the unique minimal Markov basis for the quasi-independence model of* $R \times C$ *contingency tables with structural zeros.*

Proof. We have already seen that the set of loops forms a Markov basis. We have also seen that every df 1 loop is indispensable. Therefore it suffices to show that the set of the df 1 loops is itself a Markov basis.

Table 10.1 An example of a block triangular table: Initial and final ratings on disability of stroke patients

	Final state				
Initial state	A	B	C	D	E
E	11	23	12	15	8
D	9	10	4	1	[0]
C	6	4	4	[0]	[0]
B	4	5	[0]	[0]	[0]
A	5	[0]	[0]	[0]	[0]

Source: Bishop and Fienberg [25]

Suppose a Markov basis contains non-df-1 loops. Without loss of generality let $\boldsymbol{z}_r([r];[r])$ be a non-df-1 loop of the highest degree and $(1,a) \in S, 3 \le \exists a \le r$. Then this loop is decomposed as (10.6). The two loops overlap (i.e., have nonzero element in a common position) only at $(1,a) \in S$. The $(1,a)$ elements of these loops are -1 and $+1$, thus we can add or subtract these loops in an appropriate order to/from $\boldsymbol{x} \in \mathscr{F}_{\boldsymbol{t}}(S)$ without forcing negative entries on the way, instead of adding or subtracting $\boldsymbol{z}_r([r];[r])$ to/from $\boldsymbol{x}$. Therefore $\boldsymbol{z}_r([r];[r])$ can be removed from the Markov basis and the remaining set is still a Markov basis. □

10.1.3 Enumerating Elements of the Minimal Markov Basis

In this section we discuss how to list all the elements of the unique minimal Markov basis for some specific quasi-independence models. By considering the structure discussed in Lemma 10.2, we can obtain an explicit form of the minimal basis for some typical situations, which play important roles in applications.

Block Triangular Tables

For a row index i, define $\mathscr{J}(i) := \{j \mid (i,j) \in S\}$. An incomplete table is called of *block triangular* form if, for every pair i and i', either $\mathscr{J}(i) \subset \mathscr{J}(i')$ or $\mathscr{J}(i) \supset \mathscr{J}(i')$ holds ([26,66]). Table 10.1 shows an example of a block triangle contingency table from Bishop and Fienberg [25]. In this case, the unique minimal Markov basis is the set of basic moves on S.

Square Tables with Diagonal Elements Being Structural Zeros

There are many situations that contingency tables are square and all the diagonal elements are structural zeros. Table 10.2 is an example of such tables. It is obvious

Table 10.2 An example of a square table with diagonal elements being structural zeros

	Passive participant					
Active participant	R	S	T	U	V	W
R	[0]	1	5	8	9	0
S	29	[0]	14	46	4	0
T	0	0	[0]	0	0	0
U	2	3	1	[0]	28	2
V	0	0	0	0	[0]	1
W	9	25	4	6	13	[0]

Source: Ploog [121]

that the unique minimal Markov basis for such tables contains degree 3 loops which correspond to every triple of the structural zeros. For examples, degree 3 loops such as

$$\begin{array}{|cccccc|}\hline [0] & -1 & +1 & 0 & 0 & 0 \\ +1 & [0] & -1 & 0 & 0 & 0 \\ -1 & +1 & [0] & 0 & 0 & 0 \\ 0 & 0 & 0 & [0] & 0 & 0 \\ 0 & 0 & 0 & 0 & [0] & 0 \\ 0 & 0 & 0 & 0 & 0 & [0] \\ \hline \end{array} \quad \text{or} \quad \begin{array}{|cccccc|}\hline [0] & 0 & 0 & 0 & 0 & 0 \\ 0 & [0] & 0 & 0 & +1 & -1 \\ 0 & 0 & [0] & 0 & 0 & 0 \\ 0 & 0 & 0 & [0] & 0 & 0 \\ 0 & -1 & 0 & 0 & [0] & +1 \\ 0 & +1 & 0 & 0 & -1 & [0] \\ \hline \end{array}$$

are needed to construct a connected Markov chain. It is seen that for $I \times I$ contingency tables, there are $\binom{I}{2}\binom{I-2}{2}$ degree 2 moves and $\binom{I}{3}$ df 1 degree 3 loops in the unique minimal Markov basis.

General Incomplete Tables

In general, we can use the following recursive algorithm to list all the elements in the minimal basis.

Algorithm 10.1
Input: $\mathscr{I} = \{1,\ldots,R\}, \mathscr{J} = \{1,\ldots,C\}, S$
Output: elements of the unique minimal Markov basis
ListMoves($\mathscr{I}$; $\mathscr{J}$)
{
  Choose $i^* \in \mathscr{I}$ and $\mathscr{J}(i^*) = \{j \mid (i^*, j) \in S\}$;
  List df 1 moves that have ± 1 elements in $\mathscr{R}(i^*; \mathscr{J}(i^*))$;
  $\mathscr{I} \leftarrow \mathscr{I} \setminus \{i^*\}$;
  if $\mathscr{I} \neq \emptyset$
    ListMoves($\mathscr{I}$; $\mathscr{J}$);
  else exit;
}

Table 10.3 Classification of Purum marriages

	Sib of husband				
Sib of wife	Marrim	Makan	Parpa	Thao	Kheyang
Marrim	[0]	5	17	[0]	6
Makan	5	[0]	0	16	2
Parpa	[0]	2	[0]	10	11
Thao	10	[0]	[0]	[0]	9
Kheyang	6	20	8	0	1

Source: White [150]

Example 10.5. To illustrate the algorithm, we list the elements of the unique minimal Markov basis for the incomplete table of the form in Table 10.3.

According to the algorithm, we first choose $i^* = 1$ and hence $\mathscr{J}(i^*) = \{2,3,5\}$. We also denote $\widetilde{\mathscr{I}} = \mathscr{I} - \{i^*\} = \{2,3,4,5\}$, and $\widetilde{\mathscr{J}} = \mathscr{J} \setminus \mathscr{J}(i^*) = \{1,4\}$.

$$
\begin{array}{rr|cc|ccc|}
 & & \multicolumn{2}{c|}{\widetilde{\mathscr{J}}} & \multicolumn{3}{c}{\mathscr{J}(i^*)} \\
 & & 1 & 4 & 2 & 3 & 5 \\
\cline{3-7}
i^* & 1 & [0] & [0] & x_{12} & x_{13} & x_{15} \\
\hline
 & 2 & x_{21} & x_{24} & [0] & x_{23} & x_{25} \\
 & 3 & [0] & x_{34} & x_{32} & [0] & x_{35} \\
\widetilde{\mathscr{I}} & 4 & x_{41} & [0] & [0] & [0] & x_{45} \\
 & 5 & x_{51} & x_{54} & x_{52} & x_{53} & x_{55} \\
\hline
\end{array}
$$

Next step of the algorithm is to list all df 1 loops that have ± 1 elements in $\mathscr{R}(i^*; \mathscr{J}(i^*))$. To perform this step, we can make use of the fact that such a loop has exactly one $+1$ and one -1 both in $\mathscr{R}(i^*; \mathscr{J}(i^*))$ and $\mathscr{R}(\widetilde{\mathscr{I}}; \mathscr{J}(i^*))$. For example, if we select $\{2,3\}$ from $\mathscr{J}(i^*)$ and $\{2,3\}$ from $\widetilde{\mathscr{I}}$, we can ignore column 5 and row 5. We can also ignore column 1 because this column has only one cell in S when we ignore the rows 4 and 5. Then the table is reduced to the following.

$$
\begin{array}{rr|cc|cc|}
 & & \multicolumn{2}{c|}{\widetilde{\mathscr{J}}} & \multicolumn{2}{c}{\mathscr{J}(i^*)} \\
 & & 1 & 4 & 2 & 3 \\
\cline{3-6}
i^* & 1 & [0] & [0] & x_{12} & x_{13} \\
\hline
\widetilde{\mathscr{I}} & 2 & x_{21} & x_{24} & [0] & x_{23} \\
 & 3 & [0] & x_{34} & x_{32} & [0] \\
 & 4 & x_{41} & [0] & [0] & [0] \\
\hline
\end{array}
$$

This subtable contains $\mathbf{z}_3(1,2,3;2,3,4)$. Similarly we can list all loops that have exactly one $+1$ and one -1 both in $\mathscr{R}(i^*; \mathscr{J}(i^*))$ and $\mathscr{R}(\widetilde{\mathscr{I}}; \mathscr{J}(i^*))$ by listing all pairs of columns in $\mathscr{J}(i^*)$. In this case,

- If we select $\{2,3\}$ from $\mathscr{J}(i^*)$, $\mathbf{z}_3(1,2,3;2,3,4)$ and $\mathbf{z}_2(1,5;2,3)$ are listed.
- If we select $\{2,5\}$ from $\mathscr{J}(i^*)$, $\mathbf{z}_2(1,3;2,5)$ and $\mathbf{z}_2(1,5;2,5)$ are listed.
- If we select $\{3,5\}$ from $\mathscr{J}(i^*)$, $\mathbf{z}_2(1,2;3,5)$ and $\mathbf{z}_2(1,5;3,5)$ are listed.

Table 10.4 Effects of decision alternatives on the verdicts and social perceptions of simulated jurors

	Condition						
Alternative	1	2	3	4	5	6	7
First-degree	11	[0]	[0]	2	7	[0]	2
Second-degree	[0]	20	[0]	22	[0]	11	15
Manslaughter	[0]	[0]	22	[0]	16	13	5
Not guilty	13	4	2	0	1	0	2

Source: Vidmar (1972)

Table 10.5 Maximum likelihood estimate for Table 10.6

	Condition						
Alternative	1	2	3	4	5	6	7
First-degree	14.05	[0]	[0]	2.61	3.64	[0]	1.70
Second-degree	[0]	21.93	[0]	19.55	[0]	13.75	12.77
Manslaughter	[0]	[0]	20.95	[0]	17.78	8.95	8.32
Not guilty	9.95	2.07	3.05	1.84	2.58	1.30	1.21

Substitute $\mathscr{I} \leftarrow \widetilde{\mathscr{I}}$ and iterate a similar procedure until $\mathscr{I} = \emptyset$. Then we can see that basic moves and a degree 3 loop $\mathbf{z}_3(1,2,3;2,3,4)$ form the unique minimal Markov basis.

Example 10.6. As seen in Example 10.3, the unique minimal Markov basis for the separable table (10.3) is the union of the unique minimal Markov bases for two subtables (10.4). By using Algorithm 10.1, we easily see that it is $\{\mathbf{z}_2(1,3;1,2), \mathbf{z}_3(1,2,3;1,3,4), \mathbf{z}_3(1,2,3;2,3,4), \mathbf{z}_3(4,5,6;5,6,7)\}$.

10.1.4 Numerical Example of a Quasi-Independence Model

In this section we give an example of testing the hypothesis of quasi-independence for a given data set via the MCMC method. Table 10.4 shows a data collected by Vidmar [147] for discovering the possible effects on decision making of limiting the number of alternatives available to the members of a jury panel. This is a 4×7 contingency table that has 9 structural zero cells. The degree of freedom for testing quasi-independence is 9. The maximum likelihood estimate under the hypothesis of quasi-independence is calculated by an iterative method as displayed in Table 10.5. See Bishop et al. [26] for maximum likelihood estimation of incomplete tables.

As a test statistic, we use the (twice log) likelihood ratio statistic

$$G^2 = 2\sum_S x_{ij} \log \frac{x_{ij}}{\hat{m}_{ij}},$$

where $\hat{m}_{ij}$ is the MLE of the expectation parameter m_{ij}. The observed value of G^2 is 18.816 and the corresponding asymptotic p-value is 0.0268 from the asymptotic distribution χ^2_9.

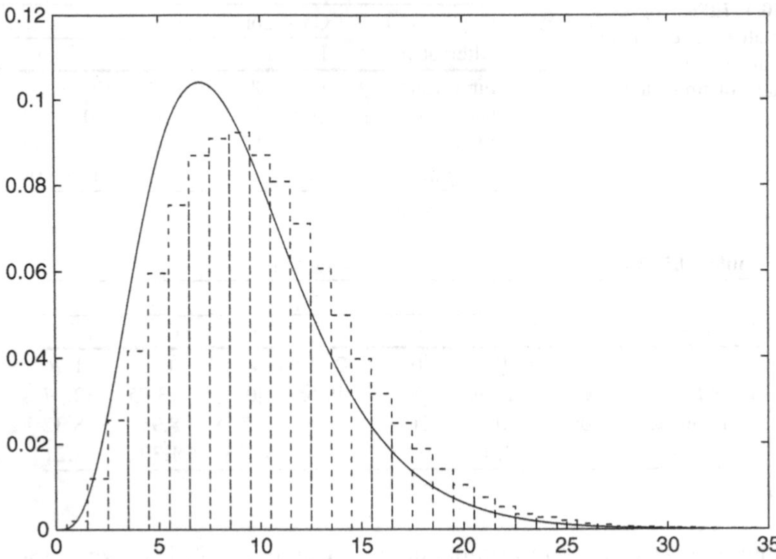

Fig. 10.1 Asymptotic and exact distributions for G^2 under the quasi-independence model

To perform the Markov chain Monte Carlo method, first we obtain the unique minimal Markov basis. From the considerations in the above sections, we easily see that a set of basic moves and a degree 3 loop $z_3(1,2,3;5,4,6)$ constitute the unique minimal Markov basis. The estimated exact p-value is 0.0444, with estimated standard deviation 0.00052. Figure 10.1 shows a histogram of the Monte Carlo sampling generated from the exact distribution of the likelihood ratio statistic under the quasi-independence hypothesis, along with the corresponding asymptotic distribution χ^2_9. We see that the asymptotic distribution understates the probability that the test statistic is greater than the observed value, and overemphasizes the significance.

10.2 Markov Bases for Subtable Sum Problem

10.2.1 Introduction of Subtable Sum Problem

Let $\boldsymbol{x} = \{x_{ij}\}$ be an $R \times C$ table and let $S \subset \mathscr{I}$ be a subset of cells of $\boldsymbol{x}$. Consider the following model for cell probabilities $\{p_{ij}\}$,

$$\log p_{ij} = \begin{cases} \mu + \alpha_i + \beta_j + \gamma, & (i,j) \in S, \\ \mu + \alpha_i + \beta_j, & \text{otherwise,} \end{cases} \tag{10.8}$$

where two-way interactions exist only on S.

The model (10.8) includes some practical models. When S is rectangular, that is, $S = \{(i,j) \mid 1 \le i \le r, 1 \le j \le c\}$ for $r \le R, c \le C$, (10.8) coincides with the block interaction model or two-way change point model ([87, 107]). For a square table such that frequencies along the diagonal cells are relatively larger (or smaller) compared to off-diagonal cells, the following model is often used

$$\log p_{ij} = \mu + \alpha_i + \beta_j + \gamma_i \delta_{ij}, \tag{10.9}$$

where δ_{ij} is Kronecker's delta. The model (10.9) is also called the quasi-independence model for a square table. We can consider the null hypothesis $\gamma_1 = \cdots = \gamma_R = \gamma$:

$$\log p_{ij} = \mu + \alpha_i + \beta_j + \gamma \delta_{ij}. \tag{10.10}$$

Then (10.10) belongs to (10.8) with $S = \{(i,j) \mid i = j\}$. We call the model (10.10) a *common diagonal effect model (CDEM)* and discuss it again in Sect. 10.2.3.

Let $x(S)$ denote the sum of cell counts in a subtable S,

$$x(S) = \sum_{(i,j)\in S} x_{ij}.$$

Then a sufficient statistic $\boldsymbol{t}$ for (10.8) is the set of row sums, column sums, and $x(S)$

$$\boldsymbol{t} = \{x_{1+}, \ldots, x_{R+}, x_{+1}, \ldots, x_{+C}, x(S)\}. \tag{10.11}$$

For $S = \emptyset$ or $S = \mathscr{I}$, we have $x(\emptyset) \equiv 0$ or $x(\mathscr{I}) = x_{++} = n$. In these cases $x(S)$ is redundant and the model reduces to the two-way complete independence model. Therefore in the following, we consider S which is a nonempty proper subset of $\mathscr{I}$. We note that $x(S^C) = x_{++} - x(S)$, where S^C is the complement of S. Therefore fixing $x(S)$ is equivalent to fixing $x(S^C)$.

In the following section we discuss Markov bases for the model (10.8). We call the problem the *subtable sum problem.* We note that if $x(S^C) = 0$ (which is equivalent to $x(S) = x_{++}$), the fiber $\mathscr{F}_{\boldsymbol{t}}$ is the one of the quasi-independence model with structural zeros in S^c. Hence the unique minimal Markov basis for a quasi-independence model is a subset of the Markov basis for subtable sum problems.

10.2.2 Markov Bases Consisting of Basic Moves

In this section we denote by $\mathscr{B}(S)$ the set of moves for (10.8); that is,

$$\mathscr{B}(S) := \{\boldsymbol{z} = \{z_{ij}\} \mid z_{i+} = 0,\ z_{+j} = 0,\ z(S) = 0\}.$$

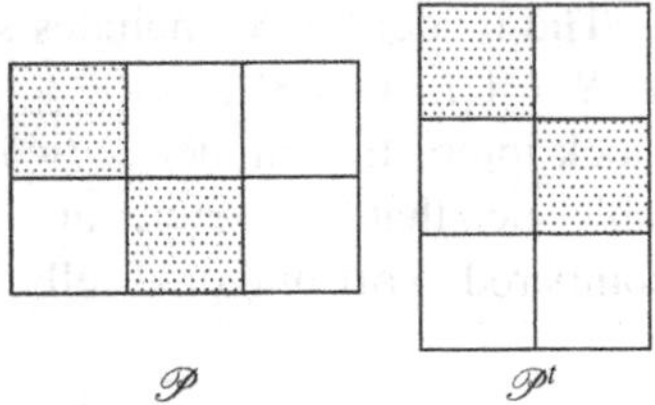

Fig. 10.2 The pattern $\mathscr{P}$ and $\mathscr{P}^t$

We note that $z(S) = z(S^C) = 0$. Therefore $\mathscr{B}(S)$ is equivalent to $\mathscr{B}(S^C)$. As shown in Chap. 8 the set $\mathscr{B}_0$ of all basic moves forms the unique minimal Markov basis for the two-way complete independence model. Define the set of basic moves for (10.8) by

$$\mathscr{B}_0(S) := \mathscr{B}_0 \cap \mathscr{B}(S).$$

$\mathscr{B}_0(S)$ does not always form a Markov basis for (10.8).

Figure 10.2 shows patterns of 2×3 and 3×2 tables. A shaded area represents a cell belonging to S. In the following, let a shaded area represent a cell belonging to S or a rectangular block of cells belonging to S. We call these two patterns in Fig. 10.2 the pattern $\mathscr{P}$ and $\mathscr{P}^t$, respectively. Then a necessary and sufficient condition on S that $\mathscr{B}_0(S)$ forms a Markov basis for (10.8) is given by the following theorem.

Theorem 10.2 (Hara et al. [79]). *$\mathscr{B}_0(S)$ is a Markov basis for* (10.8) *if and only if there exist no patterns of the form $\mathscr{P}$ or $\mathscr{P}^t$ in any 2×3 and 3×2 subtable of S or S^C after any interchange of rows and columns.*

Note that if $\mathscr{B}_0(S)$ is a Markov basis for (10.8), it is the unique minimal Markov basis, because basic moves in $\mathscr{B}_0(S)$ are all indispensable.

The proof of necessity is easy and is given in the following Proposition 10.1. The proof of sufficiency is given by the distance reducing method. However, the proof is complicated and is omitted here. For details, see Sect. 3 of Hara et al. [79]. Gröbner bases for the subtable sum problem are studied in Ohsugi and Hibi [112].

Proposition 10.1. *If there exists a pattern of $\mathscr{P}$ or $\mathscr{P}^t$ in any 2×3 and 3×2 subtable after any interchange of rows and columns, $\mathscr{B}_0(S)$ is not a Markov basis for (10.8).*

Proof. Assume that S has the pattern $\mathscr{P}$. Without loss of generality we can assume that $\mathscr{P}$ belongs to $\{(i,j) \mid i = 1,2,\ j = 1,2,3\}$. Consider a fiber such that

- $x_{1+} = x_{2+} = 2, x_{+1} = x_{+2} = 1, x_{+3} = 2,$
- $x_{i+} = 0$ and $x_{+j} = 0$ for all $(i,j) \notin \{(i,j) \mid i = 1,2,\ j = 1,2,3\},$
- $\sum_{(i,j)\in S} x_{ij} = 1.$

Then it is easy to check that this fiber contains only the following two elements

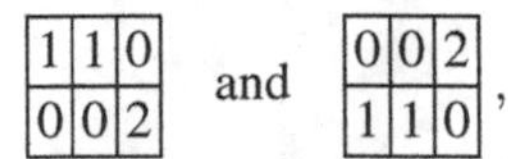

1	1	0
0	0	2

and

0	0	2
1	1	0

,

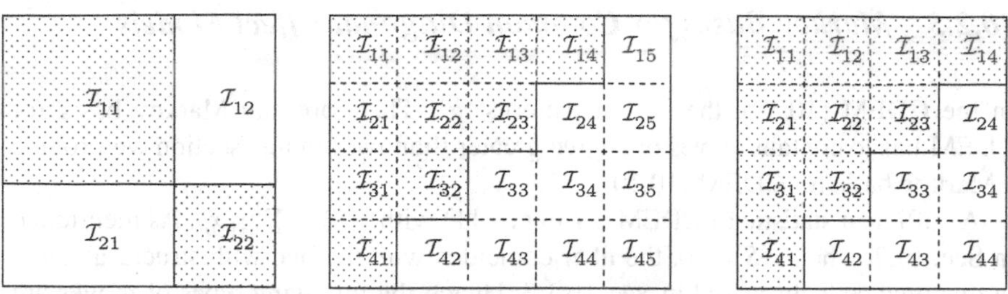

Fig. 10.3 2×2 block diagonal set and triangular sets

which implies that

$$z = \begin{array}{|c|c|c|}\hline 1 & 1 & -2 \\ \hline -1 & -1 & 2 \\ \hline \end{array} \tag{10.12}$$

is an indispensable move. Therefore if S has the pattern $\mathscr{P}$, there does not exist a Markov basis consisting of basic moves. When S has the pattern $\mathscr{P}^t$, a proof is similar. □

After an appropriate interchange of rows and columns, if S satisfies that

$$S = \{(i,j) \mid i \le r, j \le c\} \cup \{(i,j) \mid i > r, j > c\}$$

for some $r < R$ and $c < C$, we say that S is equivalent to a 2×2 block diagonal set. Figure 10.3(i) shows a 2×2 block diagonal set. A 2×2 block diagonal set is decomposed into four blocks consisting of one or more cells.

As in Sect. 10.1.3, we say that S is equivalent to a *triangular set* if, for every pair i and i', either $\mathscr{J}(i) \subset \mathscr{J}(i')$ or $\mathscr{J}(i) \supset \mathscr{J}(i')$. A triangular set is expressed as in Fig. 10.3(ii) after an appropriate interchange of rows and columns.

Proposition 10.2. *There exist no patterns of the form $\mathscr{P}$ or $\mathscr{P}^t$ in any 2×3 and 3×2 subtable of S after any interchange of rows and columns if and only if S is equivalent to a 2×2 block diagonal set or a triangular set.*

For the proof of Proposition 10.2, see Sect. 3.2 in Hara et al. [79]. From Proposition 10.2, Theorem 10.2 is rewritten as follows.

Corollary 10.1. *$\mathscr{B}_0(S)$ is a Markov basis for* (10.8) *if and only if S is equivalent to a 2×2 block diagonal set or a triangular set.*

The block interaction model ([87, 107]) is equivalent to a triangular set. Therefore, from Corollary 10.1, $\mathscr{B}_0(S)$ forms the unique minimal Markov basis for the block interaction model.

10.2.3 Markov Bases for Common Diagonal Effect Models

In the CDEM (10.10), there exist patterns $\mathscr{P}$. Therefore any Markov basis for CDEM has to contain moves of degree greater than two. In this section we provide a Markov basis for CDEM (10.10).

A sufficient statistic of CDEM is $\boldsymbol{t}$ in (10.11) with $x(S) = \sum_{i=1}^{R} x_{ii}$. As mentioned in Sec. 10.2, when $x(S) = 0$, the fiber coincides with the one with structural zeros in diagonal cells discussed in Sec. 10.1.3. Hence the following types of moves are required in a Markov basis.

- Type I : $z_2(i,i';j,j')$, where i,i',j,j' are all distinct.
- Type II : $z_3(i,i',i'';j,j',j'')$, where i,i',i'',j,j',j'' are all distinct.

In addition to these moves, we introduce the following four types of moves.

- Type III (dispensable moves of degree 3 for $\min(R,C) \geq 3$):

$$\begin{array}{cccc} & i & i' & i'' \\ i & +1 & 0 & -1 \\ i' & 0 & -1 & +1 \\ i'' & -1 & +1 & 0 \end{array}.$$

Note that given three distinct indices i,i',i'', there are three moves in the same fiber:

$$\begin{array}{ccc} +1 & 0 & -1 \\ 0 & -1 & +1 \\ -1 & +1 & 0 \end{array} \qquad \begin{array}{ccc} +1 & -1 & 0 \\ -1 & 0 & +1 \\ 0 & +1 & -1 \end{array} \qquad \begin{array}{ccc} 0 & -1 & +1 \\ -1 & +1 & 0 \\ +1 & 0 & -1 \end{array}.$$

Any two of these suffice for the connectivity of the fiber. Therefore we can choose any two moves in this fiber for minimality of Markov basis.

- Type IV (indispensable moves of degree 3 for $\max(R,C) \geq 4$):

$$\begin{array}{cccc} & i & i' & j \\ i & +1 & 0 & -1 \\ i' & 0 & -1 & +1 \\ j' & -1 & +1 & 0 \end{array},$$

where i, i', j, j' are all distinct. We note that Type IV is similar to Type III but unlike the moves in Type III, the moves of Type IV are indispensable.

- Type V (indispensable moves of degree 4 which are non-square-free):

$$\begin{array}{cccc} & j & j' & j'' \\ i & +1 & +1 & -2 \\ i' & -1 & -1 & +2 \end{array},$$

where $i = j$ and $i' = j'$; that is, two cells are on the diagonal. Note that we also include the transpose of this type as Type V moves.

- Type VI: (square-free indispensable moves of degree 4 for $\max(R,C) \geq 4$):

$$\begin{array}{ccccc} & j & j' & j'' & j'''' \\ i & +1 & +1 & -1 & -1 \\ i' & -1 & -1 & +1 & +1 \end{array},$$

where $i = j$ and $i' = j'$. Type VI includes the transpose of this type.

Theorem 10.3 (Hara et al. [80]). *The above moves of Types I–VI form a Markov basis for the CDEM with* $\min(R,C) \geq 3$ *and* $\max(R,C) \geq 4$.

Proof. Let $\mathbf{x}$ and $\mathbf{y}$ be two tables in the same fiber. If

$$x_{ii} = y_{ii}, \quad \forall i = 1, \ldots, \min(R,C),$$

then the problem reduces to the structural zero problem in Sect. 10.1.3. Therefore we only need to consider the difference

$$\mathbf{y} - \mathbf{x} = \mathbf{z} = \{z_{ij}\},$$

where there exists at least one i such that $z_{ii} \neq 0$. Note that in this case there are two indices $i \neq i'$ such that

$$z_{ii} > 0, \qquad z_{i'i'} < 0,$$

because the diagonal sum of $\mathbf{z}$ is zero. Without loss of generality we let $i = 1$, $i' = 2$. We prove the theorem by exhausting various sign patterns of the differences in other cells and confirming the distance reduction by the moves of Types I–VI. We distinguish two cases: $z_{12}z_{21} \geq 0$ and $z_{12}z_{21} < 0$.

Case 1 ($z_{12}z_{21} \geq 0$): In this case without loss of generality assume that $z_{12} \geq 0$, $z_{21} \geq 0$. Let $0+$ denote a cell with a nonnegative value of $\mathbf{z}$ and let $*$ denote a cell with an arbitrary value of $\mathbf{z}$. Then $\mathbf{z}$ looks like

$$\begin{array}{cccc} + & 0+ & * & \cdots \\ 0+ & - & * & \cdots \\ * & * & * & \cdots \\ \vdots & \vdots & \vdots & \ddots \end{array}.$$

Note that there has to be a negative cell on the first row and on the first column. Let $z_{1j} < 0$, $z_{j'1} < 0$. Then $\mathbf{z}$ looks like

$$\begin{array}{cccccc} & 1 & 2 & \cdots & j & \cdots \\ 1 & + & 0+ & \cdots & - & \cdots \\ 2 & 0+ & - & \cdots & * & \cdots \\ & \vdots & \vdots & \vdots & \vdots & \cdots \\ j' & - & * & \cdots & * & \cdots \\ & \vdots & \vdots & \vdots & \vdots & \ddots \end{array}.$$

If $j = j'$, we can apply a Type III move to reduce the 1-norm. If $j \neq j'$, we can apply a Type IV move to reduce the 1-norm. This takes care of the case $z_{12}z_{21} \geq 0$.

Case 2 ($z_{12}z_{21} < 0$): Without loss of generality assume that $z_{12} > 0$, $z_{21} < 0$. Then $\boldsymbol{z}$ looks like

$$\begin{matrix} + & + & * & \cdots \\ - & - & * & \cdots \\ * & * & * & \cdots \\ \vdots & \vdots & \vdots & \ddots \end{matrix}.$$

There has to be a negative cell on the first row and there has to be a positive cell on the second row. Without loss of generality we can let $z_{13} < 0$ and at least one of z_{23}, z_{24} is positive. Therefore $\boldsymbol{z}$ looks like

$$\begin{matrix} + & + & - & * & * & \cdots \\ - & - & * & + & * & \cdots \\ * & * & * & * & * & \cdots \\ \vdots & \vdots & \vdots & \vdots & \vdots & \ddots \end{matrix} \quad \text{or} \quad \begin{matrix} + & + & - & * & \cdots \\ - & - & + & * & \cdots \\ * & * & * & * & \cdots \\ \vdots & \vdots & \vdots & \vdots & \ddots \end{matrix}. \tag{10.13}$$

These two cases are not mutually exclusive. We look at $\boldsymbol{z}$ as the left pattern whenever possible. Namely, whenever we can find two different columns $j, j' \geq 3$, $j \neq j'$ such that $z_{1j}z_{2j'} < 0$, then we consider $\boldsymbol{z}$ to be of the left pattern. We first take care of the case where $\boldsymbol{z}$ does not look like the left pattern of (10.13); that is, there are no $j, j' \geq 3$, $j \neq j'$, such that $z_{1j}z_{2j'} < 0$.

Case 2–1 ($\boldsymbol{z}$ does not look like the left pattern of (10.13)): If there exists some $j \geq 4$ such that $z_{1j} < 0$, then in view of $z_{23} > 0$ we have $z_{1j}z_{23} < 0$ and $\boldsymbol{z}$ looks like the left pattern of (10.13). Therefore we can assume

$$z_{1j} \geq 0, \quad \forall j \geq 4.$$

Similarly

$$z_{2j} \leq 0, \quad \forall j \geq 4$$

and $\boldsymbol{z}$ looks like

$$\begin{matrix} + & + & - & 0+ & \cdots & 0+ \\ - & - & + & 0- & \cdots & 0- \\ * & * & * & * & \cdots & * \\ \vdots & \vdots & \vdots & \vdots & \vdots & \vdots \end{matrix}$$

Because the first row and the second row sum to zero, we have

$$z_{13} \leq -2, \quad z_{23} \geq 2.$$

However then we can apply Type V move to reduce the 1-norm.

Case 2–2 ($\boldsymbol{z}$ looks like the left pattern of (10.13)): Suppose that there exists some $i \geq 3$ such that $z_{i3} > 0$. If $z_{33} > 0$, then $\boldsymbol{z}$ looks like

$$\begin{array}{cccccc} + & + & - & * & * & \cdots \\ - & - & * & + & * & \cdots \\ * & * & + & * & * & \cdots \\ * & * & * & * & * & \cdots \\ \vdots & \vdots & \vdots & \vdots & \vdots & \ddots \end{array}.$$

Then we can apply a type III move involving

$$z_{12} > 0,\ z_{13} < 0,\ z_{22} < 0,\ z_{24} > 0,\ z_{33} > 0,\ z_{34} : \text{arbitrary}$$

and reduce the 1-norm. On the other hand if $z_{i3} > 0$ for $i \geq 4$, then $\boldsymbol{z}$ looks like

$$\begin{array}{cccccc} + & + & - & * & * & \cdots \\ - & - & * & + & * & \cdots \\ * & * & * & * & * & \cdots \\ * & * & + & * & * & \cdots \\ * & * & * & * & * & \cdots \\ \vdots & \vdots & \vdots & \vdots & \vdots & \ddots \end{array}.$$

Then we can apply a type IV move involving

$$z_{11} > 0,\ z_{13} < 0,\ z_{21} < 0,\ z_{24} > 0,\ z_{i3} > 0,\ z_{ii} : \text{arbitrary}$$

and reduce the 1-norm. Therefore we only need to consider $\boldsymbol{z}$ that looks like

$$\begin{array}{cccccc} + & + & - & * & * & \cdots \\ - & - & * & + & * & \cdots \\ * & * & 0- & * & * & \cdots \\ \vdots & \vdots & \vdots & \vdots & \vdots & \cdots \\ * & * & 0- & * & * & \cdots \end{array}.$$

Similar consideration for the fourth column of $\boldsymbol{z}$ forces

$$\begin{array}{cccccc} + & + & - & * & * & \cdots \\ - & - & * & + & * & \cdots \\ * & * & 0- & 0+ & * & \cdots \\ \vdots & \vdots & \vdots & \vdots & \vdots & \cdots \\ * & * & 0- & 0+ & * & \cdots \end{array}.$$

However, because the third and fourth column's sum to zero, we have $z_{23} > 0$ and $z_{14} < 0$ and $\boldsymbol{z}$ looks like

$$\begin{array}{cccccc} + & + & - & - & * & \cdots \\ - & - & + & + & * & \cdots \\ * & * & 0- & 0+ & * & \cdots \\ \vdots & \vdots & \vdots & \vdots & \vdots & \cdots \\ * & * & 0- & 0+ & * & \cdots \end{array}.$$

Then we apply a Type VI move to reduce the 1-norm.

Now we have exhausted all possible sign patterns of $\boldsymbol{z}$ and shown that the 1-norm can always be decreased by some move of Types I–VI. □

Because moves of Type I, II, IV, V, and VI are indispensable, we have the following corollary.

Corollary 10.2. *A minimal Markov basis for the diagonal sum problem with* $\min(R,C) \geq 3$ *and* $\max(R,C) \geq 4$ *consists of moves of Types I, II, IV, V, VI and two moves of Type III for each given triple* (i,i',i'').

10.2.4 Numerical Examples of Common Diagonal Effect Models

In this section we give examples of testing the null hypothesis of CDEM (10.10) against the alternative hypothesis of the quasi-independence model (10.9) for two real data sets via the MCMC method. Denote expected cell frequencies under the quasi-independence model and CDEM by

$$\hat{m}_{ij}^{QI} = n\hat{p}_{ij}^{QI}, \qquad \hat{m}_{ij}^{S} = n\hat{p}_{ij}^{S},$$

respectively. These expected cell frequencies can be computed via the iterative proportional fitting (IPF). IPF for the quasi-independence model is explained in Chap. 5 of [26]. IPF for the common diagonal effect model is given as follows. The superscript k denotes the step count.

1. Set $m_{ij}^{S,k} = m_{ij}^{S,k-1} x_{i+}/m_{i+}^{S,k-1}$ for all i, j and set $k = k+1$. Then go to Step 2.
2. Set $m_{ij}^{S,k} = m_{ij}^{S,k-1} x_{+j}/m_{+j}^{S,k-1}$ for all i, j and set $k = k+1$. Then go to Step 3.
3. Set $m_{ii}^{S,k} = m_{ii}^{S,k-1} x(S)/m^{S,k-1}(S)$ for all $i = 1,\ldots,\min(R,C)$ and $m_{ij}^{S,k} = m_{ij}^{S,k-1}(n - m^{S,k-1}(S))/(n - x(S))$ for all $i \neq j$, where $m^{S,k-1}(S)$ is the sum of fitted diagonal frequencies. Then set $k = k+1$ and go to Step 1.

After convergence we set

$$\hat{m}_{ij}^{S} = m_{ij}^{S,k} \quad \text{for all } i,\ j.$$

Table 10.6 Married couples in Arizona

	Never/occasionally	Fairly often	Very often	almost always
Never/occasionally	7	7	2	3
Fairly often	2	8	3	7
Very often	1	5	4	9
Almost always	2	8	9	14

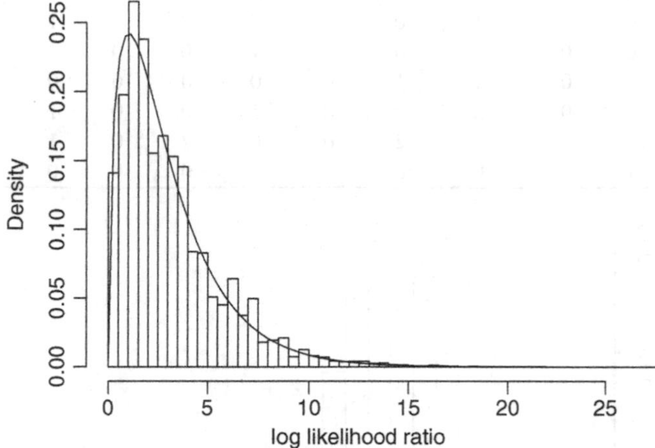

Fig. 10.4 A histogram of sampled tables via MCMC with a Markov basis computed for Table 10.6. The solid line shows the asymptotic distribution χ_3^2

We can initialize $m^{S,0}$ by

$$m_{ij}^{S,0} = n/(R \cdot C) \quad \text{for all } i, j.$$

As the discrepancy measure from the hypothesis of the common diagonal model, we calculate (twice) the log likelihood ratio statistic

$$G^2 = 2\sum_i \sum_j x_{ij} \log \frac{\hat{m}_{ij}^{QI}}{\hat{m}_{ij}^{S}}$$

for each sampled table $\boldsymbol{x} = \{x_{ij}\}$. In all experiments we sampled 10,000 tables after 8,000 burn-in steps.

The first example is Table 2.12 from [4]. Table 10.6 summarizes responses of 91 married couples in Arizona about how often sex is fun. Columns represent wives' responses and rows represent husbands' responses.

The value of G^2 for the observed table in Table 10.6 is 6.18159 and the corresponding asymptotic p-value is 0.1031 from the asymptotic distribution χ_3^2.

A histogram of sampled tables via MCMC with a Markov basis for Table 10.6 is shown in Fig.10.4. We estimated the p-value 0.12403 via MCMC with the Markov

Table 10.7 Relationship between birthday and death day

	Jan	Feb	March	April	May	June	July	Aug	Sep	Oct	Nov	Dec
Jan	1	0	0	0	1	2	0	0	1	0	1	0
Feb	1	0	0	1	0	0	0	0	0	1	0	2
March	1	0	0	0	2	1	0	0	0	0	0	1
April	3	0	2	0	0	0	1	0	1	3	1	1
May	2	1	1	1	1	1	1	1	1	1	1	0
June	2	0	0	0	1	0	0	0	0	0	0	0
July	2	0	2	1	0	0	0	0	1	1	1	2
Aug	0	0	0	3	0	0	1	0	0	1	0	2
Sep	0	0	0	1	1	0	0	0	0	0	1	0
Oct	1	1	0	2	0	0	1	0	0	1	1	0
Nov	0	1	1	1	2	0	0	2	0	1	1	0
Dec	0	1	1	0	0	0	1	0	0	0	0	0

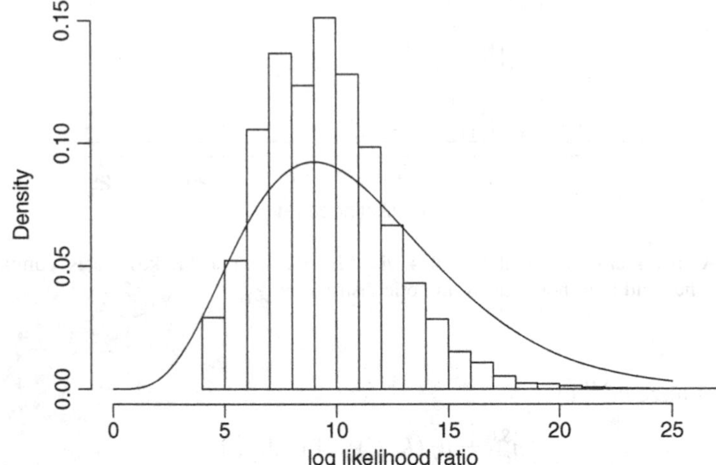

Fig. 10.5 A histogram of sampled tables via MCMC with a Markov basis computed for Table 10.7. The solid line shows the asymptotic distribution χ^2_{11}

basis defined in Theorem 10.3. Therefore the CDEM model is accepted at the significance level of 5%. We also see that χ^2_3 approximates these observed data well.

The second example is Table 1 from [50]. Table 10.7 shows data gathered to test the hypothesis of an association between birthday and death day. The table records the month of the birth and death for 82 descendants of Queen Victoria ([50]). A widely stated claim is that (birthday, death day) pairs are associated. Columns represent the month of the birthday and rows represent the month of the death day. As discussed in [50], Pearson's χ^2 statistic for the usual independence model is 115.6 with 121 degrees of freedom. Therefore the usual independence model is accepted for these data. However, when the CDEM is fitted, the Pearson's χ^2 becomes 111.5 with 120 degrees of freedom. Therefore the fit of CDEM is better than the usual independence model.

We now test the CDEM against the quasi-independence model. The value of G^2 for the observed table in Table 10.7 is 6.18839 and the corresponding asymptotic p-value is 0.860503 from the asymptotic distribution χ^2_{11}.

A histogram of sampled tables via MCMC with a Markov basis for Table 10.7 is shown in Fig. 10.5. We estimated the p-value 0.89454 via MCMC with the Markov basis in Theorem 10.3. There exists a large discrepancy between the asymptotic distribution and the distribution estimated by MCMC due to the sparsity of the table. This result indicates that the exact test via Markov basis technology is effective.

Chapter 11
Regular Factorial Designs with Discrete Response Variables

11.1 Conditional Tests for Designed Experiments with Discrete Observations

11.1.1 Conditional Tests for Log-Linear Models of Poisson Observations

First we investigate the case where the observations are counts of some events. In this case, it is natural to consider a Poisson model. To clarify the procedures of conditional tests, we take a close look at an example of *fractional factorial design* with count observations. Table 11.1 is a $\frac{1}{8}$ fraction of a full factorial design, that is, a 2^{7-3} regular fractional factorial design, defined from the *aliasing relation*

$$\text{ABDE} = \text{ACDF} = \text{BCDG} = \text{I},$$

and response data analyzed in [40] and [70]. There are $16 = 2^{7-3}$ runs in the whole experiment.

In Table 11.1, the observation x is the number of defects arising in a wave-soldering process in attaching components to an electronic circuit card. In Chap. 7 of [40], the following seven factors of a wave-soldering process are considered: (A) prebake condition, (B) flux density, (C) conveyer speed, (D) preheat condition, (E) cooling time, (F) ultrasonic solder agitator, and (G) solder temperature, each at two levels with three boards from each run being assessed for defects. Here we code the two levels as $\{0, 1\}$. The aim of this experiment is to decide which levels for each factor are desirable to reduce solder defects.

In this chapter, we only consider designs with a single observation for each run. This is natural for the settings of Poisson models, because the set of the total counts for each run is a sufficient statistic for the parameters. The same argument also holds for the settings of binomial models in Sect. 11.1.3. In our example, we focus on the totals for all runs in Table 11.1. We also ignore the second observation in run 11,

S. Aoki et al., *Markov Bases in Algebraic Statistics*, Springer Series in Statistics 199, DOI 10.1007/978-1-4614-3719-2_11,

Table 11.1 Design and number of defects x for the wave-solder experiment

	Factor							x		
Run	A	B	C	D	E	F	G	1	2	3
1	0	0	0	0	0	0	0	13	30	26
2	0	0	0	1	1	1	1	4	16	11
3	0	0	1	0	0	1	1	20	15	20
4	0	0	1	1	1	0	0	42	43	64
5	0	1	0	0	1	0	1	14	15	17
6	0	1	0	1	0	1	0	10	17	16
7	0	1	1	0	1	1	0	36	29	53
8	0	1	1	1	0	0	1	5	9	16
9	1	0	0	0	1	1	0	29	0	14
10	1	0	0	1	0	0	1	10	26	9
11	1	0	1	0	1	0	1	28	173	19
12	1	0	1	1	0	1	0	100	129	151
13	1	1	0	0	0	1	1	11	15	11
14	1	1	0	1	1	0	0	17	2	17
15	1	1	1	0	0	0	0	53	70	89
16	1	1	1	1	1	1	1	23	22	7

which is an obvious outlier as pointed out in [70]. We use the weighted total of run 11 as $(28+19)\times 3/2 = 70.5 \simeq 71$. Hence we have the η-dimensional column vector of frequencies as

$$\boldsymbol{x} = (69,31,55,149,46,43,118,30,43,45,71,380,37,36,212,52)'.$$

In this chapter, η, the dimension of the frequency vector $\boldsymbol{x}$ defined in Chap. 4, is the number of runs and the sample space is written as $\mathscr{I} = \{1,\ldots,\eta\}$. For this frequency vector $\boldsymbol{x}$, we can define the conditional sampling space in a similar way to the previous chapters. A slight difference is that a natural sampling model for this type of data is the Poisson model rather than the multinomial model described in (4.3) of Chap. 4.

We adopt the theory of *generalized linear models* [104] as follows. Assume that the observations $\boldsymbol{x} = \{x(\boldsymbol{i}),\ \boldsymbol{i} \in \mathscr{I}\} = \{x(1),\ldots,x(\eta)\}$ are mutually independently distributed as Poisson distributions with the mean parameters $\{\mu(\boldsymbol{i}),\ \boldsymbol{i} \in \mathscr{I}\}$. Because the canonical link function for a Poisson distribution is $\log(\cdot)$, we express the mean parameter $\mu(\boldsymbol{i})$ as

$$\log \mu(\boldsymbol{i}) = \sum_{j=0}^{\nu-1} \theta_j a_j(\boldsymbol{i}). \tag{11.1}$$

Note that we express the ν-dimensional parameter as $\{\theta_0,\ldots,\theta_{\nu-1}\}$ instead of $\{\theta_1,\ldots,\theta_\nu\}$ in this chapter, because it is more traditional in the theory of the (generalized) linear models to include the intercept term. The joint probability function of $\boldsymbol{x}$ is written as

$$p(\boldsymbol{x}) = \prod_{\boldsymbol{i}=1}^{\eta} \frac{\mu(\boldsymbol{i})^{x(\boldsymbol{i})}}{x(\boldsymbol{i})!} e^{-\mu(\boldsymbol{i})} = \left(\prod_{\boldsymbol{i}=1}^{\eta} \frac{e^{-\mu(\boldsymbol{i})}}{x(\boldsymbol{i})!} \right) \exp \left(\sum_{j=0}^{\nu-1} \theta_j \sum_{\boldsymbol{i}=1}^{\eta} a_j(\boldsymbol{i}) x(\boldsymbol{i}) \right).$$

Then we have a sufficient statistic for the parameter $\{\theta_0,\ldots,\theta_{\nu-1}\}$ as $\{t_0,\ldots,t_{\nu-1}\}$ where $t_j = \sum_{\boldsymbol{i}=1}^{\eta} a_j(\boldsymbol{i})x(\boldsymbol{i})$. We write this relation $\boldsymbol{t} = A\boldsymbol{x}$ as we have seen in (4.4) of Chap. 4.

The conditional distribution of $\boldsymbol{x}$ given $\boldsymbol{t}$, the hypergeometric distribution (4.7), is written as

$$p(\boldsymbol{x} \mid \boldsymbol{t}) = c \times \frac{1}{\prod_{\boldsymbol{i}=1}^{\eta} x(\boldsymbol{i})!}, \quad \boldsymbol{x} \in \mathscr{F}_{\boldsymbol{t}}, \quad c = \left[\sum_{\boldsymbol{x} \in \mathscr{F}_{\boldsymbol{t}}} \frac{1}{\prod_{\boldsymbol{i}=1}^{\eta} x(\boldsymbol{i})!} \right]^{-1}, \tag{11.2}$$

where $\mathscr{F}_{\boldsymbol{t}} = \{\boldsymbol{x} \geq 0 \mid A\boldsymbol{x} = \boldsymbol{t}\}$ is the fiber.

To define conditional tests, we specify the *null model* and the *alternative model* in terms of the parameter θ. Suppose the null model is ν-dimensional and expressed as (11.1). Then the null model is regarded as a subspace of some high-dimensional model if $\nu < \eta$. For example, the highest-dimensional model is the saturated model, which is written as

$$\log(\mu(\boldsymbol{i})) = \sum_{j=0}^{\eta-1} \theta_j a_j(\boldsymbol{i}).$$

If we consider various goodness-of-fit tests, the alternative model is the saturated model and the hypotheses are written as

$$H_0 : (\theta_\nu,\ldots,\theta_{\eta-1}) = (0,\ldots,0),$$
$$H_1 : (\theta_\nu,\ldots,\theta_{\eta-1}) \neq (0,\ldots,0).$$

On the other hand, if we consider the significance test of some additional individual effects, the alternative model is written in the form of

$$H_1 : (\theta_\nu,\ldots,\theta_{\nu+m-1}) \neq (0,\ldots,0),$$

where $\theta_\nu,\ldots,\theta_{\nu+m-1}$ express the additional effects to the null model with m degrees of freedom. In the two-level case, a single effect is expressed as a single parameter. On the other hand, for the three-level case, a single effect has more than one degree of freedom. We see how to specify models in the form of (11.1) in Sect. 11.1.2.

Depending on the hypotheses, we also specify the appropriate test statistic $T(\boldsymbol{x})$. For example, the likelihood ratio test statistic or Pearson's chi-square test statistic is frequently used. Once we specify the null model and the test statistic, our purpose is to calculate the p-value. Similarly to the context of the analysis of the contingency tables, the Markov chain Monte Carlo procedure is a valuable tool, especially when the traditional large-sample approximation is inadequate and the exact calculation of the p-value is infeasible.

11.1.2 Models and Aliasing Relations

Now we consider how to define models in terms of θ. In other words, we have to define a $\nu \times \eta$ configuration matrix A with the $(j,\boldsymbol{i})$ element $a_j(\boldsymbol{i})$ to define the sufficient statistic $\boldsymbol{t} = A\boldsymbol{x}$ and fiber $\mathscr{F}_{\boldsymbol{t}}$. In the literature of designed experiments, A', the transpose of A, is usually called a *design matrix*. It is also called a *covariate matrix* or a *model matrix*. Unlike the other literature of designed experiments, we call a matrix A (not A') a design matrix or a configuration matrix, which is consistent with the other chapters in this book. We illustrate how to define A corresponding to the main and interaction effects we want to consider in the cases of two-level and three-level regular fractional factorial designs see [16] for two-level case and [17] for three-level case for detail. See also the literature on designed experiments such as [151] for detail.

First we define a regular fractional factorial design. The theories of the regular fractional factorial designs with two or three levels are well developed and elegantly written in the literature dealing with theoretical aspects of the designed experiments. See [123], for example. In this section we first consider the two-level case and then consider the three-level case.

11.1.2.1 Two-Level Case

Suppose there are s controllable factors, $\mathrm{Y}_1,\ldots,\mathrm{Y}_s$, with two levels. Let $\mathscr{D}$ be a 2^s full factorial design with levels being 0 and 1 as in Table 11.1. Therefore

$$\mathscr{D} = \{(y_1,\ldots,y_s) \mid y_i \in \{0,1\},\ i = 1,\ldots,s\}.$$

In most of the literature considering designed experiments from an algebraic viewpoint, two levels are coded as $\{-1,1\}$ rather than $\{0,1\}$. There is no essential difference between them. In this section we use the coding $\{0,1\}$, because it generalizes to the three-level case somewhat more easily. In Chap. 15 we use the coding $\{-1,1\}$.

A fractional factorial design $\mathscr{F}$ is a subset of $\mathscr{D}$. $\mathscr{F}$ is a regular fractional factorial design if there are some $k > 0$ and $c_{ij} \in \{0,1\}, i = 1,\ldots,k,\ j = 0,\ldots,s$ satisfying

$$\mathscr{F} = \mathscr{D} \cap \left\{ (y_1,\ldots,y_s) \;\middle|\; \sum_{j=1}^{s} c_{ij} y_j \equiv c_{i0} \pmod 2,\ \ i = 1,\ldots,k \right\}.$$

The k relations

$$\sum_{j=1}^{s} c_{ij} y_j \equiv c_{i0} \pmod 2,\quad i = 1,\ldots,k$$

are called *defining relations* or *aliasing relations*. Without loss of generality, we assume that k relations are linearly independent over the finite field $\mathrm{GF}(2) = \{0,1\}$, where the addition is carried out modulo 2. For example, three relations

$$y_1 + y_2 + y_3 + y_4 \equiv 0,\ y_1 + y_2 + y_4 + y_5 \equiv 0,\ y_3 + y_5 \equiv 0$$

are linearly dependent in GF(2) inasmuch as

$$(y_1+y_2+y_3+y_4)+(y_1+y_2+y_4+y_5) \equiv 2y_1+2y_2+y_3+2y_4+y_5 \equiv y_3+y_5 \pmod 2.$$

Considering the change of levels $0 \leftrightarrow 1$ for each factor, we can also assume that $c_{i0}=0,\ i=1,\dots,k$ without loss of generality.

Denote the observation at level $(y_1,\dots,y_s)$ as $x_{y_1\cdots y_s}$. A simple way of modeling is to treat the elements of $\boldsymbol{\theta}$ as a *parameter contrast* of the main and the interaction effects. Note that the main effect of Y_1 is given by

$$\frac{1}{2^{s-1}}\left(\sum_{y_2,\dots,y_s} x_{0y_2\cdots y_s} - \sum_{y_2,\dots,y_s} x_{1y_2\cdots y_s}\right), \tag{11.3}$$

whereas the interaction effect of Y_1 and Y_2 is given by

$$\frac{1}{2^{s-2}}\left(\sum_{y_3,\dots,y_s} (x_{00y_3\cdots y_s}+x_{11y_3\cdots y_s}) - \sum_{y_3,\dots,y_s} (x_{01y_3\cdots y_s}+x_{10y_3\cdots y_s})\right).$$

We construct a design matrix A so that each element of $A\boldsymbol{x}$ corresponds to the sufficient statistic for the parameter contrast of the main and interaction effect as follows.

Definition 11.1. For models of regular fractional factorial design $\mathscr{F}$ with two levels, a design matrix $A=\{a_j(\boldsymbol{i})\}$ is an $\nu \times \eta$ matrix satisfying

- The first row of A is $(1,\dots,1)$.
- If the model includes the main effect of the factor Y_p, there is j such that the row j of A is

$$a_j(\boldsymbol{i}) = \begin{cases} 1 \text{ for } y_p = 0, \\ 0 \text{ for } y_p = 1. \end{cases}$$

- If the model includes the m-factor interaction effect $\mathrm{Y}_{p_1}\times\cdots\times\mathrm{Y}_{p_m}$, there is j such that the row j of A is

$$a_j(\boldsymbol{i}) = \begin{cases} 1 \text{ for } y_{p_1}+\cdots+y_{p_m} \equiv 0 \pmod 2, \\ 0 \text{ for } y_{p_1}+\cdots+y_{p_m} \equiv 1 \pmod 2. \end{cases}$$

Note that we define A as the simplest form in the above definition. For example, to reflect the main effect of Y_p, we only use the sum $\sum_{y_p=0} x_{y_1\cdots y_s}$ instead of (11.3). This simplification is valid because we consider the intercept term. The constant $\frac{1}{2^{s-1}}$ also can be ignored because we consider the same conditional sample space $\mathscr{F}_{\boldsymbol{t}}$. These simplifications allow us to regard the design matrix A as the configuration of the toric ideal that we consider in this book.

Example 11.1. We construct a design matrix A for the wave-soldering data given in Table 11.1. For the simple main effect model, A is constructed as

$$A = \begin{pmatrix} 1&1&1&1&1&1&1&1&1&1&1&1&1&1&1&1 \\ 1&1&1&1&1&1&1&1&0&0&0&0&0&0&0&0 \\ 1&1&1&1&0&0&0&0&1&1&1&1&0&0&0&0 \\ 1&1&0&0&1&1&0&0&1&1&0&0&1&1&0&0 \\ 1&0&1&0&1&0&1&0&1&0&1&0&1&0&1&0 \\ 1&0&1&0&0&1&0&1&0&1&0&1&1&0&1&0 \\ 1&0&0&1&1&0&0&1&0&1&1&0&0&1&1&0 \\ 1&0&0&1&0&1&1&0&1&0&0&1&0&1&1&0 \end{pmatrix}.$$

If we include the interaction effect of $\mathrm{Y}_1 \times \mathrm{Y}_2$ (A $\times$ B in Table 11.1), the row

$$(1,1,1,1,0,0,0,0,0,0,0,0,1,1,1,1)$$

is added to A. Similarly, if we include the three-factor interaction $\mathrm{Y}_1 \times \mathrm{Y}_2 \times \mathrm{Y}_3$, the row

$$(1,1,0,0,0,0,1,1,0,0,1,1,1,1,0,0)$$

is added to A. The design matrix for the saturated model has $\eta = 16 (= \nu)$ rows, which is the Hadamard matrix of order 16, when 0 is replaced by -1.

Note that we can only consider models consistent with the aliasing relations. For example, if the aliasing relation

$$y_1 + y_2 + y_3 + y_4 \equiv 0 \pmod 2$$

exists, the two two-factor interaction effects, $\mathrm{Y}_1 \times \mathrm{Y}_2$ and $\mathrm{Y}_3 \times \mathrm{Y}_4$ are not simultaneously identifiable. In this case, at most one of $\mathrm{Y}_1 \times \mathrm{Y}_2$ and $\mathrm{Y}_3 \times \mathrm{Y}_4$ can be included in the model. Mathematically, this corresponds to the singularity of the matrix AA' in GF(2). See [14] for detail.

11.1.2.2 Three-Level Case

Next we consider three-level designs. We code the three levels of s controllable factors $\mathrm{Y}_1, \ldots, \mathrm{Y}_s$ as $\{0,1,2\}$. Then the 3^s full factorial design is

$$\mathscr{D} = \{(y_1, \ldots, y_s) \mid y_i \in \{0,1,2\},\ i = 1, \ldots, s\},$$

and $\mathscr{F} \subset \mathscr{D}$ is a fractional factorial design. $\mathscr{F}$ is a regular fractional factorial design if there are some $k > 0$ and $c_{ij} \in \{0,1,2\}, i = 1, \ldots, k, j = 0, \ldots, s$ satisfying

$$\mathscr{F} = \mathscr{D} \cap \left\{ (y_1 \ldots, y_s) \;\middle|\; \sum_{j=1}^{s} c_{ij} y_j \equiv c_{i0} \pmod 3,\ i = 1, \ldots, k \right\}.$$

Similarly to the two-level case, we assume that the k relations

$$\sum_{j=1}^{s} c_{ij} y_j \equiv c_{i0} \pmod 3, \qquad i = 1, \ldots, k$$

are independent over GF(3). Without loss of generality, we also assume $c_{i0} = 0$ for $i = 1, \ldots, k$. In addition, we assume that *the coefficient for the first nonzero factor is* 1, that is,

$$c_{ij^*} = 1, \qquad j^* = \min\{j \mid c_{ij} \neq 0\}$$

for $i = 1, \ldots, k$ without loss of generality. This notational convention is presented in [151].

To define a design matrix A, we also consider the sufficient statistics for the parameter contrast of the main and the interaction effects. The difference from the two-level case is that there is more than one degree of freedom in the three-level case. For example, to consider the main effect of Y_1, we might be interested in pairwise comparison of the average responses of three sets, giving

$$\frac{1}{3^{s-1}} \left(\sum_{y_2, \cdots, y_s} x_{0y_2 \cdots y_s} - \sum_{y_2, \cdots, y_s} x_{1y_2 \cdots y_s} \right) \tag{11.4}$$

which compares responses to level 0 and level 1,

$$\frac{1}{3^{s-1}} \left(\sum_{y_2, \cdots, y_s} x_{0y_2 \cdots y_s} - \sum_{y_2, \cdots, y_s} x_{2y_2 \cdots y_s} \right) \tag{11.5}$$

which compares responses to level 0 and level 2, and

$$\frac{1}{3^{s-1}} \left(\sum_{y_2, \cdots, y_s} x_{1y_2 \cdots y_s} - \sum_{y_2, \cdots, y_s} x_{2y_2 \cdots y_s} \right)$$

which compares responses to level 1 and level 2. However, these three comparisons are not independent inasmuch as we can calculate the third comparison from the other two comparisons.

In this sense, the degree of freedom for the main effect is two. We express the main effect of each factor as two parameters, which correspond to the two comparisons (11.4) and (11.5). Similarly, there are 2^m degrees of freedom for the m-factor interaction. For example, two-factor interaction $Y_1 \times Y_2$ is decomposed into two components, namely, Y_1Y_2 and $Y_1Y_2^2$. Y_1Y_2 expresses the group satisfying

$$y_1 + y_2 \equiv 0, 1, 2 \pmod 3,$$

whereas $Y_1Y_2^2$ expresses the group satisfying

$$y_1 + 2y_2 \equiv 0, 1, 2 \pmod 3.$$

Each group has two degrees of freedom as we have seen. Similarly, three-factor interaction $\mathrm{Y}_1 \times \mathrm{Y}_2 \times \mathrm{Y}_3$ is decomposed into four components, $\mathrm{Y}_1\mathrm{Y}_2\mathrm{Y}_3$, $\mathrm{Y}_1\mathrm{Y}_2\mathrm{Y}_3^2$, $\mathrm{Y}_1\mathrm{Y}_2^2\mathrm{Y}_3$, $\mathrm{Y}_1\mathrm{Y}_2^2\mathrm{Y}_3^2$, and so on.

Now we give a definition.

Definition 11.2. For models of a regular fractional factorial design $\mathscr{F}$ with three levels, a design matrix $A = \{a_j(\boldsymbol{i})\}$ is a $\nu \times \eta$ matrix satisfying

- The first row of A is $(1,\ldots,1)$,
- If the model includes the main effect of the factor Y_p, there are j_1 and j_2 such that the row j_1 of A is
$$a_{j_1}(\boldsymbol{i}) = \begin{cases} 1 & \text{for } y_p = 0, \\ 0 & \text{for } y_p = 1,2 \end{cases}$$
and the row j_2 of A is
$$a_{j_2}(\boldsymbol{i}) = \begin{cases} 1 & \text{for } y_p = 1, \\ 0 & \text{for } y_p = 0,2. \end{cases}$$
- If the model includes the m-factor interaction effect $\mathrm{Y}_{p_1} \times \cdots \times \mathrm{Y}_{p_m}$, there are 2^m distinct js such that the row j of A is
$$a_j(\boldsymbol{i}) = \begin{cases} 1 & \text{for } y_{p_1} + c_{p_2}y_{p_2} + \cdots + c_{p_m}y_{p_m} \equiv c_0 \pmod 3, \\ 0 & \text{for } y_{p_1} + c_{p_2}y_{p_2} + \cdots + c_{p_m}y_{p_m} \equiv 1 - c_0, 2 \pmod 3, \end{cases}$$
for $c_0 = 0,1$, $c_{p_r} = 1,2$, $r = 2,\ldots,m$.

Similarly to the two-level cases, we can only consider a model that is consistent with the aliasing relations. Because this point is somewhat complicated in the three-level case, we illustrate it by an example.

Example 11.2. Table 11.2 shows a 3^{4-1} fractional factorial design defined by

$$y_1 + y_2 + y_3 + 2y_4 \equiv 0 \pmod 3. \tag{11.6}$$

In the traditional expression of designed experiments, this design is written as $\mathrm{Y}_4 = \mathrm{Y}_1\mathrm{Y}_2\mathrm{Y}_3$.

For this design, the model consistent with the aliasing relation is specified as follows. From the relation (11.6), for example, we see that the components expressed as $\mathrm{Y}_1\mathrm{Y}_2$, $\mathrm{Y}_3\mathrm{Y}_4^2$ and $\mathrm{Y}_1\mathrm{Y}_2\mathrm{Y}_3^2\mathrm{Y}_4$ are mutually confounded with each other (in other words, linearly dependent over GF(3)). In fact, by adding $y_1 + y_2$ to both side of (11.6), we have

$$y_1 + y_2 \equiv 2y_1 + 2y_2 + y_3 + 2y_4 \equiv y_1 + y_2 + 2y_3 + y_4 \pmod 3,$$

which means that the three groups defined by $y_1 + y_2 \equiv 0,1,2 \pmod 3$ are identical to the three groups defined by $y_1 + y_2 + 2y_3 + y_4 \equiv 0,1,2 \pmod 3$. Similarly, by adding $2(y_1 + y_2)$ to both side of (11.6), we have

$$2y_1 + 2y_2 \equiv 3y_1 + 3y_2 + y_3 + 2y_4 \equiv y_3 + 2y_4 \pmod 3,$$

Table 11.2 Design and observations for a 3^{4-1} fractional factorial design

Run	Factor Y_1	Y_2	Y_3	Y_4	x
1	0	0	0	0	x_1
2	0	0	1	1	x_2
3	0	0	2	2	x_3
4	0	1	0	1	x_4
5	0	1	1	2	x_5
6	0	1	2	0	x_6
7	0	2	0	2	x_7
8	0	2	1	0	x_8
9	0	2	2	1	x_9
10	1	0	0	1	x_{10}
11	1	0	1	2	x_{11}
12	1	0	2	0	x_{12}
13	1	1	0	2	x_{13}
14	1	1	1	0	x_{14}
15	1	1	2	1	x_{15}
16	1	2	0	0	x_{16}
17	1	2	1	1	x_{17}
18	1	2	2	2	x_{18}
19	2	0	0	2	x_{19}
20	2	0	1	0	x_{20}
21	2	0	2	1	x_{21}
22	2	1	0	0	x_{22}
23	2	1	1	1	x_{23}
24	2	1	2	2	x_{24}
25	2	2	0	1	x_{25}
26	2	2	1	2	x_{26}
27	2	2	2	0	x_{27}

which means that the three groups defined by $2y_1+2y_2 \equiv 0,1,2 \pmod 3$, or equivalently by $y_1+y_2 \equiv 0,1,2 \pmod 3$, are also identical to the three groups defined by $y_3+2y_4 \equiv 0,1,2 \pmod 3$. Following the usual notational convention, we write this relation as

$$\mathrm{Y}_1\mathrm{Y}_2 = \mathrm{Y}_3\mathrm{Y}_4^2 = \mathrm{Y}_1\mathrm{Y}_2\mathrm{Y}_3^2\mathrm{Y}_4.$$

By the similar modulus 3 calculus, we can derive all the aliasing relations as follows.

$$\begin{array}{ll}
\mathrm{Y}_1 = \mathrm{Y}_2\mathrm{Y}_3\mathrm{Y}_4^2 = \mathrm{Y}_1\mathrm{Y}_2^2\mathrm{Y}_3^2\mathrm{Y}_4 & \mathrm{Y}_2 = \mathrm{Y}_1\mathrm{Y}_3\mathrm{Y}_4^2 = \mathrm{Y}_1\mathrm{Y}_2^2\mathrm{Y}_3\mathrm{Y}_4^2 \\
\mathrm{Y}_3 = \mathrm{Y}_1\mathrm{Y}_2\mathrm{Y}_4^2 = \mathrm{Y}_1\mathrm{Y}_2\mathrm{Y}_3^2\mathrm{Y}_4^2 & \mathrm{Y}_4 = \mathrm{Y}_1\mathrm{Y}_2\mathrm{Y}_3 = \mathrm{Y}_1\mathrm{Y}_2\mathrm{Y}_3\mathrm{Y}_4 \\
\mathrm{Y}_1\mathrm{Y}_2 = \mathrm{Y}_3\mathrm{Y}_4^2 = \mathrm{Y}_1\mathrm{Y}_2\mathrm{Y}_3^2\mathrm{Y}_4 & \mathrm{Y}_1\mathrm{Y}_2^2 = \mathrm{Y}_1\mathrm{Y}_3^2\mathrm{Y}_4 = \mathrm{Y}_2\mathrm{Y}_3^2\mathrm{Y}_4 \\
\mathrm{Y}_1\mathrm{Y}_3 = \mathrm{Y}_2\mathrm{Y}_4^2 = \mathrm{Y}_1\mathrm{Y}_2^2\mathrm{Y}_3\mathrm{Y}_4 & \mathrm{Y}_1\mathrm{Y}_3^2 = \mathrm{Y}_1\mathrm{Y}_2^2\mathrm{Y}_4 = \mathrm{Y}_2\mathrm{Y}_3^2\mathrm{Y}_4^2. \\
\mathrm{Y}_1\mathrm{Y}_4 = \mathrm{Y}_1\mathrm{Y}_2^2\mathrm{Y}_3^2 = \mathrm{Y}_2\mathrm{Y}_3\mathrm{Y}_4 & \mathrm{Y}_1\mathrm{Y}_4^2 = \mathrm{Y}_2\mathrm{Y}_3 = \mathrm{Y}_1\mathrm{Y}_2^2\mathrm{Y}_3^2\mathrm{Y}_4^2 \\
\mathrm{Y}_2\mathrm{Y}_3^2 = \mathrm{Y}_1\mathrm{Y}_2^2\mathrm{Y}_4^2 = \mathrm{Y}_1\mathrm{Y}_3^2\mathrm{Y}_4^2 & \mathrm{Y}_2\mathrm{Y}_4 = \mathrm{Y}_1\mathrm{Y}_2^2\mathrm{Y}_3 = \mathrm{Y}_1\mathrm{Y}_3\mathrm{Y}_4 \\
\mathrm{Y}_3\mathrm{Y}_4 = \mathrm{Y}_1\mathrm{Y}_2\mathrm{Y}_3^2 = \mathrm{Y}_1\mathrm{Y}_2\mathrm{Y}_4 &
\end{array} \quad (11.7)$$

From the above relations, we can clarify models for which all the effects are simultaneously estimable for the design (11.6). For example, the model of the main effects for the factors $\mathrm{Y}_1, \mathrm{Y}_2, \mathrm{Y}_3, \mathrm{Y}_4$ and the two-factor interaction effects $\mathrm{Y}_1 \times \mathrm{Y}_2$ is estimable, because the two components $\mathrm{Y}_1\mathrm{Y}_2$, $\mathrm{Y}_1\mathrm{Y}_2^2$ of $\mathrm{Y}_1 \times \mathrm{Y}_2$ are not confounded with any main effect. Among the models of the main effects and two two-factor interaction effects, the model with $\mathrm{Y}_1 \times \mathrm{Y}_2$ and $\mathrm{Y}_1 \times \mathrm{Y}_3$ is estimable, whereas the model with $\mathrm{Y}_1 \times \mathrm{Y}_2$ and $\mathrm{Y}_3 \times \mathrm{Y}_4$ is not estimable because the components $\mathrm{Y}_1\mathrm{Y}_2$ and $\mathrm{Y}_3\mathrm{Y}_4^2$ are confounded. In [151], main effects or components of two-factor interaction effects are called *clear* if they are not confounded with any other main effects or components of two-factor interaction effects. Moreover, a two-factor interaction effect, say $\mathrm{Y}_1 \times \mathrm{Y}_2$ is called *clear* if both of its components, $\mathrm{Y}_1\mathrm{Y}_2$ and $\mathrm{Y}_1\mathrm{Y}_2^2$, are clear. Therefore (11.7) implies that each of the main effect and the components, $\mathrm{Y}_1\mathrm{Y}_2^2, \mathrm{Y}_1\mathrm{Y}_3^2, \mathrm{Y}_1\mathrm{Y}_4, \mathrm{Y}_2\mathrm{Y}_3^2, \mathrm{Y}_2\mathrm{Y}_4, \mathrm{Y}_3\mathrm{Y}_4$ are clear, and there is no clear two-factor interaction effect.

We give a design matrix for Table 11.2. The design matrix for the main effect model is given as

$$A = \begin{pmatrix} 1\,1\,1\,1\,1\,1\,1\,1\,1\,1\,1\,1\,1\,1\,1\,1\,1\,1\,1\,1\,1\,1\,1\,1\,1\,1\,1 \\ 1\,1\,1\,1\,1\,1\,1\,1\,1\,0\,0\,0\,0\,0\,0\,0\,0\,0\,0\,0\,0\,0\,0\,0\,0\,0\,0 \\ 0\,0\,0\,0\,0\,0\,0\,0\,0\,1\,1\,1\,1\,1\,1\,1\,1\,1\,0\,0\,0\,0\,0\,0\,0\,0\,0 \\ 1\,1\,1\,0\,0\,0\,0\,0\,0\,1\,1\,1\,0\,0\,0\,0\,0\,0\,1\,1\,1\,0\,0\,0\,0\,0\,0 \\ 0\,0\,0\,1\,1\,1\,0\,0\,0\,0\,0\,0\,1\,1\,1\,0\,0\,0\,0\,0\,0\,1\,1\,1\,0\,0\,0 \\ 1\,0\,0\,1\,0\,0\,1\,0\,0\,1\,0\,0\,1\,0\,0\,1\,0\,0\,1\,0\,0\,1\,0\,0\,1\,0\,0 \\ 0\,1\,0\,0\,1\,0\,0\,1\,0\,0\,1\,0\,0\,1\,0\,0\,1\,0\,0\,1\,0\,0\,1\,0\,0\,1\,0 \\ 1\,0\,0\,0\,0\,1\,0\,1\,0\,0\,0\,1\,0\,1\,0\,1\,0\,0\,0\,1\,0\,1\,0\,0\,0\,0\,1 \\ 0\,1\,0\,1\,0\,0\,0\,0\,1\,1\,0\,0\,0\,0\,1\,0\,1\,0\,0\,0\,1\,0\,1\,0\,1\,0\,0 \end{pmatrix}.$$

If we include the two-factor interaction $\mathrm{Y}_1 \times \mathrm{Y}_2$, the four rows

$$\begin{pmatrix} 1\,1\,1\,0\,0\,0\,0\,0\,0\,0\,0\,0\,0\,0\,0\,1\,1\,1\,0\,0\,0\,1\,1\,1\,0\,0\,0 \\ 0\,0\,0\,1\,1\,1\,0\,0\,0\,1\,1\,1\,0\,0\,0\,0\,0\,0\,0\,0\,0\,0\,0\,0\,1\,1\,1 \\ 1\,1\,1\,0\,0\,0\,0\,0\,0\,0\,0\,0\,1\,1\,1\,0\,0\,0\,0\,0\,0\,0\,0\,0\,1\,1\,1 \\ 0\,0\,0\,0\,0\,0\,1\,1\,1\,1\,1\,1\,0\,0\,0\,0\,0\,0\,0\,0\,0\,1\,1\,1\,0\,0\,0 \end{pmatrix}$$

are added to A. If we want to include another two-factor interaction, $\mathrm{Y}_3 \times \mathrm{Y}_4$ cannot be estimated because $\mathrm{Y}_1\mathrm{Y}_2$ and $\mathrm{Y}_3\mathrm{Y}_4^2$ are confounded.

On the other hand, the models with two two-factor interactions $\mathrm{Y}_1 \times \mathrm{Y}_2$ and $\mathrm{Y}_1 \times \mathrm{Y}_3$ are estimable. In this case, the four rows

$$\begin{pmatrix} 1\,0\,0\,1\,0\,0\,1\,0\,0\,0\,0\,1\,0\,0\,1\,0\,0\,1\,0\,1\,0\,0\,1\,0\,0\,1\,0 \\ 0\,1\,0\,0\,1\,0\,0\,1\,0\,1\,0\,0\,1\,0\,0\,1\,0\,0\,0\,0\,1\,0\,0\,1\,0\,0\,1 \\ 1\,0\,0\,1\,0\,0\,1\,0\,0\,0\,1\,0\,0\,1\,0\,0\,1\,0\,0\,0\,1\,0\,0\,1\,0\,0\,1 \\ 0\,0\,1\,0\,0\,1\,0\,0\,1\,1\,0\,0\,1\,0\,0\,1\,0\,0\,0\,1\,0\,0\,1\,0\,0\,1\,0 \end{pmatrix}$$

are also added to A. In addition, we can include the three-factor interaction $Y_1 \times Y_2 \times Y_3$ inasmuch as none of four components, $Y_1Y_2Y_3, Y_1Y_2Y_3^2, Y_1Y_2^2Y_3, Y_1Y_2^2Y_3^2$, is confounded with the four main effects and the components of the two-factor interaction effects. In this case, the eight rows

$$\begin{pmatrix}
1\,0\,0\,0\,0\,1\,0\,1\,0\,0\,0\,1\,0\,1\,0\,1\,0\,0\,0\,1\,0\,1\,0\,0\,0\,0\,1 \\
0\,1\,0\,1\,0\,0\,0\,0\,1\,1\,0\,0\,0\,0\,1\,0\,1\,0\,0\,0\,1\,0\,1\,0\,1\,0\,0 \\
1\,0\,0\,0\,1\,0\,0\,0\,1\,0\,1\,0\,0\,0\,1\,1\,0\,0\,0\,0\,1\,1\,0\,0\,0\,1\,0 \\
0\,0\,1\,1\,0\,0\,0\,1\,0\,1\,0\,0\,0\,1\,0\,0\,0\,1\,0\,1\,0\,0\,0\,1\,1\,0\,0 \\
1\,0\,0\,0\,1\,0\,0\,0\,1\,0\,0\,1\,1\,0\,0\,0\,1\,0\,0\,1\,0\,0\,0\,1\,1\,0\,0 \\
0\,1\,0\,0\,0\,1\,1\,0\,0\,1\,0\,0\,0\,1\,0\,0\,0\,1\,0\,0\,1\,1\,0\,0\,0\,1\,0 \\
1\,0\,0\,0\,0\,1\,0\,1\,0\,0\,1\,0\,1\,0\,0\,0\,0\,1\,0\,0\,1\,0\,1\,0\,1\,0\,0 \\
0\,0\,1\,0\,1\,0\,1\,0\,0\,1\,0\,0\,0\,0\,1\,0\,1\,0\,0\,1\,0\,1\,0\,0\,0\,0\,1
\end{pmatrix}$$

are further added to A.

11.1.3 Conditional Tests for Logistic Models of Binomial Observations

Next we consider the case that the observation for each run is a ratio of counts. The arguments, especially the relations between the identifiable models and aliasing relations, are almost the same as the Poisson case. Therefore we give a brief consideration on the sufficient statistics and the design matrix here. Table 11.3 is a $1/2$ fraction of a full factorial design (that is, a 2^{4-1} fractional factorial design) defined from the relation

$$\mathrm{ACD} = \mathrm{I} \tag{11.8}$$

and response data given by [103] and reanalyzed in [70]. In Table 11.3, the observation x is the number of good parts out of 1,000 during the stamping process in manufacturing windshield modeling. The purpose of Martin et al. [103] is to decide the levels for four factors, (A) poly-film thickness, (B) oil mixture, (C) gloves, and (D) metal blanks, which most improve the slugging condition.

As for a statistical model for this type of data, it is natural to suppose that the distribution of the observation $x(\boldsymbol{i})$ is the mutually independent binomial distribution $\mathrm{Bin}(\mu(\boldsymbol{i}), n_{\boldsymbol{i}})$, $\boldsymbol{i} = 1, \ldots, \eta$, where $n_{\boldsymbol{i}} = 1{,}000$, $i = 1, \ldots, \eta (= 8)$ for this example. Following the theory of generalized linear models, we consider the logit link, which is the canonical link for the binomial distribution. It expresses the relation between the mean parameter $\mu(\boldsymbol{i})$ and the systematic part as

$$\mathrm{logit}(\mu(\boldsymbol{i})) = \log \frac{\mu(\boldsymbol{i})}{1 - \mu(\boldsymbol{i})} = \sum_{j=0}^{\nu-1} \theta_j a_j(\boldsymbol{i}).$$

Table 11.3 Design and number of good parts x out of 1,000 for the windshield molding slugging experiment

	Factor				
Run	A	B	C	D	x
1	0	0	0	0	338
2	0	0	1	1	826
3	0	1	0	0	350
4	0	1	1	1	647
5	1	0	0	1	917
6	1	0	1	0	977
7	1	1	0	1	953
8	1	1	1	0	972

The joint probability function in this case is written as

$$\begin{aligned}
&\prod_{i=1}^{\eta}\binom{n_i}{x(\boldsymbol{i})}\mu(\boldsymbol{i})^{x(\boldsymbol{i})}(1-\mu(\boldsymbol{i}))^{n_i-x(\boldsymbol{i})}\\
&=\prod_{i=1}^{\eta}\binom{n_i}{x(\boldsymbol{i})}(1-\mu(\boldsymbol{i}))^{n_i}\left(\frac{\mu(\boldsymbol{i})}{1-\mu(\boldsymbol{i})}\right)^{x(\boldsymbol{i})}\\
&=\prod_{i=1}^{\eta}\binom{n_i}{x(\boldsymbol{i})}(1-\mu(\boldsymbol{i}))^{n_i}\exp\left(\sum_{j=0}^{\nu-1}\theta_j\sum_{i=1}^{\eta}a_j(\boldsymbol{i})x(\boldsymbol{i})\right),
\end{aligned}$$

which implies that a sufficient statistic for the parameter θ is $t_j=\sum_{i=1}^{\eta}a_j(\boldsymbol{i})x(\boldsymbol{i})$ and $n_1,\ldots,n_\eta$. Consequently, the exact conditional tests are based on the conditional distribution,

$$p(\boldsymbol{x}\mid \boldsymbol{t},n_1,\ldots,n_\eta)=c\times\frac{1}{\prod_{i=1}^{\eta}x(\boldsymbol{i})!(n_i-x(\boldsymbol{i}))!}, \tag{11.9}$$

where c is the normalizing constant determined from $\boldsymbol{t}$ and $n_1,\ldots,n_\eta$ written as

$$c=\left[\sum_{\boldsymbol{x}\in\mathscr{F}_{\boldsymbol{t}}}\frac{1}{\prod_{i=1}^{\eta}x(\boldsymbol{i})!(n_i-x(\boldsymbol{i}))!}\right]^{-1}$$

and

$$\mathscr{F}_{\boldsymbol{t}}=\{\boldsymbol{x}\mid A\boldsymbol{x}=\boldsymbol{t},\ x(\boldsymbol{i})\in\{0,\ldots,n_i\},\ \boldsymbol{i}=1,\ldots,\eta\}$$

is the fiber.

For notational convenience, we extend $\boldsymbol{x}$ to

$$\tilde{\boldsymbol{x}}=(x(1),\ldots,x(\eta),n_1-x(1),\ldots,n_\eta-x(\eta))'$$

for the binomial model. Corresponding to this $\tilde{\boldsymbol{x}}$, we also extend the $\nu\times\eta$ matrix A to

$$\widetilde{A}=\begin{pmatrix}A & 0\\ E_\eta & E_\eta\end{pmatrix}, \tag{11.10}$$

where 0 is the $\nu \times \eta$ zero matrix and E_η is the identity matrix of the order η. $\widetilde{A}$ is the Lawrence lifting of the configuration A. See (4.24) in Sect. 4.6. Using $\tilde{\boldsymbol{x}}$ and $\widetilde{A}$, the condition that $A\boldsymbol{x}$ and $n_1, \ldots, n_\eta$ are fixed is simply written that $\widetilde{A}\tilde{\boldsymbol{x}}$ is fixed.

Once the configuration is given as (11.10), the procedure for conducting the exact test by the Markov chain Monte Carlo method is the same as in the Poisson case.

11.1.4 Example: Wave-Soldering Data

We give an example of calculating the p-value for the wave-soldering data in Table 11.1. First we have to define a null model in which we are interested. Following [70], we focus on the model of seven main effects and two two-factor interactions, A×C and B×D. Note that the parameters for this model are simultaneously identifiable. The dimension of the parameter of this null model is $\nu = 10$, therefore the residual has $\eta - \nu = 16 - 10 = 6$ degrees of freedom.

We now consider goodness-of-fit tests. Traditional χ^2 tests evaluate the upper probability for some discrepancy measures such as the deviance, the likelihood ratio, or Pearson's chi-square, based on the asymptotic distribution, $\chi^2_{\eta-\nu}$. Here we use the (twice log) likelihood ratio statistic

$$G^2(\boldsymbol{x}) = 2 \sum_{\boldsymbol{i}=1}^{\eta} x(\boldsymbol{i}) \log \frac{x(\boldsymbol{i})}{\widehat{\mu(\boldsymbol{i})}},$$

where $\widehat{\mu(\boldsymbol{i})}$ is the maximum likelihood estimate for $\mu(\boldsymbol{i})$ under the null model (that is, fitted value), given by

$$\begin{aligned}\widehat{\mu} = (&64.53, 47.25, 53.15, 151.08, 30.43, 46.79, 115.24, 32.53,\\ &49.42, 46.13, 70.90, 360.54, 35.19, 30.26, 232.14, 51.42)'\end{aligned}$$

for our example. For the observed data $\boldsymbol{x}^o$, $G^2(\boldsymbol{x}^o)$ is calculated as $G^2(\boldsymbol{x}^o) = 19.096$ and the corresponding asymptotic p-value is 0.0040 from the asymptotic distribution χ^2_6. This result tells us that the null hypothesis is highly significant and is rejected.

Next we calculate the same p-value by the Markov chain Monte Carlo method. We use a Markov basis as a minimal Markov basis obtained by 4ti2 [1]. After 100,000 burn-in steps, we construct 1,000,000 Monte Carlo samples. In contrast to the asymptotic p-value 0.0040, the estimated p-value is 0.032, with estimated standard deviation 0.0045, where we use a batching method to obtain an estimate of variance; see [82] and [128]. Figure 11.1 shows a histogram of the Monte Carlo sampling generated from the conditional distribution of the likelihood ratio statistic under the null hypothesis, along with the corresponding asymptotic distribution χ^2_6.

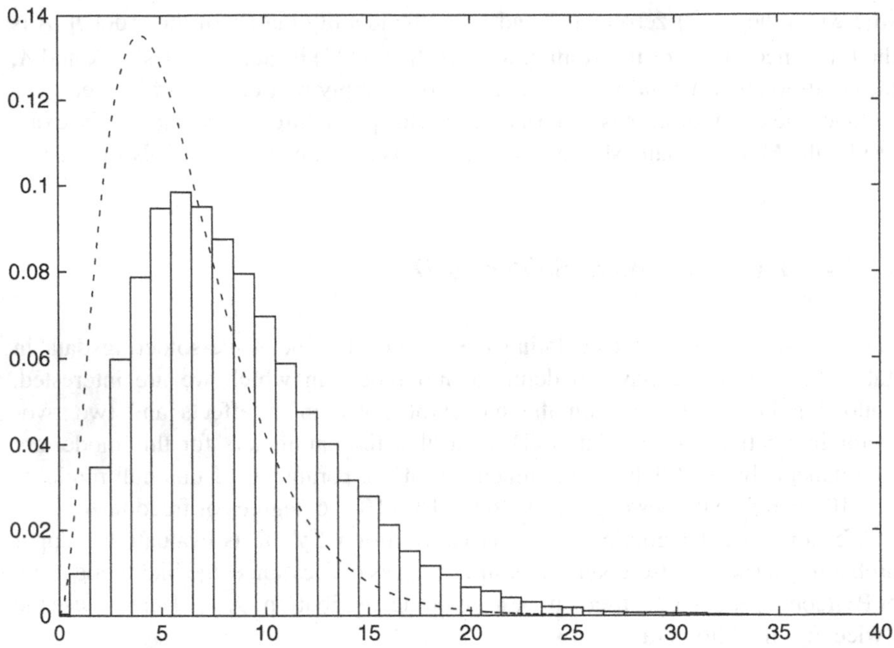

Fig. 11.1 Asymptotic and Monte Carlo estimated conditional distribution

11.2 Markov Bases and Corresponding Models for Contingency Tables

Now we investigate relationships between contingency tables and regular fractional factorial designs of two or three levels. As is shown in the previous chapters of this book, Markov bases have been mainly considered in the context of contingency tables. For example, minimal Markov bases of the decomposable models of contingency tables are considered in Chap. 8. In this chapter, considering the fractional factorial designs, we encounter some new models and Markov bases, that do not correspond to hierarchical models of contingency tables.

11.2.1 Rewriting Observations as Frequencies of a Contingency Table

The arguments of this section are very simple; that is, *we rewrite observations as if they were the frequencies of a contingency table with minimal support size*. We explain this idea by considering the relation between the fractional factorial designs with eight runs and 2^3 (Poisson model) and 2^4 (logistic model) contingency tables.

Table 11.4 Eight-run 2^{s-k} fractional factorial designs ($s-k=3$)

Number of factors s	Resolution	Design Generators
4	IV	$Y_4 = Y_1Y_2Y_3$
5	III	$Y_4 = Y_1Y_2,\ Y_5 = Y_1Y_3$
6	III	$Y_4 = Y_1Y_2,\ Y_5 = Y_1Y_3,\ Y_6 = Y_2Y_3$
7	III	$Y_4 = Y_1Y_2,\ Y_5 = Y_1Y_3,\ Y_6 = Y_2Y_3, Y_7 = Y_1Y_2Y_3$

Recall that there are s controllable factors, $Y_1,\dots,Y_s$ assigned to some regular fractional factorial design, which is defined by k linearly independent defining relations. Because we consider the factors with two levels, there are $\eta = 2^{s-k}$ runs in the design. Here we consider the case that $s-k=3$. We first show the list of the most frequently used designs with eight runs in Table 11.4. Here we use the expression such as $Y_4 = Y_1Y_2Y_3$ to define design, which is given by

$$\{0,1\}^4 \cap \{(y_1,y_2,y_3,y_4) \mid y_4 \equiv y_1+y_2+y_3 \pmod 2\}.$$

Such an expression is standard in the context of the designed experiments.

We clarify the relationships between these designs and the models of 2^3 contingency tables $\boldsymbol{x} = \{x(i_1i_2i_3)\}$, $1 \le i_1,i_2,i_3 \le 2$, for Poisson observations, and the models of 2^4 contingency tables $\boldsymbol{x} = \{x(i_1i_2i_3i_4)\}$, $1 \le i_1,i_2,i_3,i_4 \le 2$, for the binomial observations. We also write indices as subscripts for the rest of this chapter: we write $x_{i_1i_2i_3i_4}$ instead of $x(i_1i_2i_3i_4)$, for example. In the case of Poisson observations, we write eight observations as if they are the frequencies of a 2^3 contingency table;

$$\boldsymbol{x} = (x_{111},x_{112},x_{121},x_{122},x_{211},x_{212},x_{221},x_{222})'.$$

In the case of $s=5$, for example, the design and the observations are given as follows.

	Factor					
Run	Y_1	Y_2	Y_3	Y_4	Y_5	$\boldsymbol{x}$
1	0	0	0	0	0	x_{111}
2	0	0	1	0	1	x_{112}
3	0	1	0	1	0	x_{121}
4	0	1	1	1	1	x_{122}
5	1	0	0	1	1	x_{211}
6	1	0	1	1	0	x_{212}
7	1	1	0	0	1	x_{221}
8	1	1	1	0	0	x_{222}

For this type of data, we define a ν-dimensional parameter θ and the design matrix A according to an appropriate model we consider, as explained in Sect. 11.1.2.1. First consider the simple main effect model $Y_1/Y_2/Y_3/Y_4/Y_5$ ($\nu = 6$). To test this model against various alternatives, the Markov chain Monte Carlo testing procedure needs a Markov basis for the configuration

$$A = \begin{pmatrix} 1\,1\,1\,1\,1\,1\,1\,1 \\ 1\,1\,1\,1\,0\,0\,0\,0 \\ 1\,1\,0\,0\,1\,1\,0\,0 \\ 1\,0\,1\,0\,1\,0\,1\,0 \\ 1\,1\,0\,0\,0\,0\,1\,1 \\ 1\,0\,1\,0\,0\,1\,0\,1 \end{pmatrix}.$$

Note that each component of $A\boldsymbol{x}$ corresponds to a sufficient statistic under the null model $\mathrm{Y}_1/\mathrm{Y}_2/\mathrm{Y}_3/\mathrm{Y}_4/\mathrm{Y}_5$. In this case, a sufficient statistic is given as

$$\begin{gathered} x_{\cdots},\ x_{1\cdot\cdot},\ x_{2\cdot\cdot},\ x_{\cdot 1\cdot},\ x_{\cdot 2\cdot},\ x_{\cdot\cdot 1},\ x_{\cdot\cdot 2}, \\ x_{11\cdot}+x_{22\cdot},\ x_{12\cdot}+x_{21\cdot},\ x_{1\cdot 1}+x_{2\cdot 2},\ x_{1\cdot 2}+x_{2\cdot 1}, \end{gathered} \tag{11.11}$$

where we use the notations such as

$$x_{\cdots} = \sum_{i_1=1}^{2}\sum_{i_2=1}^{2}\sum_{i_3=1}^{2} x_{i_1 i_2 i_3}, \qquad x_{i_1\cdot\cdot} = \sum_{i_2=1}^{2}\sum_{i_3=1}^{2} x_{i_1 i_2 i_3}, \qquad x_{i_1 i_2\cdot} = \sum_{i_3=1}^{2} x_{i_1 i_2 i_3}$$

for marginal frequencies. Here we see that the sufficient statistic (11.11) is nothing but a sufficient statistic for the *conditional independence model* $\mathrm{Y}_1\mathrm{Y}_2/\mathrm{Y}_1\mathrm{Y}_3$, given as

$$\{x_{i_1 i_2\cdot}\},\ \{x_{i_1\cdot i_3}\},\ i_1,i_2,i_3 = 1,2. \tag{11.12}$$

The one-to-one relation between (11.11) and (11.12) is easily shown as

$$x_{i_1 i_2\cdot} = \frac{x_{i_1\cdot\cdot}+x_{\cdot i_2\cdot}-(x_{i_1 i_2^*\cdot}+x_{i_1^* i_2\cdot})}{2}, \qquad x_{i_1\cdot i_3} = \frac{x_{i_1\cdot\cdot}+x_{\cdot\cdot i_3}-(x_{i_1\cdot i_3^*}+x_{i_1^*\cdot i_3})}{2}, \tag{11.13}$$

where $\{i_1,i_1^*\},\{i_2,i_2^*\}$, and $\{i_3,i_3^*\}$ are distinct indices, respectively. This correspondence is, of course, due to the aliasing relations $\mathrm{Y}_4=\mathrm{Y}_1\mathrm{Y}_2,\ \mathrm{Y}_5=\mathrm{Y}_1\mathrm{Y}_3$.

We consider another model. Because there are eight observations, we can estimate eight parameters at most (in the saturated model). Because the saturated model cannot be tested, let us consider the models of $\nu = 7$ parameters. If we restrict our attention to hierarchical models, five main effects and one of the two-factor interaction effects, $\mathrm{Y}_2\mathrm{Y}_3, \mathrm{Y}_2\mathrm{Y}_5, \mathrm{Y}_3\mathrm{Y}_4, \mathrm{Y}_4\mathrm{Y}_5$, can be included in the models, inasmuch as the aliasing relation is given as

$$\begin{aligned} &\mathrm{Y}_1=\mathrm{Y}_2\mathrm{Y}_4=\mathrm{Y}_3\mathrm{Y}_5,\ \mathrm{Y}_2=\mathrm{Y}_1\mathrm{Y}_4,\ \mathrm{Y}_3=\mathrm{Y}_1\mathrm{Y}_5,\ \mathrm{Y}_4=\mathrm{Y}_1\mathrm{Y}_2,\ \mathrm{Y}_5=\mathrm{Y}_1\mathrm{Y}_3, \\ &\mathrm{Y}_2\mathrm{Y}_3=\mathrm{Y}_4\mathrm{Y}_5,\ \mathrm{Y}_2\mathrm{Y}_5=\mathrm{Y}_3\mathrm{Y}_4=\mathrm{Y}_1\mathrm{Y}_2\mathrm{Y}_3. \end{aligned}$$

If our null model includes $\mathrm{Y}_2\mathrm{Y}_3$ or $\mathrm{Y}_4\mathrm{Y}_5$ (i.e., if our null model is written as $\mathrm{Y}_1/\mathrm{Y}_2\mathrm{Y}_3/\mathrm{Y}_4/\mathrm{Y}_5$ or $\mathrm{Y}_1/\mathrm{Y}_2/\mathrm{Y}_3/\mathrm{Y}_4\mathrm{Y}_5$), we add the row

$$(1\,0\,0\,1\,1\,0\,0\,1)$$

to the design matrix A. In this case, a sufficient statistic under the null model includes $x_{\cdot 11}+x_{\cdot 22}$ and $x_{\cdot 12}+x_{\cdot 21}$ in addition to (11.11), which is nothing but a well-known sufficient statistic for the *no-three-factor interaction model*, $\mathrm{Y}_1\mathrm{Y}_2/\mathrm{Y}_1\mathrm{Y}_3/\mathrm{Y}_2\mathrm{Y}_3$,

$$\{x_{i_1 i_2 \cdot}\},\ \{x_{i_1 \cdot i_3}\},\ \{x_{\cdot i_2 i_3}\},\ i_1,i_2,i_3 = 1,2,$$

by the similar relations to (11.13).

On the other hand, if our null model includes $\mathrm{Y}_2\mathrm{Y}_5$ or $\mathrm{Y}_3\mathrm{Y}_4$, that is, if our null model is written as $\mathrm{Y}_1/\mathrm{Y}_2\mathrm{Y}_5/\mathrm{Y}_3/\mathrm{Y}_4$ or $\mathrm{Y}_1/\mathrm{Y}_2/\mathrm{Y}_3\mathrm{Y}_4/\mathrm{Y}_5$, we have to add the row

$$(1\,0\,0\,1\,0\,1\,1\,0)$$

to the design matrix A. In this case, a sufficient statistic under the null model includes $x_{111}+x_{122}+x_{212}+x_{221}$ and $x_{112}+x_{121}+x_{211}+x_{222}$ in addition to (11.11). This is one of the models that do not have corresponding models in the hierarchical models of three-way contingency tables. We write this new model as

$$\mathrm{Y}_1\mathrm{Y}_2/\mathrm{Y}_1\mathrm{Y}_3+(\mathrm{Y}_1\mathrm{Y}_2\mathrm{Y}_3).$$

A sufficient statistic for this model is

$$\{x_{i_1 i_2 \cdot}\},\ \{x_{i_1 \cdot i_3}\},\ i_1,i_2,i_3 = 1,2,$$
$$x_{111}+x_{122}+x_{212}+x_{221},\ x_{112}+x_{121}+x_{211}+x_{222}.$$

Similarly, we can specify the corresponding models of three-way contingency tables (for the factors $\mathrm{Y}_1,\mathrm{Y}_2,\mathrm{Y}_3$) to all the possible models for the designs of Table 11.4, as if the observations were the frequencies of a 2^3 contingency table. The result is summarized in Table 11.5. In Table 11.5, we use the notation $(\mathrm{Y}_1\mathrm{Y}_2\mathrm{Y}_3)$ for the models where the sufficient statistic contains $\{x_{111}+x_{122}+x_{212}+x_{221},\ x_{112}+x_{121}+x_{211}+x_{222}\}$.

In the case of binomial observations, there are 16 observations. Similarly to the Poisson case, we treat the observations as if they were the frequencies of a 2^4 contingency table. In the case of $s=5$, for example, the design and the observations are given as follows.

	Factor						
Run	Y_1	Y_2	Y_3	Y_4	Y_5	$\boldsymbol{x}$	
1	0	0	0	0	0	x_{1111}	x_{1112}
2	0	0	1	0	1	x_{1121}	x_{1122}
3	0	1	0	1	0	x_{1211}	x_{1212}
4	0	1	1	1	1	x_{1221}	x_{1222}
5	1	0	0	1	1	x_{2111}	x_{2112}
6	1	0	1	1	0	x_{2121}	x_{2122}
7	1	1	0	0	1	x_{2211}	x_{2212}
8	1	1	1	0	0	x_{2221}	x_{2222}

Table 11.5 Eight-run 2^{s-k} fractional factorial designs and the corresponding models of three-way contingency tables ($s-k=3$)

Design : $s=4$, $Y_4=Y_1Y_2Y_3$		
v	Null model	Corresponding model of 2^3 table
5	$Y_1/Y_2/Y_3/Y_4$	$Y_1/Y_2/Y_3+(Y_1Y_2Y_3)$
6	$Y_1Y_2/Y_3/Y_4$	$Y_1Y_2/Y_3+(Y_1Y_2Y_3)$
7	$Y_1Y_2/Y_1Y_3/Y_4$	$Y_1Y_2/Y_1Y_3+(Y_1Y_2Y_3)$
Design : $s=5$, $Y_4=Y_1Y_2,\ Y_5=Y_1Y_3$		
v	Null model	Corresponding model of 2^3 table
6	$Y_1/Y_2/Y_3/Y_4/Y_5$	Y_1Y_2/Y_1Y_3
7	$Y_1/Y_2Y_3/Y_4/Y_5$	$Y_1Y_2/Y_1Y_3/Y_2Y_3$
	$Y_1/Y_2Y_5/Y_3/Y_4$	$Y_1Y_2/Y_1Y_3+(Y_1Y_2Y_3)$
Design : $s=6$, $Y_4=Y_1Y_2,\ Y_5=Y_1Y_3,\ Y_6=Y_2Y_3$		
v	Null model	Corresponding model of 2^3 table
7	$Y_1/Y_2/Y_3/Y_4/Y_5/Y_6$	$Y_1Y_2/Y_1Y_3/Y_2Y_3$

For this type of data, we also specify parameter θ and the design matrix according to the appropriate models, by replacing A by $\widetilde{A}$ of (11.10). Note that the elements of $\boldsymbol{x}$ are ordered as

$$\boldsymbol{x}=(x_{1111},x_{1121},\ldots,x_{2211},x_{2221},x_{1112},x_{1122},\ldots,x_{2212},x_{2222})'.$$

Accordingly, correspondences to the models of 2^4 contingency tables are easily obtained and the result is given in Table 11.6. In Table 11.6, we use the notations $(Y_1Y_2Y_3)$ and $(Y_1Y_2Y_3Y_4)$ for the models where a sufficient statistic contains $\{x_{i_1i_2i_3\cdot}\}$, $i_1,i_2,i_3=1,2$, and $\{x_{111\ell}+x_{122\ell}+x_{212\ell}+x_{221\ell},x_{112\ell}+x_{121\ell}+x_{211\ell}+x_{222\ell}\}$, $\ell=1,2$, respectively.

Table 11.6 is automatically converted from Table 11.5 as follows. By definition, Y_4 is added to all the generating sets. Note also that the sufficient statistic for each model includes $\{x_{i_1i_2i_3\cdot}\},1\le i_1,i_2,i_3\le 2$, by definition, which yields Table 11.6. Therefore the models that do not include all of Y_1Y_2,Y_1Y_3 and Y_2Y_3 do not correspond to hierarchical models.

In (11.13), we see that the sufficient statistic $A\boldsymbol{x}$ for the main effect model of 2^{5-2} fractional factorial design is equivalent to the two-dimensional marginals of 2^3 contingency tables. This correspondence is due to the aliasing relations $Y_4=Y_1Y_2,Y_5=Y_1Y_3$. In fact, such a correspondence holds in general. We now state a proposition for the general two-level and three-level regular fractional factorial designs.

Proposition 11.1. *For 2^s and 3^s full factorial designs, write observations as $\boldsymbol{x}=\{x_{i_1\cdots i_s}\}$. Then the necessary and the sufficient condition that the $\{i_1,\ldots,i_n\}$-marginal n-dimensional table ($n\le s$) is uniquely determined from $A\boldsymbol{x}$ is that the design matrix A includes the contrasts for all (the components of) m-factor interaction effects $Y_{j_1}\times\cdots\times Y_{j_m}$ for all $\{j_1,\ldots,j_m\}\subset\{i_1,\ldots,i_n\}$, $m\le n$.*

Table 11.6 Eight-run 2^{s-k} fractional factorial designs and the corresponding models of three-way contingency tables ($s-k=3$)

Design : $s=4$, $Y_4=Y_1Y_2Y_3$		
v	Null model	Corresponding model of 2^4 table
5	$Y_1/Y_2/Y_3/Y_4$	$Y_1Y_4/Y_2Y_4/Y_3Y_4+(Y_1Y_2Y_3)+(Y_1Y_2Y_3Y_4)$
6	$Y_1Y_2/Y_3/Y_4$	$Y_1Y_2Y_4/Y_3Y_4+(Y_1Y_2Y_3)+(Y_1Y_2Y_3Y_4)$
7	$Y_1Y_2/Y_1Y_3/Y_4$	$Y_1Y_2Y_4/Y_1Y_3Y_4+(Y_1Y_2Y_3)+(Y_1Y_2Y_3Y_4)$
Design : $s=5$, $Y_4=Y_1Y_2$, $Y_5=Y_1Y_3$		
v	Null model	Corresponding model of 2^4 table
6	$Y_1/Y_2/Y_3/Y_4/Y_5$	$Y_1Y_2Y_4/Y_1Y_3Y_4+(Y_1Y_2Y_3)$
7	$Y_1/Y_2Y_3/Y_4/Y_5$	$Y_1Y_2Y_4/Y_1Y_3Y_4/Y_2Y_3Y_4/Y_1Y_2Y_3$
	$Y_1/Y_2Y_5/Y_3/Y_4$	$Y_1Y_2Y_4/Y_1Y_3Y_4+(Y_1Y_2Y_3)+(Y_1Y_2Y_3Y_4)$
Design : $s=6$, $Y_4=Y_1Y_2$, $Y_5=Y_1Y_3$, $Y_6=Y_2Y_3$		
v	Null model	Corresponding model of 2^4 table
7	$Y_1/Y_2/Y_3/Y_4/Y_5/Y_6$	$Y_1Y_2Y_4/Y_1Y_3Y_4/Y_2Y_3Y_4/Y_1Y_2Y_3$

Proof. We just count the degrees of freedom. The saturated model for the 2^n full factorial design is expressed as the contrast for the total mean, n contrasts for the main effects, $\binom{n}{m}$ contrasts for the m-factor interaction effects for $m=2,\ldots,n$, because they are linearly independent and

$$1+n+\sum_{m=2}^{n}\binom{n}{m}=2^n.$$

Similarly, the saturated model for the 3^n full factorial design is expressed as the contrast for the total mean, $2\times n$ contrasts for the main effects, $2^m\times\binom{n}{m}$ contrasts for the m-factor interaction effects for $m=2,\ldots,n$, inasmuch as they are linearly independent and

$$1+2n+\sum_{m=2}^{n}2^m\binom{n}{m}=(1+2)^n=3^n. \qquad \square$$

Proposition 11.1 states that the hierarchical models for the controllable factors in the full factorial designs just correspond to the hierarchical models for the contingency table. On the other hand, hierarchical models for the controllable factors in the 2^{s-k} and 3^{s-k} fractional factorial designs do not correspond to the hierarchical models for the 2^s and 3^s contingency tables in general. This is because A contains only part of the contrasts of interaction elements in the case of fractional factorial designs, especially for the cases of three-level designs. Consequently, many interesting structures appear in considering Markov bases for the fractional factorial designs.

Table 11.7 Sixteen-run 2^{s-k} fractional factorial designs ($s-k=4$)

Number of Factors s	Resolution	Design Generators
5	V	$Y_5 = Y_1Y_2Y_3Y_4$
6	IV	$Y_5 = Y_1Y_2Y_3, Y_6 = Y_1Y_2Y_4$
7	IV	$Y_5 = Y_1Y_2Y_3, Y_6 = Y_1Y_2Y_4,\ Y_7 = Y_1Y_3Y_4$
8	IV	$Y_5 = Y_1Y_2Y_3, Y_6 = Y_1Y_2Y_4,\ Y_7 = Y_1Y_3Y_4Y_8 = Y_2Y_3Y_4$
9	III	$Y_5 = Y_1Y_2Y_3, Y_6 = Y_1Y_2Y_4,\ Y_7 = Y_1Y_3Y_4Y_8 = Y_2Y_3Y_4,$ $Y_9 = Y_1Y_2Y_3Y_4$
10	III	$Y_5 = Y_1Y_2Y_3, Y_6 = Y_1Y_2Y_4,\ Y_7 = Y_1Y_3Y_4\ Y_8 = Y_2Y_3Y_4,$ $Y_9 = Y_1Y_2Y_3Y_4,\ Y_{10} = Y_3Y_4$

11.2.2 Models for the Two-Level Regular Fractional Factorial Designs with 16 Runs

Next we consider fractional factorial designs with 16 runs, that is, the case of $s-k=4$. Table 11.7 is a list of 16-run 2^{s-k} fractional factorial designs ($s-k=4, s \leq 10$) from Sect. 4 of [151].

By similar considerations to the 8-run cases, we can seek the corresponding models of 2^4 contingency tables for Poisson observations, and models of 2^5 contingency tables for the binomial observations. Modeling for binomial observations can be easily obtained from the Poisson case as we have seen, therefore we only consider the Poisson case here.

Because at most 16 parameters are estimable for the 16-run designs, we can consider various models of main effects and interaction effects. For example, the saturated model of the $s=5$ design, $Y_5 = Y_1Y_2Y_3Y_4$, can include all the main effects and two-factor interactions,

$$Y_1Y_2/Y_1Y_3/Y_1Y_4/Y_1Y_5/Y_2Y_3/Y_2Y_4/Y_2Y_5/Y_3Y_4/Y_3Y_5/Y_4Y_5.$$

Note that for the models of $s=5,6,7,8$ in Table 11.7, each main effect and two-factor interaction is simultaneously estimable. (On the other hand, for the resolution III models of $s=9,10$, some of the two-factor interactions are not simultaneously estimable.) Among the models that include all the main effects and some of the two-factor interaction effects, some models have the corresponding hierarchical model in the 2^4 contingency tables when we write the 16 observations as $\boldsymbol{x} = \{x_{i_1i_2i_3i_4}\}, i_1,i_2,i_3,i_4 = 1,2$.

For example, for the $s=6$ design of $Y_5 = Y_1Y_2Y_3, Y_6 = Y_1Y_2Y_4$, the model of 6 main effects and 5 two-factor interaction effects,

$$Y_1Y_2/Y_1Y_3/Y_1Y_4/Y_2Y_3/Y_2Y_4/Y_5/Y_6,$$

Table 11.8 Sixteen-run 2^{s-k} fractional factorial designs and the corresponding hierarchical models of 2^4 contingency tables ($s-k=4$)

Design : $s=6$, $Y_5=Y_1Y_2Y_3, Y_6=Y_1Y_2Y_4$	
$\nu=12$	
Representative null model	$Y_1Y_2/Y_1Y_3/Y_1Y_4/Y_2Y_3/Y_2Y_4/Y_5/Y_6$
Num. of the null models	48
Corresponding model of 2^4 table	$Y_1Y_2Y_3/Y_1Y_2Y_4$
Design : $s=6$, $Y_5=Y_1Y_2Y_3, Y_6=Y_1Y_2Y_4$	
$\nu=13$	
Representative null model	$Y_1Y_2/Y_1Y_3/Y_1Y_4/Y_2Y_3/Y_2Y_4/Y_3Y_4/Y_5/Y_6$
Num. of the null models	96
Corresponding model of 2^4 table	$Y_1Y_2Y_3/Y_1Y_2Y_4/Y_3Y_4$
Design : $s=7$, $Y_5=Y_1Y_2Y_3, Y_6=Y_1Y_2Y_4, Y_7=Y_1Y_3Y_4$	
$\nu=12$	
Representative null model	$Y_1Y_2/Y_1Y_3/Y_1Y_4/Y_2Y_3/Y_2Y_4/Y_3Y_4/Y_5/Y_6/Y_7$
Num. of the null models	$3^6=729$
Corresponding model of 2^4 table	$Y_1Y_2Y_3/Y_1Y_2Y_4/Y_1Y_3Y_4$
Design : $s=8$, $Y_5=Y_1Y_2Y_3, Y_6=Y_1Y_2Y_4, Y_7=Y_1Y_3Y_4, Y_8=Y_2Y_3Y_4$	
$\nu=12$	
Representative null model	$Y_1Y_2/Y_1Y_3/Y_1Y_4/Y_2Y_3/Y_2Y_4/Y_3Y_4/Y_5/Y_6/Y_7$
Num. of the null models	$4^6=4096$
Corresponding model of 2^4 table	$Y_1Y_2Y_3/Y_1Y_2Y_4/Y_1Y_3Y_4/Y_2Y_3Y_4$

has a corresponding model of $Y_1Y_2Y_3/Y_1Y_2Y_4$ for the 2^4 contingency tables. By the aliasing relations

$$Y_1Y_2=Y_3Y_5=Y_4Y_6,\ Y_1Y_3=Y_2Y_5,\ Y_1Y_4=Y_2Y_6,\ Y_1Y_5=Y_2Y_3,$$
$$Y_1Y_6=Y_2Y_4,\ Y_3Y_4=Y_5Y_6,\ Y_3Y_6=Y_4Y_5=Y_1Y_2Y_3Y_4,$$

it is seen that there are $3\cdot 2\cdot 2\cdot 2\cdot 2=48$ distinct models such as

$$\begin{array}{c} Y_1Y_2/Y_1Y_3/Y_1Y_4/Y_1Y_5/Y_1Y_6/Y_5/Y_6,\\ Y_1Y_2/Y_1Y_3/Y_1Y_4/Y_1Y_5/Y_2Y_4/Y_5/Y_6,\\ Y_1Y_2/Y_1Y_3/Y_1Y_4/Y_2Y_3/Y_1Y_6/Y_5/Y_6,\\ Y_1Y_2/Y_1Y_3/Y_1Y_4/Y_2Y_3/Y_2Y_4/Y_5/Y_6,\\ \vdots\\ Y_4Y_6/Y_2Y_5/Y_2Y_6/Y_2Y_3/Y_1Y_6/Y_5/Y_6,\\ Y_4Y_6/Y_2Y_5/Y_2Y_6/Y_2Y_3/Y_2Y_4/Y_5/Y_6, \end{array}$$

which correspond to the model of $Y_1Y_2Y_3/Y_1Y_2Y_4$ in the 2^4 contingency tables. By similar considerations, we can specify all the models for the designs of Table 11.7, which correspond to some hierarchical models in the 2^4 contingency tables. The result is shown in Table 11.8.

One of the merits of specifying corresponding hierarchical models of contingency tables is a possibility to make use of already known general results on Markov bases of contingency tables. For example, we see in Chap. 8 that a Markov basis can be constructed by degree 2 basic moves for the decomposable models in the contingency tables. In our designed experiments, therefore, the Markov basis for the models that correspond to decomposable models of contingency tables can be constructed by basic moves (square-free moves of degree 2) only. Among the results of Tables 11.5, 11.6, and 11.8, there are two models that correspond to decomposable models in the contingency tables. We can confirm that minimal Markov bases for these models consist of basic moves as follows.

- 2^{5-2} fractional factorial design of $\mathrm{Y}_4 = \mathrm{Y}_1\mathrm{Y}_2, \mathrm{Y}_5 = \mathrm{Y}_1\mathrm{Y}_3$:
  The main effects model $\mathrm{Y}_1/\mathrm{Y}_2/\mathrm{Y}_3/\mathrm{Y}_4/\mathrm{Y}_5$ corresponds to the decomposable model $\mathrm{Y}_1\mathrm{Y}_2/\mathrm{Y}_1\mathrm{Y}_3$ of the 2^3 contingency tables. This is a conditional independence model between Y_2 and Y_3 given Y_1 and a minimal Markov basis is constructed by basic moves as
  $$(111)(122) - (112)(121),\ (211)(222) - (212)(221).$$
- 2^{6-2} fractional factorial design of $\mathrm{Y}_5 = \mathrm{Y}_1\mathrm{Y}_2\mathrm{Y}_3, \mathrm{Y}_6 = \mathrm{Y}_1\mathrm{Y}_2\mathrm{Y}_4$:
  The model $\mathrm{Y}_1\mathrm{Y}_2/\mathrm{Y}_1\mathrm{Y}_3/\mathrm{Y}_1\mathrm{Y}_4/\mathrm{Y}_2\mathrm{Y}_3/\mathrm{Y}_2\mathrm{Y}_4/\mathrm{Y}_5/\mathrm{Y}_6$ corresponds to the decomposable model $\mathrm{Y}_1\mathrm{Y}_2\mathrm{Y}_3/\mathrm{Y}_1\mathrm{Y}_2\mathrm{Y}_4$ of the 2^4 contingency tables. This is a conditional independence model between Y_3 and Y_4 given $\{\mathrm{Y}_1, \mathrm{Y}_2\}$ and a minimal Markov basis is again constructed by basic moves as
  $$\begin{array}{l}(1111)(1122) - (1112)(1121),\ (1211)(1222) - (1212)(1221),\\ (2111)(2122) - (2112)(2121),\ (2211)(2222) - (2212)(2221).\end{array}$$

For the other designs of Table 11.7 ($p = 5, 9, 10$), all the models include the sufficient statistic

$$\begin{array}{l}x_{1111} + x_{1122} + x_{1212} + x_{1221} + x_{2112} + x_{2121} + x_{2211} + x_{2222},\\ x_{1112} + x_{1121} + x_{1211} + x_{1222} + x_{2111} + x_{2122} + x_{2212} + x_{2221},\end{array}$$

and therefore have no corresponding hierarchical models in the 2^4 contingency tables. For example, a sufficient statistic of the main effect models for the 2^{5-1} design of $\mathrm{Y}_5 = \mathrm{Y}_1\mathrm{Y}_2\mathrm{Y}_3\mathrm{Y}_4$ is

$$\begin{array}{l}\{x_{i_1\cdots}\},\ \{x_{\cdot i_2\cdot\cdot}\},\ \{x_{\cdot\cdot i_3\cdot}\},\ \{x_{\cdots i_4}\},\ i_1, i_2, i_3, i_4 = 1, 2,\\ x_{1111} + x_{1122} + x_{1212} + x_{1221} + x_{2112} + x_{2121} + x_{2211} + x_{2222},\\ x_{1112} + x_{1121} + x_{1211} + x_{1222} + x_{2111} + x_{2122} + x_{2212} + x_{2221},\end{array}$$

and a sufficient statistic of the main effect models for the 2^{10-6} design of

$$\begin{array}{l}\mathrm{Y}_5 = \mathrm{Y}_1\mathrm{Y}_2\mathrm{Y}_3,\ \mathrm{Y}_6 = \mathrm{Y}_1\mathrm{Y}_2\mathrm{Y}_4,\ \mathrm{Y}_7 = \mathrm{Y}_1\mathrm{Y}_3\mathrm{Y}_4,\ \mathrm{Y}_8 = \mathrm{Y}_2\mathrm{Y}_3\mathrm{Y}_4,\ \mathrm{Y}_9 = \mathrm{Y}_1\mathrm{Y}_2\mathrm{Y}_3\mathrm{Y}_4,\\ \mathrm{Y}_{10} = \mathrm{Y}_3\mathrm{Y}_4\end{array}$$

is

$$\{x_{i_1i_2i_3\cdot}\},\ \{x_{i_1i_2\cdot i_4}\},\ \{x_{i_1\cdot i_3i_4}\},\ \{x_{\cdot i_2i_3i_4}\},\ i_1,i_2,i_3,i_4=1,2,$$
$$x_{1111}+x_{1122}+x_{1212}+x_{1221}+x_{2112}+x_{2121}+x_{2211}+x_{2222},$$
$$x_{1112}+x_{1121}+x_{1211}+x_{1222}+x_{2111}+x_{2122}+x_{2212}+x_{2221}.$$

11.2.3 Three-Level Regular Fractional Factorial Designs and $3^{s\text{-}k}$ Continent Tables

Next we consider the three-level designs. As the simplest example, we first consider a design with 9 runs for three controllable factors, that is, 3^{3-1} fractional factorial design. Write three controllable factors as $\mathrm{Y}_1,\mathrm{Y}_2,\mathrm{Y}_3$, and define $\mathrm{Y}_3=\mathrm{Y}_1\mathrm{Y}_2$. In this design, the design matrix for the main effects model of $\mathrm{Y}_1,\mathrm{Y}_2,\mathrm{Y}_3$ is defined as

$$A=\begin{pmatrix} 1\,1\,1\,1\,1\,1\,1\,1\,1 \\ 1\,1\,1\,0\,0\,0\,0\,0\,0 \\ 0\,0\,0\,1\,1\,1\,0\,0\,0 \\ 1\,0\,0\,1\,0\,0\,1\,0\,0 \\ 0\,1\,0\,0\,1\,0\,0\,1\,0 \\ 1\,0\,0\,0\,0\,1\,0\,1\,0 \\ 0\,1\,0\,1\,0\,0\,0\,0\,1 \end{pmatrix}.$$

To investigate the structure of the fiber, write the observation as a frequency of the 3×3 contingency table, $x_{11},\ldots,x_{33}$. Then the fiber is the set of tables with the same row sums $\{x_{i_1\cdot}\}$, column sums $\{x_{\cdot i_2}\}$, and the contrast displayed as

$$\boxed{\begin{matrix} 0 & 1 & 2 \\ 1 & 2 & 0 \\ 2 & 0 & 1 \end{matrix}}\,.$$

Concerning a minimal Markov basis, we see that the moves to connect the following three-element fiber are sufficient,

$$\left\{ \boxed{\begin{matrix} 1 & 0 & 0 \\ 0 & 1 & 0 \\ 0 & 0 & 1 \end{matrix}},\ \boxed{\begin{matrix} 0 & 1 & 0 \\ 0 & 0 & 1 \\ 1 & 0 & 0 \end{matrix}},\ \boxed{\begin{matrix} 0 & 0 & 1 \\ 1 & 0 & 0 \\ 0 & 1 & 0 \end{matrix}} \right\}.$$

Therefore any two moves from the following three moves,

$$(11)(22)(33)-(12)(23)(31),$$
$$(11)(22)(33)-(13)(21)(32),$$
$$(12)(23)(31)-(13)(21)(32),$$

is a minimal Markov basis.

For the rest of this chapter, we consider three types of fractional factorial designs with 27 runs, which are important for practical applications. We investigate the relations between various models for the fractional factorial designs and the $3 \times 3 \times 3$ contingency table. In the context of the Markov basis for the contingency tables, the Markov basis for the $3 \times 3 \times 3$ contingency tables has been investigated by many researchers, especially for the no-three-factor interaction model in Chap. 9. In the following, we investigate Markov bases for some models; we are especially concerned with their minimality, unique minimality, and indispensability of their elements (cf. Sect. 5.2). Similarly to Chap. 9, we write three 3×3 slices to display $3 \times 3 \times 3$ moves of higher degrees.

11.2.3.1 3_{IV}^{4-1} Fractional Factorial Design Defined from $\mathrm{Y}_4 = \mathrm{Y}_1\mathrm{Y}_2\mathrm{Y}_3$

In the case of four controllable factors for design with 27 runs, we have a resolution IV design by setting $\mathrm{Y}_4 = \mathrm{Y}_1\mathrm{Y}_2\mathrm{Y}_3$. As seen in Sect. 11.1.2.2, all the main effects are clear, whereas all the two-factor interactions are not clear in this design.

For the main effect model in this design, the sufficient statistic is written as

$$\{x_{i_1\cdot\cdot}\},\ \{x_{\cdot i_2 \cdot}\},\ \{x_{\cdot\cdot i_3}\}$$

and

$$\begin{aligned}
&x_{111} + x_{123} + x_{132} + x_{213} + x_{222} + x_{231} + x_{312} + x_{321} + x_{333},\\
&x_{112} + x_{121} + x_{133} + x_{211} + x_{223} + x_{232} + x_{313} + x_{322} + x_{331},\\
&x_{113} + x_{122} + x_{131} + x_{212} + x_{221} + x_{233} + x_{311} + x_{323} + x_{332}.
\end{aligned}$$

By 4ti2 [1], the minimal Markov basis for this model consists of 54 degree 2 moves and 24 degree 3 moves. All the elements of the same degrees are on the same orbit (see Chap. 7).

The moves of degree 2 connect three-element fibers such as

$$\{(112)(221),\ (121)(212),\ (122)(211)\} \tag{11.14}$$

into a tree, and the moves of degree 3 connect three-element fibers such as

$$\{(111)(122)(133),\ (112)(123)(131),\ (113)(121)(132)\} \tag{11.15}$$

into a tree. For the fiber (11.14), for example, two moves such as

$$(121)(212) - (112)(221),\ (122)(211) - (112)(221)$$

are needed for a Markov basis.

Considering the aliasing relations given by (11.7), we can consider models with interaction effects. We see by using 4ti2 that the structures of the minimal Markov bases for each model are given as follows.

- For the model of the main effects and the interaction effect $Y_1 \times Y_2$, 27 indispensable moves of degree 2 such as $(113)(321) - (111)(323)$ and 54 dispensable moves of degree 3 constitute a minimal Markov basis. The degree 3 elements are on two orbits; one connects 9 three-element fibers such as (11.15) and the other connects 18 three-element fibers such as

$$\{(111)(133)(212), (112)(131)(213), (113)(132)(211)\}.$$

- For the model of the main effects and the interaction effects $Y_1 \times Y_2, Y_1 \times Y_3$, 6 dispensable moves of degree 3, 81 indispensable moves of degree 4 such as

$$\begin{array}{|ccc|}\hline -1 & +1 & 0 \\ +1 & -1 & 0 \\ 0 & 0 & 0 \\ \hline\end{array}\quad \begin{array}{|ccc|}\hline -1 & 0 & +1 \\ +1 & 0 & -1 \\ 0 & 0 & 0 \\ \hline\end{array}\quad \begin{array}{|ccc|}\hline 0 & 0 & 0 \\ 0 & 0 & 0 \\ 0 & 0 & 0 \\ \hline\end{array} \tag{11.16}$$

and 171 indispensable moves of degree 6, 63 moves such as

$$\begin{array}{|ccc|}\hline -1 & +1 & 0 \\ +1 & 0 & -1 \\ 0 & -1 & +1 \\ \hline\end{array}\quad \begin{array}{|ccc|}\hline -1 & 0 & +1 \\ 0 & +1 & -1 \\ +1 & -1 & 0 \\ \hline\end{array}\quad \begin{array}{|ccc|}\hline 0 & 0 & 0 \\ 0 & 0 & 0 \\ 0 & 0 & 0 \\ \hline\end{array} \tag{11.17}$$

and 108 moves such as

$$\begin{array}{|ccc|}\hline -1 & +1 & 0 \\ +1 & -1 & 0 \\ 0 & 0 & 0 \\ \hline\end{array}\quad \begin{array}{|ccc|}\hline -1 & 0 & +1 \\ 0 & 0 & 0 \\ +1 & 0 & -1 \\ \hline\end{array}\quad \begin{array}{|ccc|}\hline +1 & 0 & -1 \\ -1 & 0 & +1 \\ 0 & 0 & 0 \\ \hline\end{array}$$

constitute a minimal Markov basis. The degree 3 elements connect three-element fibers such as (11.15).
- For the model of the main effects and the interaction effects $Y_1 \times Y_2, Y_1 \times Y_3, Y_2 \times Y_3$, 27 indispensable moves of degree 6 such as (11.17) and 27 indispensable moves of degree 8 such as

$$\begin{array}{|ccc|}\hline +2 & -1 & -1 \\ -1 & +1 & 0 \\ -1 & 0 & +1 \\ \hline\end{array}\quad \begin{array}{|ccc|}\hline -1 & +1 & 0 \\ +1 & -1 & 0 \\ 0 & 0 & 0 \\ \hline\end{array}\quad \begin{array}{|ccc|}\hline -1 & 0 & +1 \\ 0 & 0 & 0 \\ +1 & 0 & -1 \\ \hline\end{array}$$

constitute a unique minimal Markov basis.
- For the model of the main effects and the interaction effects $Y_1 \times Y_2, Y_1 \times Y_3, Y_1 \times Y_4$, 6 dispensable moves of degree 3 constitute a minimal Markov basis, which connects three-element fibers such as (11.15).

11.2.3.2 3_{III}^{5-2} Fractional Factorial Design Defined from $Y_4 = Y_1Y_2, Y_5 = Y_1Y_2^2Y_3$

In the case of five controllable factors for designs with 27 runs, the parameter contrasts for the two main factors are allocated by two aliasing relations.

In this section, we consider two designs from Table 5A.2 of [151]. First we consider the 3_{III}^{5-2} fractional factorial design defined by $Y_4 = Y_1Y_2, Y_5 = Y_1Y_2^2Y_3$. For this design, we can consider the following nine distinct hierarchical models (except for the saturated model). Minimal Markov bases for these models are calculated by 4ti2 as follows.

- For the model of the main effects of Y_1, Y_2, Y_3, Y_4, Y_5, 27 indispensable moves of degree 2 such as $(112)(221) - (111)(222)$, 56 dispensable moves of degree 3, 54 indispensable moves of degree 4 such as

$$\begin{array}{|ccc|}\hline +1 & 0 & 0 \\ +1 & -1 & 0 \\ -1 & 0 & 0 \\ \hline\end{array}\ \begin{array}{|ccc|}\hline -1 & 0 & 0 \\ 0 & +1 & 0 \\ 0 & 0 & 0 \\ \hline\end{array}\ \begin{array}{|ccc|}\hline 0 & 0 & 0 \\ 0 & -1 & 0 \\ 0 & +1 & 0 \\ \hline\end{array}$$

and 9 indispensable moves of degree 6 such as

$$\begin{array}{|ccc|}\hline +2 & 0 & 0 \\ 0 & -1 & 0 \\ -1 & 0 & 0 \\ \hline\end{array}\ \begin{array}{|ccc|}\hline -1 & 0 & 0 \\ +1 & +1 & 0 \\ 0 & -1 & 0 \\ \hline\end{array}\ \begin{array}{|ccc|}\hline 0 & -1 & 0 \\ -1 & 0 & 0 \\ 0 & +2 & 0 \\ \hline\end{array}$$

constitute a minimal Markov basis. The degree three moves are in three orbits, which connect three types of three-element fibers, that is,

18 moves for 9 fibers: $\{(111)(123)(132), (113)(122)(131), (112)(121)(133)\}$,
36 moves for 18 fibers: $\{(111)(123)(212), (113)(122)(211), (112)(121)(213)\}$,
2 moves for the fiber: $\{(112)(223)(331), (131)(212)(323), (121)(232)(313)\}$.

- For the model of the main effects and the interaction effect $Y_1 \times Y_3$, 18 dispensable moves of degree 3, 162 indispensable moves of degree 4 such as (11.16), 81 indispensable moves of degree 5 such as

$$\begin{array}{|ccc|}\hline -1 & +1 & +1 \\ +1 & -1 & -1 \\ 0 & 0 & 0 \\ \hline\end{array}\ \begin{array}{|ccc|}\hline 0 & 0 & 0 \\ +1 & 0 & 0 \\ -1 & 0 & 0 \\ \hline\end{array}\ \begin{array}{|ccc|}\hline -1 & 0 & 0 \\ 0 & 0 & 0 \\ +1 & 0 & 0 \\ \hline\end{array}$$

and 54 indispensable moves of degree 5 such as

$$\begin{array}{|ccc|}\hline -1 & +2 & 0 \\ +1 & -1 & 0 \\ 0 & -1 & 0 \\ \hline\end{array}\ \begin{array}{|ccc|}\hline -1 & 0 & 0 \\ +1 & 0 & 0 \\ 0 & 0 & 0 \\ \hline\end{array}\ \begin{array}{|ccc|}\hline 0 & 0 & 0 \\ -1 & 0 & 0 \\ +1 & 0 & 0 \\ \hline\end{array}$$

and 54 indispensable moves of degree 6 such as (11.17) constitute a minimal Markov basis. The degree 3 moves connect three-element fibers such as

$$\{(111)(123)(132),\ (112)(121)(133),\ (113)(122)(131)\}. \tag{11.18}$$

- For the model of the main effects and the interaction effect $Y_3 \times Y_5$, 27 indispensable moves of degree 2 such as $(112)(221) - (111)(222)$ constitute the unique minimal Markov basis.
- For the model of the main effects and the interaction effects $Y_1 \times Y_3, Y_1 \times Y_5$, 6 dispensable moves of degree 3 and 81 indispensable moves of degree 6 such as (11.17) constitute a minimal Markov basis. The degree 3 moves connect three-element fibers such as (11.18).
- For the model of the main effects and the interaction effects $Y_1 \times Y_3, Y_2 \times Y_3$, 27 indispensable moves of degree 4 such as (11.16) and 54 indispensable moves of degree 6 such as

$$\begin{array}{|ccc|}\hline -1 & +1 & 0 \\ +1 & 0 & -1 \\ 0 & -1 & +1 \\ \hline \end{array}\ \begin{array}{|ccc|}\hline +1 & -1 & 0 \\ -1 & 0 & +1 \\ 0 & +1 & -1 \\ \hline \end{array}\ \begin{array}{|ccc|}\hline 0 & 0 & 0 \\ 0 & 0 & 0 \\ 0 & 0 & 0 \\ \hline \end{array} \tag{11.19}$$

  constitute the unique minimal Markov basis.
- For the model of the main effects and the interaction effects $Y_1 \times Y_3, Y_3 \times Y_5$, 27 indispensable moves of degree 4 such as (11.16) and 54 indispensable moves of degree 6 such as (11.17) constitute the unique minimal Markov basis.
- For the model of the main effects and the interaction effects $Y_1 \times Y_3, Y_1 \times Y_5, Y_3 \times Y_5$, 9 indispensable moves of degree 6 such as (11.17) constitute the unique minimal Markov basis.
- For the model of the main effects and the interaction effects $Y_1 \times Y_3, Y_2 \times Y_3, Y_3 \times Y_4$, 9 indispensable moves of degree 6 such as (11.17) constitute the unique minimal Markov basis.
- For the model of the main effects and the interaction effects $Y_1 \times Y_3, Y_2 \times Y_3, Y_3 \times Y_5$, 9 indispensable moves of degree 6 such as (11.19) constitute the unique minimal Markov basis.

11.2.3.3 $\mathbf{3^{5-2}_{III}}$ Fractional Factorial Design Defined from $\mathbf{Y_4 = Y_1Y_2, Y_5 = Y_1Y_2^2}$

Next we consider 3^{5-2}_{III} fractional factorial design defined from $Y_4 = Y_1Y_2, Y_5 = Y_1Y_2^2$. For this design, we can consider the following four distinct hierarchical models (except for the saturated model). Minimal Markov bases for these models are calculated by 4ti2 as follows.

- For the model of the main effects of Y_1, Y_2, Y_3, Y_4, Y_5, 108 indispensable moves of degree 2 such as $(112)(121) - (111)(122)$ constitute the unique minimal Markov basis.
- For the model of the main effects and the interaction effect $Y_1 \times Y_3$, 27 indispensable moves of degree 2 such as $(112)(121) - (111)(122)$ constitute the unique minimal Markov basis.
- For the model of the main effects and the interaction effects $Y_1 \times Y_3, Y_2 \times Y_3$, 27 indispensable moves of degree 4 such as

$$\begin{array}{|ccc|}\hline -1 & +1 & 0 \\ +1 & -1 & 0 \\ 0 & 0 & 0 \\ \hline \end{array}\ \begin{array}{|ccc|}\hline +1 & -1 & 0 \\ -1 & +1 & 0 \\ 0 & 0 & 0 \\ \hline \end{array}\ \begin{array}{|ccc|}\hline 0 & 0 & 0 \\ 0 & 0 & 0 \\ 0 & 0 & 0 \\ \hline \end{array}$$

  and 54 indispensable moves of degree 6 such as (11.19) constitute the unique minimal Markov basis.
- For the model of the main effects and the interaction effects $Y_1 \times Y_3, Y_2 \times Y_3, Y_3 \times Y_4$, 9 indispensable moves of degree 6 such as (11.19) constitute the unique minimal Markov basis.

Chapter 12
Groupwise Selection Models

12.1 Examples of Groupwise Selections

First we introduce two data sets from the viewpoint of the groupwise selection. In Sect. 12.1.1, we take a close look at patterns of subject selections in the National Center Test for university entrance examinations in Japan. In Sect. 12.1.2, we illustrate an important problem of population genetics from the viewpoint of groupwise selection.

12.1.1 The Case of National Center Test in Japan

One important practical problem of groupwise selections is the entrance examination for universities in Japan. In Japan, as the common first-stage screening process, most students applying for universities take the National Center Test (NCT hereafter) for university entrance examinations administered by the National Center for University Entrance Examinations (NCUEE). Basic information on the NCT is available on the NCUEE website ([106] in the references.

After obtaining the NCT score, students apply to departments of individual universities and take second-stage examinations administered by the universities. Due to time constraints of the NCT schedule, there are rather complicated restrictions on possible combinations of subjects. Furthermore, each department of each university can impose different additional requirements on the combinations of subjects of NCT to students applying to the department.

In NCT, students, or examinees, can choose subjects in mathematics, social studies, and science. These three major subjects are divided into subcategories. For example, mathematics is divided into Mathematics 1 and Mathematics 2 and these are then composed of individual subjects. In the test carried out in 2006, examinees could select two mathematics subjects, two social studies subjects, and three science subjects at most as shown below. The details of the subjects can be found on

Table 12.1 Number of examinees who take social studies subjects

	Geography and History						Civics		
	WHA	WHB	JHA	JHB	GeoA	GeoB	ContS	Ethics	P&E
1 subject	496	29,108	1,456	54,577	1,347	27,152	40,677	16,607	25,321
2 subjects	1,028	61,132	3,386	90,427	5,039	83,828	180,108	27,064	37,668

Table 12.2 Number of examinees who select two social studies subjects

	Geography and History					
Civics	WHA	WHB	JHA	JHB	GeoA	GeoB
ContSoc	687	39,913	2,277	62,448	3,817	70,966
Ethics	130	10,966	409	10.482	405	4,672
P&E	211	10,253	700	17,497	817	8,190

Table 12.3 Number of examinees who take science subjects

	Science 1				Science 2			Science 3			
	CSciB	BioI	ISci	BioIA	CSciA	ChemI	ChemIA	PhysI	EarthI	PhysIA	EarthIA
1 subject	2,558	80,385	511	1,314	1,569	19,616	717	14,397	10,788	289	236
2 subjects	6,878	79,041	523	1,195	26,848	158,027	2,777	106,822	6,913	905	259
3 subjects	7,942	18,519	728	490	6,838	20,404	437	18,451	8,423	361	444

web pages and publications of NCUEE. We omit mathematics for simplicity, and only consider selections in social studies and science. In parentheses we show our abbreviations for the subjects in this chapter.

- Social Studies:
  - Geography and History: One subject from {World History A (WHA), World History B (WHB), Japanese History A (JHA), Japanese History B (JHB), Geography A (GeoA), Geography B (GeoB)}
  - Civics: One subject from {Contemporary Society (ContSoc), Ethics, Politics and Economics (P&E)}
- Science:
  - Science 1: One subject from {Comprehensive Science B (CSciB), Biology I (BioI), Integrated Science (IntegS), Biology IA (BioIA)}
  - Science 2: One subject from {Comprehensive Science A (CSciA), Chemistry I (ChemI), Chemistry IA (ChemIA)}
  - Science 3: One subject from {Physics I (PhysI), Earth Science I (EarthI), Physics IA (PhysIA), Earth Science IA (EarthIA)}

Frequencies of the examinees selecting each combination of subjects in 2006 are given on the NCUEE website. Part of them are reproduced in [8], which we show in Tables 12.1–12.5.

As seen in these tables, examinees may select or not select these subjects. For example, one examinee may select two subjects from social studies and three subjects from science, whereas another examinee may select only one subject from science and none from social studies. Hence each examinee is categorized into one

Table 12.4 Number of examinees who select two science subjects

		Science 2			Science 3			
		CSciA	ChemI	ChemIA	PhysI	EarthI	PhysIA	EarthIA
Science 1	CSciB	1,501	1,334	23	120	3,855	1	44
	BioI	21,264	54,412	244	1,366	1,698	5	52
	ISci	147	165	50	43	92	5	21
	BioIA	128	212	715	16	33	29	62
Science 3	PhysI	3,243	101,100	934	–	–	–	–
	EarthI	485	730	20	–	–	–	–
	PhysIA	43	54	768	–	–	–	–
	EarthIA	37	20	23	–	–	–	–

Table 12.5 Number of examinees who select three science subjects

Science 3		PhysI			EarthI		
Science 2		CSciA	ChemI	ChemIA	CSciA	ChemI	ChemIA
Science 1	CSciB	1,155	5,152	17	1,201	317	7
	BioI	553	10,901	31	3,386	3,342	16
	ISci	80	380	23	62	34	4
	BioIA	6	114	39	22	22	10
Science 3		PhysIA			EarthIA		
Science 2		CSciA	ChemI	ChemIA	CSciA	ChemI	ChemIA
Science 1	CSciB	16	5	16	48	5	3
	BioI	30	35	19	130	56	20
	ISci	32	13	27	48	14	11
	BioIA	12	6	150	57	8	44

of the $(6+1) \times \cdots \times (4+1) = 2{,}800$ combinations of individual subjects. Here 1 is added for not choosing from the subcategory. As mentioned above, individual departments of universities impose different additional requirements on the choices of NCT subjects. For example, many science or engineering departments of national universities ask the students to take two subjects from science and one subject from social studies.

Let us observe some tendencies of the selections by the examinees to illustrate what kind of statistical questions one might ask concerning the data in Tables 12.1–12.5.

(i) The most frequent triple of science subjects is {BioI, ChemI, PhysI} in Table 12.5, which seems to be consistent with Table 12.3 because these three subjects are the most frequently selected subjects in Science 1, Science 2 and Science 3, respectively. However in Table 12.4, although the pairs {BioI, ChemI} and {ChemI, PhysI} are the most frequently selected pairs in {Science 1, Science2} and {Science 2, Science 3}, respectively, the pair {BioI, PhysI} is not the first choice in {Science 1, Science 3}. This fact indicates differences in the selection of science subjects between the examinees selecting two subjects and those selecting three subjects.

(ii) In Table 12.2 the most frequent pair is {GeoB, ContSoc}. However, the most frequent single subject from geography and history is JHB both in Tables 12.1 and 12.2. This fact indicates the interaction effect in selecting pairs of social studies.

These observations lead to many interesting statistical questions. However Tables 12.1–12.5 only give frequencies of choices separately for social studies and science; that is, they are the marginal tables for these two major subjects. In this chapter we are interested in independence across these two subjects, such as "are the selections on social studies and science related or not?" We give various models for NCT data in Sect. 12.2.1 and numerical analysis in Sect. 12.5.1.

12.1.2 The Case of Hardy–Weinberg Models for Allele Frequency Data

We also consider problems of population genetics. This is another important application of the methodology of this chapter.

The allele frequency data are usually given as genotype frequencies. For multi-allele locus with alleles $A_1, A_2, \ldots, A_m$, the probability of the genotype A_iA_j in an individual from a randomly breeding population is q_i^2 $(i = j)$ or $2q_iq_j$ $(i \neq j)$, where q_i is the proportion of the allele A_i. These are known as the Hardy–Weinberg equilibrium probabilities as we have seen in Sect. 6.2.2. The Hardy–Weinberg law plays an important role in the field of population genetics and often serves as a basis for genetic inference, therefore much attention has been paid to tests of the hypothesis that a population being sampled is in the Hardy–Weinberg equilibrium against the hypothesis that disturbing forces cause some deviation from the Hardy–Weinberg ratio. See [43] and [67] for example. Although Guo and Thompson [67] consider the exact test of the Hardy–Weinberg equilibrium for multiple loci, the exact procedure becomes infeasible if the data size or the number of alleles is moderately large. Therefore MCMC is also useful for this problem. In Sect. 6.2.2, we have considered minimal Markov bases for the conditional tests of the Hardy–Weinberg model by using MCMC.

Due to the rapid progress of sequencing technology, more and more information is available on the combination of alleles on the same chromosome. A combination of alleles at more than one locus on the same chromosome is called a haplotype and data on haplotype counts are called haplotype frequency data. The haplotype analysis has gained increasing attention in the mapping of complex disease genes, because of the limited power of conventional single-locus analyses.

Haplotype data may come with or without pairing information on homologous chromosomes. It is technically more difficult to determine pairs of haplotypes of the corresponding loci on a pair of homologous chromosomes. A pair of haplotypes on homologous chromosomes is called a diplotype. In this chapter we are interested in diplotype frequency data, because haplotype frequency data on individual chromosomes without pairing information are standard contingency table data and can be analyzed by statistical methods for usual contingency tables. For the diplotype frequency data, the null model we want to consider is the independence model that the probability for each diplotype is expressed by the product of probabilities for each genotype.

We consider the models for genotype frequency data and diplotype frequency data in Sect. 12.2.2. Note that the availability of haplotype data or diplotype data requires a separate treatment in our arguments. Finally we give numerical examples of the analysis of diplotype frequency data in Sect. 12.5.2.

12.2 Conditional Tests for Groupwise Selection Models

In the context of selection problems, a finite sample space $\mathscr{I}$ is the space of possible selections and each element $\boldsymbol{i} \in \mathscr{I}$ represents a combination of choices. We also call each $\boldsymbol{i} \in \mathscr{I}$ a cell, following the terminology of contingency tables, It should be noted that unlike the case of standard multiway contingency tables, our index set $\mathscr{I}$ cannot be written as a direct product in general. We show the structures of $\mathscr{I}$ for NCT data and allele frequency data in Sects. 12.2.1 and 12.2.2, respectively.

Let $p(\boldsymbol{i})$ denote the probability of selecting the combination $\boldsymbol{i}$ (or the probability of cell $\boldsymbol{i}$) and write $\boldsymbol{p} = \{p(\boldsymbol{i})\}_{\boldsymbol{i}\in\mathscr{I}}$. In this chapter, we do not necessarily assume that $\boldsymbol{p}$ is normalized. In fact, in the models of this chapter, we only give an unnormalized functional specification of $p(\cdot)$. Recall that we need not calculate the normalizing constant $1/\sum_{\boldsymbol{i}\in\mathscr{I}} p(\boldsymbol{i})$ for performing an MCMC procedure (cf. Chap. 2). Denote the result of the selections by n individuals as $\boldsymbol{x} = \{x(\boldsymbol{i})\}_{\boldsymbol{i}\in\mathscr{I}}$, where $x(\boldsymbol{i})$ is the frequency of the cell $\boldsymbol{i}$.

In the models considered in this chapter, the cell probability $p(\boldsymbol{i})$ is written as some product of functions that correspond to various marginal probabilities. Let $\mathscr{J}$ denote the index set of the marginals. Then our models can be written as

$$p(\boldsymbol{i}) = h(\boldsymbol{i}) \prod_{\boldsymbol{j}\in\mathscr{J}} q(\boldsymbol{j})^{a_{\boldsymbol{j}}(\boldsymbol{i})}, \tag{12.1}$$

where $h(\boldsymbol{i})$ is a known function and the $q(\boldsymbol{j})$s are the parameters. An important point here is that the sufficient statistic $\boldsymbol{t} = \{t(\boldsymbol{j}), \boldsymbol{j} \in \mathscr{J}\}$ is written in a matrix form as

$$\boldsymbol{t} = A\boldsymbol{x}, \quad A = (a_{\boldsymbol{j}}(\boldsymbol{i}))_{\boldsymbol{j}\in\mathscr{J},\boldsymbol{i}\in\mathscr{I}}, \tag{12.2}$$

where A is a $\nu \times \eta$ matrix of nonnegative integers and $\nu = |\mathscr{J}|$, $\eta = |\mathscr{I}|$. As in Sect. 1.1 we call A a *configuration*.

As we have seen in Chap. 2, we can perform a conditional test of the model (12.1) based on the conditional distribution given the sufficient statistic $\boldsymbol{t}$. An important point in this chapter is that we can make use of the theory of the Gröbner basis for the Segre–Veronese configuration to obtain a Markov basis.

12.2.1 Models for NCT Data

Following the general formalization above, we formulate data types and their statistical models in view of NCT. Suppose that there are J different groups (or categories) and m_j different subgroups in group j for $j = 1,\dots,J$. There are m_{jk} different *items* in subgroup k of group j ($k = 1,\dots,m_j$, $j = 1,\dots,J$). In NCT, $J = 2$, $m_1 = |\{\text{Geography and History, Civics}\}| = 2$ and similarly $m_2 = 3$. The sizes of subgroups are $m_{11} = |\{\text{WHA, WHB, JHA, JHB, GeoA, GeoB}\}| = 6$ and similarly $m_{12} = 3$, $m_{21} = 4$, $m_{22} = 3$, $m_{23} = 4$.

Each individual selects c_{jk} items from the subgroup k of group j. We assume that the total number τ of items chosen is fixed and common for all individuals. In NCT c_{jk} is either 0 or 1. For example, if an examinee is required to take two science subjects in NCT, then (c_{21}, c_{22}, c_{23}) is $(1,1,0)$, $(1,0,1)$, or $(0,1,1)$. For the analysis of genotypes in Sect. 12.2.2, $c_{jk} \equiv 2$ although there is no nesting of subgroups, and the same item (allele) can be selected more than once (selection "with replacement").

We now set up our notation for indexing a combination of choices carefully. In NCT, if an examinee chooses WHA from "Geography and History" of Social Studies and PhysI from Science 3 of Science, we denote the combination of these two choices as (111)(231). In this notation, the selection of c_{jk} items from the subgroup k of group j is indexed as

$$\boldsymbol{i}_{jk} = (jkl_1)(jkl_2)\dots(jkl_{c_{jk}}), \quad 1 \le l_1 \le \dots \le l_{c_{jk}} \le m_{jk}.$$

Here $\boldsymbol{i}_{jk}$ is regarded as a string. If nothing is selected from the subgroup, we define $\boldsymbol{i}_{jk}$ to be an empty string. Now by concatenation of strings, the set $\mathscr{I}$ of combinations is written as

$$\mathscr{I} = \{\boldsymbol{i} = \boldsymbol{i}_1 \dots \boldsymbol{i}_J\}, \quad \boldsymbol{i}_j = \boldsymbol{i}_{j1} \dots \boldsymbol{i}_{jm_j}, \quad j = 1,\dots,J.$$

For example, the choice of (P&E, BioI, ChemI) in NCT is denoted by $\boldsymbol{i} = (123)(212)(222)$. In the following we denote $\boldsymbol{i}' \subset \boldsymbol{i}$ if $\boldsymbol{i}'$ appears as a substring of $\boldsymbol{i}$.

Now we consider some statistical models for $\boldsymbol{p}$. For NCT data, we consider three simple statistical models, namely *complete independence model*, *subgroupwise independence model*, and *groupwise independence model*. The complete independence model is defined as

$$p(\boldsymbol{i}) = \prod_{j=1}^{J} \prod_{\substack{k=1 \\ \boldsymbol{i}_{jk} \subset \boldsymbol{i}}}^{m_j} \prod_{t=1}^{c_{jk}} q_{jk}(l_t) \qquad (12.3)$$

for some parameters $q_{jk}(l)$, $j = 1,\dots,J$; $k = 1,\dots,m_j$; $l = 1,\dots,m_{jk}$. Note that if $c_{jk} > 1$ we need a multinomial coefficient in (12.3). The complete independence model means that each $p(\boldsymbol{i})$, the inclination of the combination $\boldsymbol{i}$, is explained by the set of inclinations $q_{jk}(l)$ of each item. Here $q_{jk}(l)$ corresponds to the marginal probability of the item (jkl). However, we do not necessarily normalize them as $1 = \sum_{l=1}^{m_{jk}} q_{jk}(l)$, because the normalization for $\boldsymbol{p}$ is not trivial anyway. The same comment applies to other models below.

Similarly, the subgroupwise independence model is defined as

$$p(\boldsymbol{i}) = \prod_{j=1}^{J} \prod_{\substack{k=1 \\ \boldsymbol{i}_{jk} \subset \boldsymbol{i}}}^{m_j} q_{jk}(\boldsymbol{i}_{jk}) \tag{12.4}$$

for some parameters $q_{jk}(\cdot)$, and the groupwise independence model is defined as

$$p(\boldsymbol{i}) = \prod_{j=1}^{J} q_j(\boldsymbol{i}_j) \tag{12.5}$$

for some parameters $q_j(\cdot)$.

In this chapter, we treat these models as the *null models* and give testing procedures to assess their fitting to observed data following the general theory in Chap. 2.

12.2.2 Models for Allele Frequency Data

Next we consider the allele frequency data. First we consider the models for the genotype frequency data. We assume that there are J distinct loci. In the locus j, there are m_j distinct alleles, $A_{j1},\ldots,A_{jm_j}$. In this case, we can imagine that each individual selects two alleles for each locus with replacement. Therefore the set of the combinations is written as

$$\mathscr{I} = \{\boldsymbol{i} = (i_{11}i_{12})(i_{21}i_{22})\ldots(i_{J1}i_{J2}) \mid 1 \le i_{j1} \le i_{j2} \le m_j,\ j = 1,\ldots,J\}.$$

For the genotype frequency data, we consider two models of hierarchical structure, namely the *genotypewise independence model*

$$p(\boldsymbol{i}) = \prod_{j=1}^{J} q_j(i_{j1}i_{j2}) \tag{12.6}$$

and the Hardy–Weinberg model

$$p(\boldsymbol{i}) = \prod_{j=1}^{J} \tilde{q}_j(i_{j1}i_{j2}), \tag{12.7}$$

where

$$\tilde{q}_j(i_{j1}i_{j2}) = \begin{cases} q_j(i_{j1})^2 & \text{if } i_{j1} = i_{j2}, \\ 2q_j(i_{j1})q_j(i_{j2}) & \text{if } i_{j1} \neq i_{j2}. \end{cases} \tag{12.8}$$

Note that for both cases the sufficient statistic $\boldsymbol{t}$ can be written as $\boldsymbol{t} = A\boldsymbol{x}$ for an appropriate matrix A as shown in Sect. 12.5.2.

Next we consider the diplotype frequency data. In order to illustrate the difference between genotype data and diplotype data, consider a simple case of $J = 2, m_1 = m_2 = 2$ and suppose that genotypes of $n = 4$ individuals are given as

$$\{A_{11}A_{11}, A_{21}A_{21}\},\ \{A_{11}A_{11}, A_{21}A_{22}\},\ \{A_{11}A_{12}, A_{21}A_{21}\},\ \{A_{11}A_{12}, A_{21}A_{22}\}.$$

In these genotype data, for an individual who has a homozygote genotype on at least one locus, the diplotypes are uniquely determined. However, for the fourth individual who has the genotype $\{A_{11}A_{12}, A_{21}A_{22}\}$, there are two possible diplotypes: $\{(A_{11}, A_{21}), (A_{12}, A_{22})\}$ and $\{(A_{11}, A_{22}), (A_{12}, A_{21})\}$.

Now suppose that information on diplotypes is available. The set of combinations for the diplotype data is given as

$$\mathscr{I} = \{\boldsymbol{i} = \boldsymbol{i}_1\boldsymbol{i}_2 = (i_{11}\cdots i_{J1})(i_{12}\cdots i_{J2}) \mid 1 \le i_{j1}, i_{j2} \le m_j,\ j = 1, \ldots, J\}.$$

In order to determine the order of $\boldsymbol{i}_1 = (i_{11}\ldots i_{J1})$ and $\boldsymbol{i}_2 = (i_{12}\ldots i_{J2})$ uniquely, we assume that these two are lexicographically ordered; that is, there exists some j such that

$$i_{11} = i_{12}, \ldots, i_{j-1,1} = i_{j-1,2},\ i_{j1} < i_{j2}$$

unless $\boldsymbol{i}_1 = \boldsymbol{i}_2$.

For the parameter $\boldsymbol{p} = \{p(\boldsymbol{i})\}$ where $p(\boldsymbol{i})$ is the probability for the diplotype $\boldsymbol{i}$, we can consider the same models as for the genotype case. Corresponding to the null hypothesis that diplotype data do not contain more information than the genotype data, we can consider the genotypewise independence model (12.6) and the Hardy–Weinberg model (12.7). A sufficient statistic for these models is the same as we have seen above.

If these models are rejected, we can further test independence in diplotype data. For example, we can consider a haplotypewise Hardy–Weinberg model

$$p(\boldsymbol{i}) = p(\boldsymbol{i}_1\boldsymbol{i}_2) = \begin{cases} q(\boldsymbol{i}_1)^2 & \text{if } \boldsymbol{i}_1 = \boldsymbol{i}_2, \\ 2q(\boldsymbol{i}_1)q(\boldsymbol{i}_2) & \text{if } \boldsymbol{i}_1 \neq \boldsymbol{i}_2. \end{cases}$$

A sufficient statistic for this model is given by the set of frequencies of each haplotype and the conditional test can be performed as in the case of the Hardy–Weinberg model for a single gene by formally identifying each haplotype as an allele.

12.3 Gröbner Basis for Segre–Veronese Configuration

In this section, we introduce toric ideals of algebras of the Segre–Veronese type [109] with a generalization to fit statistical applications in this chapter. We use the notation and the terminology of Sect. 3.

Fix integers $\tau \geq 2, M \geq 1$ and sets of integers $\mathbf{b} = \{b_1, \dots, b_M\}$, $\mathbf{c} = \{c_1, \dots, c_M\}$, $\mathbf{r} = \{r_1, \dots, r_M\}$, and $\mathbf{s} = \{s_1, \dots, s_M\}$ such that

(i) $0 \leq c_i \leq b_i$ for all $1 \leq i \leq M$,
(ii) $1 \leq s_i \leq r_i \leq \nu$ for all $1 \leq i \leq M$.

Let $A_{\tau,\mathbf{b},\mathbf{c},\mathbf{r},\mathbf{s}} \subset \mathbb{N}^\nu$ denote the configuration consisting of all nonnegative integer vectors $(f_1, f_2, \dots, f_\nu) \in \mathbb{N}^\nu$ such that

(i) $\sum_{j=1}^{\nu} f_j = \tau$.
(ii) $c_i \leq \sum_{j=s_i}^{r_i} f_j \leq b_i$ for all $1 \leq i \leq M$.

Let $k[A_{\tau,\mathbf{b},\mathbf{c},\mathbf{r},\mathbf{s}}]$ denote the semigroup ring generated by all monomials $\prod_{j=1}^{\nu} q_j{}^{f_j}$ over the field k and call it an *algebra of Segre–Veronese type*. Note that the present definition generalizes the definition in [109].

Several popular classes of semigroup rings are Segre–Veronese type algebras. If $M = 2$, $\tau = 2$, $b_1 = b_2 = c_1 = c_2 = 1$, $s_1 = 1$, $s_2 = r_1 + 1$ and $r_2 = \nu$, then the semigroup ring $k[A_{\tau,\mathbf{b},\mathbf{c},\mathbf{r},\mathbf{s}}]$ is the Segre product of polynomial rings $k[q_1, \dots, q_{r_1}]$ and $k[q_{r_1+1}, \dots, q_\nu]$. On the other hand, if $M = \nu$, $s_i = r_i = i$, $b_i = \tau$, and $c_i = 0$ for all $1 \leq i \leq M$, then the semigroup ring $k[A_{\tau,\mathbf{b},\mathbf{c},\mathbf{r},\mathbf{s}}]$ is the classical τth Veronese subring of the polynomial ring $k[q_1, \dots, q_\nu]$. Moreover, if $M = \nu$, $s_i = r_i = i$, $b_i = 1$, and $c_i = 0$ for all $1 \leq i \leq M$, then the semigroup ring $k[A_{\tau,\mathbf{b},\mathbf{c},\mathbf{r},\mathbf{s}}]$ is the τth squarefree Veronese subring of the polynomial ring $k[q_1, \dots, q_\nu]$. In addition, Veronese type algebras (i.e., $M = \nu$, $s_i = r_i = i$, and $c_i = 0$ for all $1 \leq i \leq M$) are studied in [48] and [139].

Let $k[Y]$ denote the polynomial ring with the set of variables

$$\left\{ y_{j_1 j_2 \cdots j_\tau} \;\middle|\; 1 \leq j_1 \leq j_2 \leq \cdots \leq j_\tau \leq \nu,\ \prod_{k=1}^{\tau} q_{j_k} \in \{\boldsymbol{q}^{\boldsymbol{a}_1}, \dots, \boldsymbol{q}^{\boldsymbol{a}_\eta}\} \right\},$$

where $k[A_{\tau,\mathbf{b},\mathbf{c},\mathbf{r},\mathbf{s}}] = k[\boldsymbol{q}^{\boldsymbol{a}_1}, \dots, \boldsymbol{q}^{\boldsymbol{a}_\eta}]$. The toric ideal $I_{A_{\tau,\mathbf{b},\mathbf{c},\mathbf{r},\mathbf{s}}}$ is the kernel of the surjective homomorphism $\pi : k[Y] \longrightarrow k[A_{\tau,\mathbf{b},\mathbf{c},\mathbf{r},\mathbf{s}}]$ defined by $\pi(y_{j_1 j_2 \cdots j_\tau}) = \prod_{k=1}^{\tau} q_{j_k}$. A monomial $y_{\alpha_1 \alpha_2 \cdots \alpha_\tau} y_{\beta_1 \beta_2 \cdots \beta_\tau} \cdots y_{\gamma_1 \gamma_2 \cdots \gamma_\tau}$ is called *sorted* if

$$\alpha_1 \leq \beta_1 \leq \cdots \leq \gamma_1 \leq \alpha_2 \leq \beta_2 \leq \cdots \leq \gamma_2 \leq \cdots \leq \alpha_\tau \leq \beta_\tau \leq \cdots \leq \gamma_\tau.$$

Let sort$(\cdot)$ denote the operator that takes any string over the alphabet $\{1, 2, \dots, d\}$ and sorts it into weakly increasing order. Then the quadratic Gröbner basis of toric ideal $I_{A_{\tau,\mathbf{b},\mathbf{c},\mathbf{r},\mathbf{s}}}$ is given as follows.

Theorem 12.1. *There exists a monomial order on $k[Y]$ such that the set of all binomials*

$$\{y_{\alpha_1\alpha_2\cdots\alpha_\tau}y_{\beta_1\beta_2\cdots\beta_\tau} - y_{\gamma_1\gamma_3\cdots\gamma_{2\tau-1}}y_{\gamma_2\gamma_4\cdots\gamma_{2\tau}} \mid \mathrm{sort}(\alpha_1\beta_1\alpha_2\beta_2\cdots\alpha_\tau\beta_\tau) = \gamma_1\gamma_2\cdots\gamma_{2\tau}\} \tag{12.9}$$

is the reduced Gröbner basis of the toric ideal $I_{A_{\tau,\mathbf{b},\mathbf{c},\mathbf{r},\mathbf{s}}}$. The initial ideal is generated by square-free quadratic (nonsorted) monomials.

In particular, the set of all integer vectors corresponding to the above binomials is a Markov basis. Furthermore, the set is minimal as a Markov basis.

Proof. The basic idea of the proof appears in Theorem 14.2 in [139].

Let $\mathscr{G}$ be the above set of binomials. First we show that $\mathscr{G} \subset I_{A_{\tau,\mathbf{b},\mathbf{c},\mathbf{r},\mathbf{s}}}$. Suppose that $m = y_{\alpha_1\alpha_2\cdots\alpha_\tau}y_{\beta_1\beta_2\cdots\beta_\tau}$ is not sorted and let

$$\gamma_1\gamma_2\cdots\gamma_{2\tau} = \mathrm{sort}(\alpha_1\beta_1\alpha_2\beta_2\cdots\alpha_\tau\beta_\tau).$$

Then, m is square-free because the monomial $y^2_{\alpha_1\alpha_2\cdots\alpha_\tau}$ is sorted. The binomial $y_{\alpha_1\alpha_2\cdots\alpha_\tau}y_{\beta_1\beta_2\cdots\beta_\tau} - y_{\alpha'_1\alpha'_2\cdots\alpha'_\tau}y_{\beta'_1\beta'_2\cdots\beta'_\tau} \in k[Y]$ belongs to $I_{A_{\tau,\mathbf{b},\mathbf{c},\mathbf{r},\mathbf{s}}}$ if and only if $\mathrm{sort}(\alpha_1\alpha_2\cdots\alpha_\tau\beta_1\beta_2\cdots\beta_\tau) = \mathrm{sort}(\alpha'_1\alpha'_2\cdots\alpha'_\tau\beta'_1\beta'_2\cdots\beta'_\tau)$, thus it is sufficient to show that both $y_{\gamma_1\gamma_3\cdots\gamma_{2\tau-1}}$ and $y_{\gamma_2\gamma_4\cdots\gamma_{2\tau}}$ are variables of $k[Y]$. For $1 \le i \le M$, let $\rho_i = |\{j \mid s_i \le \gamma_{2j-1} \le r_i\}|$ and $\sigma_i = |\{j \mid s_i \le \gamma_{2j} \le r_i\}|$. Because $\gamma_1 \le \gamma_2 \le \cdots \le \gamma_{2\tau}$, ρ_i and σ_i are either equal or they differ by one for each i. If $\rho_i \le \sigma_i$, then $0 \le \sigma_i - \rho_i \le 1$. Because $2c_i \le \rho_i + \sigma_i \le 2b_i$, we have $\sigma_i \le b_i + 1/2$ and $c_i - 1/2 \le \rho_i$. Thus $c_i \le \rho_i \le \sigma_i \le b_i$. If $\rho_i > \sigma_i$, then $\rho_i - \sigma_i = 1$. Because $2c_i \le \rho_i + \sigma_i \le 2b_i$, we have $\rho_i \le b_i + 1/2$ and $c_i - 1/2 \le \sigma_i$. Thus $c_i \le \sigma_i < \rho_i \le b_i$. Hence $y_{\gamma_1\gamma_3\cdots\gamma_{2\tau-1}}$ and $y_{\gamma_2\gamma_4\cdots\gamma_{2\tau}}$ are variables of $k[Y]$.

By virtue of the relation between the reduction of a monomial by $\mathscr{G}$ and sorting of the indices of a monomial, it follows that there exists a monomial order such that, for any binomial in $\mathscr{G}$, the first monomial is the initial monomial. See also Theorem 3.12 in [139].

Suppose that $\mathscr{G}$ is not a Gröbner basis. By Theorem 3.1 there exists a binomial $f \in I_{A_{\tau,\mathbf{b},\mathbf{c},\mathbf{r},\mathbf{s}}}$ such that both monomials in f are sorted. This means that $f = 0$ and f is not a binomial. Hence $\mathscr{G}$ is a Gröbner basis of $I_{A_{\tau,\mathbf{b},\mathbf{c},\mathbf{r},\mathbf{s}}}$. It is easy to see that the Gröbner basis $\mathscr{G}$ is reduced and a minimal set of generators of $I_{A_{\tau,\mathbf{b},\mathbf{c},\mathbf{r},\mathbf{s}}}$. □

Theorem 12.1 states that the minimal Markov basis for the Segre–Veronese configuration $I_{A_{\tau,\mathbf{b},\mathbf{c},\mathbf{r},\mathbf{s}}}$ is constructed as the basic moves defined by (12.9). The theory of Segre–Veronese configuration was further generalized to a class of configurations called *nested configurations*. Toric ideals for nested configurations possess many nice properties. See Aoki et al. [7], and Ohsugi and Hibi [114].

12.4 Sampling from the Gröbner Basis for the Segre–Veronese Configuration

Here we describe how to run a Markov chain using the Gröbner basis given in Theorem 12.1.

First, given a configuration A in (12.2), we check that (with appropriate reordering of rows) that A is indeed a configuration of Segre–Veronese type. It is easy to check that our models in Sects. 12.2.1 and 12.2.2 are of Segre–Veronese type, because the restrictions on choices are imposed separately for each group or each subgroup. Recall that each column of A consists of nonnegative integers whose sum τ is common.

We now associate with each column $\boldsymbol{a}(\boldsymbol{i})$ of A a set of indices indicating the rows with positive elements $a_j(\boldsymbol{i}) > 0$ and a particular index j is repeated $a_j(\boldsymbol{i})$ times. For example, if $v = 4, \tau = 3$, and $\boldsymbol{a}(\boldsymbol{i}) = (1,0,2,0)'$, then row 1 appears once and row 3 appears twice in $\boldsymbol{a}(\boldsymbol{i})$. Therefore we associate the index $(1,3,3)$ with $\boldsymbol{a}(\boldsymbol{i})$. We can consider the set of indices as $\tau \times \eta$ matrix $\tilde{A}$. Note that $\tilde{A}$ and A carry the same information.

Given $\tilde{A}$, we can choose a random element of the reduced Gröbner basis of Theorem 12.1 as follows. Choose two columns (i.e., choose two cells from $\mathscr{I}$) of $\tilde{A}$ and sort $2 \times \tau$ elements of these two columns. From the sorted elements, pick alternate elements and form two new sets of indices. For example, if $\tau = 3$ and the two chosen columns of $\tilde{A}$ are $(1,3,3)$ and $(1,2,4)$, then by sorting these six elements we obtain $(1,1,2,3,3,4)$. Picking alternate elements produces $(1,2,3)$ and $(1,3,4)$. These new sets of indices correspond to (a possibly overlapping) two columns of $\tilde{A}$, hence to two cells of $\mathscr{I}$. Now the difference of the two original columns and the two sorted columns of $\tilde{A}$ correspond to a random binomial in (12.9). It should be noted that when the sorted columns coincide with the original columns, then we discard these columns and choose two other columns. Then we can perform an MCMC procedure as explained in Chap. 2.

12.5 Numerical Examples

In this section we present numerical experiments on NCT data and diplotype frequency data.

12.5.1 *The Analysis of NCT Data*

First we consider the analysis of NCT data concerning selections in social studies and science. Because NCUEE currently does not provide cross-tabulations of frequencies of choices across the major subjects, we cannot evaluate the p-value

Table 12.6 The data set of the number of examinees in NCT in 2006 ($n = 195,094$)

	ContS	Ethics	P&E
WH	32,352	8,839	8,338
JH	51,573	8,684	14,499
Geo	59,588	4,046	7,175

	CSiA	Chem	Phys	Earth
CSiB	1,648	1,572	169	4,012
Bio	21,392	55,583	1,416	1,845
Phys	3,286	102,856	–	–
Earth	522	793	–	–

of the actual data. However, for the models in Sect. 12.2.1, the sufficient statistics (the marginal frequencies) can be obtained from Tables 12.1–12.5. Therefore in this section we evaluate the conditional null distribution of Pearson's χ^2 statistic by MCMC and compare it to the asymptotic χ^2 distribution.

In Sect. 12.2.1, we considered three models, the complete independence model, subgroupwise independence model, and groupwise independence model, for the setting of groupwise selection problems. Note, however, that the subgroupwise independence model coincides with the groupwise independence model for NCT data, because $c_{jk} \le 1$ for all j and k. Therefore we consider fitting the complete independence model and the group-wise independence model for NCT data.

As we have seen in Sect. 12.1.2, there are many kinds of choices for each examinee. However, it may be natural to treat some similar subjects as one subject. For example, WHA and WHB may well be treated as WH, ChemI and Chem IA may well be treated as Chem, and so on. As a result, we consider the following aggregation of subjects.

- In social studies: WH = {WHA, WHB}, JH = {JHA, JHB}, Geo = {GeoA, GeoB}
- In science: CSiB = {CSiB, ISci}, Bio = {BioI, BioIA}, Chem = {ChemI, ChemIA}, Phys = {PhysI, PhysIA}, Earth = {EarthI, EarthIA}

In our analysis, we take a look at examinees selecting two subjects for social studies and two subjects for science. Therefore

$$J = 2, m_1 = 2, m_2 = 3, m_{11} = m_{12} = 3, m_{21} = m_{22} = m_{23} = 2,$$
$$c_{11} = c_{12} = 1, (c_{21}, c_{22}, c_{23}) = (1,1,0) \text{ or } (1,0,1) \text{ or } (0,1,1).$$

The number of possible combinations is then $\nu = |\mathscr{I}| = 3 \cdot 3 \times 3 \cdot 2^2 = 108$. Accordingly our sample size is $n = 195,094$, which is the number of examinees selecting two subjects for science from Table 12.3. Our data set is shown in Table 12.6.

From Table 12.6, we can calculate the maximum likelihood estimates of the numbers of the examinees selecting each combination of subjects. The sufficient statistics under the complete independence model are the numbers of the examinees selecting each subject, whereas the sufficient statistics under the groupwise independence model are the numbers of the examinees selecting each combination of subjects in the same group. The maximum likelihood estimates calculated from the sufficient statistics are shown in Table 12.7. For the complete independence model the maximum likelihood estimates can be calculated as in Sect. 5.2 of [26].

The configuration A for the complete independence model is written as

$$A = \begin{pmatrix} E_3 \otimes \mathbf{1}_3' \otimes \mathbf{1}_{12}' \\ \mathbf{1}_3' \otimes E_3 \otimes \mathbf{1}_{12}' \\ \mathbf{1}_9' \quad \otimes B \end{pmatrix}$$

and the configuration A for the groupwise independence model is written as

$$A = \begin{pmatrix} E_9 \otimes \mathbf{1}_{12}' \\ \mathbf{1}_9' \otimes E_{12}' \end{pmatrix},$$

where E_n is the $n \times n$ identity matrix, $\mathbf{1}_n = (1, \ldots, 1)'$ is the $n \times 1$ column vector of 1s, $\otimes$ denotes the Kronecker product, and

$$B = \begin{pmatrix} 111100000000 \\ 000011110000 \\ 100010001100 \\ 010001000011 \\ 001000101010 \\ 000100010101 \end{pmatrix}.$$

Note that the configuration B is the vertex-edge incidence matrix of the $(2,2,2)$ complete multipartite graph. Quadratic Gröbner bases of toric ideals arising from complete multipartite graphs are studied in [109].

Given these configurations we can easily run a Markov chain as discussed in Chap. 2. After $5,000,000$ burn-in steps, we construct $10,000$ Monte Carlo samples. Two figures in Fig. 12.1 show histograms of the Monte Carlo sampling generated from the exact conditional distribution of Pearson's chi-square statistics for the NCT data under the complete independence model and the groupwise independence model along with the corresponding asymptotic distributions χ^2_{98} and χ^2_{88}, respectively.

12.5.2 The Analysis of Allele Frequency Data

Next we give a numerical example of genome data. Table 12.8 shows diplotype frequencies on the three loci, T-549C (locus 1), C-441T (locus 2), and T-197C (locus 3) in the human genome 14q22.1, which are given in [108]. Although the data are used for the genetic association studies in [108], we simply consider fitting our models. As an example, we only consider the diplotype data of patients in the population of blacks ($n = 79$).

Table 12.7 MLE of the number of the examinees selecting each combination of subjects under the complete independence model (upper) and the groupwise independence model (lower)

	WH			JH			Geo		
	ContS	Ethics	P&E	ContS	Ethics	P&E	ContS	Ethics	P&E
CSiB,CSiA	180.96	27.20	37.84	273.12	41.05	57.12	258.70	38.88	54.10
	273.28	74.66	70.43	435.65	73.36	122.48	503.35	34.18	60.61
CSiB,Chem	1,083.82	162.89	226.65	1,635.85	245.86	342.10	1,549.48	232.88	324.03
	260.68	71.22	67.18	415.56	69.97	116.83	480.14	32.60	57.81
CSiB,Phys	110.04	16.54	23.01	166.09	24.96	34.73	157.32	23.64	32.90
	28.02	7.66	7.22	44.68	7.52	12.56	51.62	3.50	6.22
CSiB,Earth	7.33	1.10	1.53	11.06	1.66	2.31	10.47	1.57	2.19
	665.30	181.77	171.47	1,060.57	178.58	298.16	1,225.39	83.20	147.55
Bio,CSiA	1,961.78	294.84	410.26	2,960.99	445.02	619.21	2,804.66	421.52	586.52
	3,547.39	969.19	914.26	5,654.96	952.20	1,589.81	6,533.81	443.64	786.74
Bio,Chem	11,749.94	1,765.93	2,457.19	17,734.63	2,665.39	3,708.74	16,798.27	2,524.66	3,512.92
	9,217.20	2,518.26	2,375.53	14,693.34	2,474.10	4,130.82	16,976.84	1,152.72	2,044.18
Bio,Phys	1,193.01	179.30	249.49	1,800.65	270.63	376.56	1,705.58	256.34	356.68
	234.81	64.15	60.52	374.32	63.03	105.23	432.49	29.37	52.08
Bio,Earth	79.43	11.94	16.61	119.88	18.02	25.07	113.55	17.07	23.75
	305.95	83.59	78.85	487.72	82.12	137.12	563.52	38.26	67.85
CSiA,Phys	2,691.94	404.58	562.95	4,063.04	610.65	849.68	3,848.52	578.41	804.82
	544.91	148.88	140.44	868.65	146.27	244.21	1,003.65	68.15	120.85
CSiA,Earth	179.22	26.94	37.48	270.50	40.65	56.57	256.22	38.51	53.58
	86.56	23.65	22.31	137.99	23.24	38.79	159.44	10.83	19.20
Bio,Phys	16,123.14	2,423.20	3,371.73	24,335.27	3,657.42	5,089.09	23,050.40	3,464.31	4,820.39
	17,056.38	4,660.03	4,395.90	27,189.93	4,578.31	7,644.05	31,415.54	2,133.10	3,782.75
Bio,Earth	1,073.41	161.33	224.48	1,620.14	243.50	338.81	1,534.60	230.64	320.92
	131.50	35.93	33.89	209.63	35.30	58.93	242.21	16.45	29.16

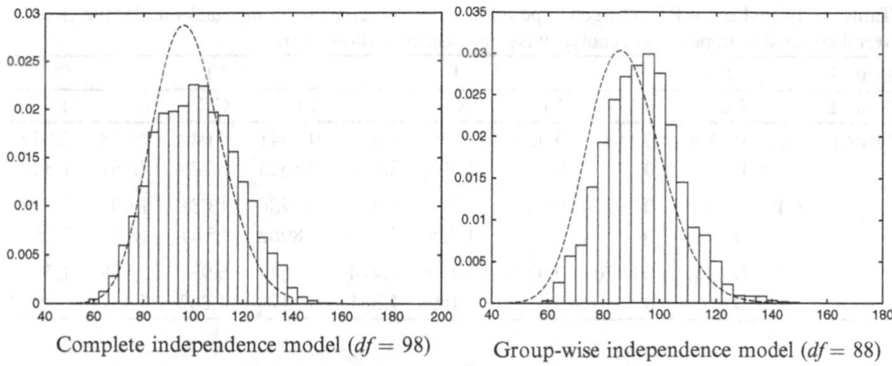

Fig. 12.1 Asymptotic and Monte Carlo sampling distributions of NCT data

Table 12.8 PTGDR diplotype frequencies among patients and controls in each population. (The order of the SNPs in the haplotype is T-549C, C-441T, and T-197C.)

	Whites		Blacks	
Diplotype	Controls	Patients	Controls	Patients
CCT/CCT	16	78	7	10
CCT/TTT	27	106	12	27
CCT/TCT	48	93	4	12
CCT/CCC	17	45	3	9
TTT/TTT	9	43	2	7
TTT/TCT	34	60	8	6
TTT/CCC	4	28	1	6
TCT/TCT	11	20	7	0
TCT/CCC	6	35	1	2
CCC/CCC	1	8	0	0

Table 12.9 The genotype frequencies for patients among blacks of PTGDR data

locus 3		CC			CT			TT		
locus 2		CC	CT	TT	CC	CT	TT	CC	CT	TT
locus 1	CC	0	0	0	9	0	0	10	0	0
	CT	0	0	0	2	6	0	12	27	0
	TT	0	0	0	0	0	0	0	6	7

First we consider the analysis of genotype frequency data. Although Table 12.8 is diplotype frequency data, here we ignore the information on the haplotypes and simply treat them as genotype frequency data. Because $n = 3$ and $m_1 = m_2 = m_3 = 2$, there are $3^3 = 27$ distinct sets of genotypes (i.e., $|\mathscr{I}| = 27$), and only 8 distinct haplotypes appear in Table 12.8. Table 12.9 is the set of genotype frequencies of patients in the population of blacks.

Under the genotypewise independence model (12.6), the sufficient statistic is the genotype frequency data for each locus. On the other hand, under the Hardy–Weinberg model (12.7), the sufficient statistic is the allele frequency data for each

Table 12.10 MLE for PTGDR genotype frequencies of patients among blacks under the Hardy–Weinberg model (upper) and genotypewise independence model (lower)

locus 3		CC			CT			TT		
locus 2		CC	CT	TT	CC	CT	TT	CC	CT	TT
locus 1	CC	0.1169	0.1180	0.0298	1.939	1.958	0.4941	8.042	8.118	2.049
		0	0	0	1.708	2.018	0.3623	6.229	7.361	1.321
	CT	0.2008	0.2027	0.0512	3.331	3.362	0.8486	13.81	13.94	3.519
		0	0	0	4.225	4.993	0.8962	15.41	18.21	3.268
	TT	0.0862	0.0870	0.0220	1.430	1.444	0.3644	5.931	5.988	1.511
		0	0	0	1.169	1.381	0.2479	4.262	5.037	0.9040

locus, and the genotype frequencies for each locus are estimated by the Hardy–Weinberg law. Accordingly, the maximum likelihood estimates for the combinations of the genotype frequencies are calculated as Table 12.10.

The configuration A for the Hardy–Weinberg model is written as

$$A = \begin{pmatrix} 222222222\ 111111111\ 000000000 \\ 000000000\ 111111111\ 222222222 \\ 222111000\ 222111000\ 222111000 \\ 000111222\ 000111222\ 000111222 \\ 210210210\ 210210210\ 210210210 \\ 012012012\ 012012012\ 012012012 \end{pmatrix}$$

and the configuration A for the genotypewise independence model is written as

$$A = \begin{pmatrix} E_3 \otimes \mathbf{1}_3' \otimes \mathbf{1}_3' \\ \mathbf{1}_3' \otimes E_3 \otimes \mathbf{1}_3' \\ \mathbf{1}_3' \otimes \mathbf{1}_3' \otimes E_3' \end{pmatrix}.$$

Because these two configurations are of the Segre–Veronese type, again we can easily perform MCMC sampling as discussed in Chap. 2. After 100,000 burn-in steps, we construct 10,000 Monte Carlo samples. Two figures in Fig. 12.2 show histograms of the Monte Carlo sampling generated from the exact conditional distribution of Pearson goodness-of-fit χ^2 statistics for the PTGDR genotype frequency data under the Hardy–Weinberg model and the genotypewise independence model along with the corresponding asymptotic distributions χ^2_{24} and χ^2_{21}, respectively.

From the Monte Carlo samples, we can also estimate the p-values for each null model. The values of Pearson goodness-of-fit χ^2 for the PTGDR genotype frequency data of Table 12.9 are $\chi^2 = 88.26$ under the Hardy–Weinberg models, whereas $\chi^2 = 103.37$ under the genotype-wise independence model. These values are highly significant ($p < 0.01$ for both models), which implies the susceptibility of the particular haplotypes.

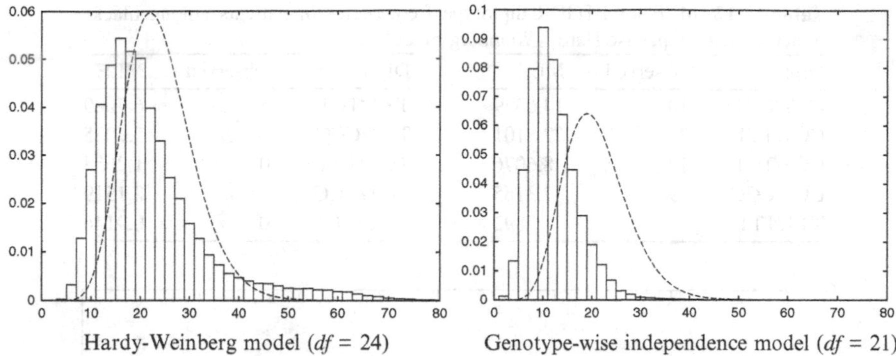

Fig. 12.2 Asymptotic and Monte Carlo sampling distributions of PTGDR genotype frequency data

Table 12.11 Observed frequency and MLE under the Hardy–Weinberg model for PTGDR haplotype frequencies of patients among blacks

Haplotype	Observed	MLE under HW	Haplotype	Observed	MLE under HW
CCC	17	6.078	TCC	0	5.220
CCT	68	50.410	TCT	20	43.293
CTC	0	3.068	TTC	0	2.635
CTT	0	25.445	TTT	53	21.853

Next we consider the analysis of the diplotype frequency data. In this case of $n = 3$ and $m_1 = m_2 = m_3 = 2$, there are $2^3 = 8$ distinct haplotypes, and there are

$$|\mathscr{I}| = 8 + \binom{8}{2} = 36$$

distinct diplotypes, whereas there are only four haplotypes and ten diplotypes in Table 12.8. The numbers of each haplotype are calculated as the second column of Table 12.11. Under the Hardy–Weinberg model, the haplotype frequencies are estimated proportionally to the allele frequencies, which are shown as the third column of Table 12.11.

The maximum likelihood estimates of the diplotype frequencies under the Hardy–Weinberg model are calculated from the maximum likelihood estimates for each haplotype. These values coincide with appropriate fractions of the values for the corresponding combinations of the genotypes in Table 12.10. For example, the MLE for the diplotype CCT/CCT coincides with the MLE for the combination of the genotypes (CC,CC,TT) in Table 12.10, whereas the MLEs for the diplotype CCC/TTT, CCT/TTC, CTC/TCT, CTT/TCC coincide with the $\frac{1}{4}$ fraction of the MLE for the combination of the genotypes (CT,CT,CT), and so on. Because we know that the Hardy–Weinberg model is highly significantly rejected, it is natural to consider the haplotypewise Hardy–Weinberg model given in Sect. 12.2.2.

Table 12.12 MLE for PTGDR diplotype frequencies of patients among blacks under the haplotypewise Hardy–Weinberg model

Diplotype	Observed	MLE	Diplotype	Observed	MLE
CCT/CCT	10	14.6329	TTT/TCT	6	6.7089
CCT/TTT	27	22.8101	TTT/CCC	6	5.7025
CCT/TCT	12	8.6076	TCT/TCT	0	1.2658
CCT/CCC	9	7.3165	TCT/CCC	2	2.1519
TTT/TTT	7	8.8892	CCC/CCC	0	0.9146

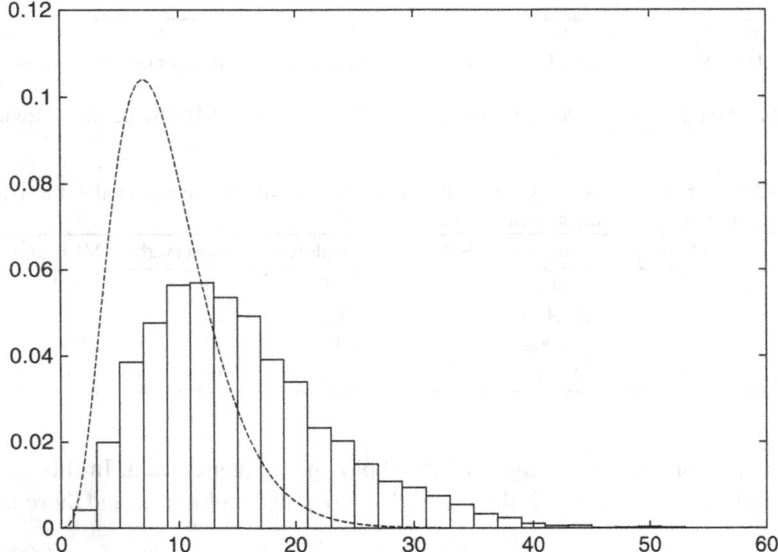

Fig. 12.3 Asymptotic and Monte Carlo sampling distributions of PTGDR diplotype frequency data under the haplotypewise Hardy–Weinberg model ($df = 9$)

Table 12.12 shows the maximum likelihood estimates under the haplotypewise Hardy–Weinberg model. It should be noted that the MLE for the other diplotypes are all zeros.

We perform the Markov chain Monte Carlo sampling for the haplotypewise Hardy–Weinberg model. The configuration A for this model is written as

$$A = \begin{pmatrix} 200000001111111000000000000000000000 \\ 020000001000000111111000000000000000 \\ 002000000100000100000111110000000000 \\ 000200000010000010000100001111000000 \\ 000020000001000001000010001000111000 \\ 000002000000100000100001000100100110 \\ 000000200000010000010000100010010101 \\ 000000020000001000001000010001001011 \end{pmatrix},$$

which is obviously of the Segre–Veronese type. We give a histogram of the Monte Carlo sampling generated from the exact conditional distribution of Pearson's chi-square statistics for the PTGDR diplotype frequency data under the haplotypewise Hardy–Weinberg model, along with the corresponding asymptotic distributions χ^2_9 in Fig. 12.3.

The p-value for this model is estimated as 0.8927 with the estimated standard deviation 0.0029. (We also discard the first 100,000 samples, and use a batching method to obtain an estimate of variance; see [82] and [128].) Note that the asymptotic p-value based on χ^2_9 is 0.6741.

Chapter 13
The Set of Moves Connecting Specific Fibers

13.1 Discrete Logistic Regression Model with One Covariate

Let $\{1,\dots,J\}$ be the set of levels of a covariate and let x_{1j} and x_{2j}, $j=1,\dots,J$, be the numbers of successes and failures for a covariate j, respectively. Let p_j be the probability for success. Assume that x_{ij} be distributed as a binomial distribution

$$x_{ij} \sim \mathrm{Bin}(x_{+j}, p_j).$$

Then the binary logistic regression model with one discrete covariate is described as

$$\mathrm{logit}(p_j) = \log \frac{p_j}{1-p_j} = \alpha + \beta j, \qquad j = 1,\dots,J. \tag{13.1}$$

A sufficient statistic for the model is $\boldsymbol{t} = (x_{1+}, \sum_{j=1}^{J} j x_{+j})$. Usually the column sums $x_{+1},\dots,x_{+J}$ are also fixed and positive by a sampling scheme. In order to perform conditional tests, we need the set of moves connecting contingency tables not only sharing $\boldsymbol{t}$ but satisfying $x_{ij} \le x_{+j}$ for $i=1,2$ and $j=1,\dots,J$.

Consider the following Poisson logistic regression model

$$x_{ij} \sim Po(\lambda_{ij}), \quad \lambda_{1j} = \lambda p_j, \quad \lambda_{2j} = \lambda(1-p_j), \tag{13.2}$$

where p_j satisfies (13.1). A sufficient statistic for the model (13.2) is

$$\boldsymbol{t}_1 := (x_{1+}, x_{+1}, \dots, x_{+J}, \sum_{j=1}^{J} j x_{+j}) = \boldsymbol{t} \cup (x_{+1}, \dots, x_{+J}).$$

We note that a Markov basis of (13.2) also connects every fiber for (13.1). In the rest of this section we discuss a Markov basis for the Poisson logistic regression model (13.2).

S. Aoki et al., *Markov Bases in Algebraic Statistics*, Springer Series in Statistics 199, DOI 10.1007/978-1-4614-3719-2_13,

Table 13.1 Maximal degrees and numbers of moves of the minimal Markov basis for the model (13.2)

J	10	11	12	13	14	15	16
Max deg	18	20	22	24	26	28	30
# of moves	1,830	3,916	8,569	16,968	34,355	66,066	123,330

Moves $\mathbf{z} = \{z_{ij}\}$ for the model satisfy $(z_{1+}, z_{+1}, \ldots, z_{+J}) = \mathbf{0}$ and

$$\sum_{j=1}^{J} j z_{+j} = 0. \tag{13.3}$$

The configuration for this model is written as a Lawrence lifting by

$$\Lambda(A) = \begin{pmatrix} A & 0 \\ E_J & E_J \end{pmatrix}, \qquad A = \begin{pmatrix} 1\,1 \ldots 1 \\ 1\,2 \ldots J \end{pmatrix}, \tag{13.4}$$

where E_J denotes the $J \times J$ identity matrix.

Table 13.1 presents maximal degrees and numbers of moves of minimal Markov bases for $\Lambda(A)$ computed by 4ti2. In general Markov bases for $\Lambda(A)$ become large and very complicated as seen from the table. As mentioned earlier, however, x_{+j} can be assumed to be positive. Actually many moves in a Markov basis for (13.2) are required only for fibers with $x_{+j} = 0$.

Now we introduce the following subset of Markov bases consisting only of degree 4 moves.

Definition 13.1. Let $\boldsymbol{e}_j$ be defined by a $2 \times J$ integer array with 1 in the $(1, j)$-cell, -1 in the $(2, j)$-cell, and 0 everywhere else. Define $\mathscr{B}_1$ by the set of moves $\mathbf{z} = (z_{ij})$ satisfying

1. $\mathbf{z} = \boldsymbol{e}_{j_1} - \boldsymbol{e}_{j_2} - \boldsymbol{e}_{j_3} + \boldsymbol{e}_{j_4}$.
2. $1 \le j_1 < j_2 \le j_3 < j_4 \le J$.
3. $j_1 - j_2 = j_3 - j_4$.

Then $\mathbf{z} \in \mathscr{B}_1$ is expressed as

$$\mathbf{z} = \begin{array}{|cccc|} \multicolumn{1}{c}{j_1} & j_2 & j_3 & \multicolumn{1}{c}{j_4} \\ \hline 1 & -1 & -1 & 1 \\ -1 & 1 & 1 & -1 \\ \hline \end{array}\,.$$

Proposition 13.1 (Hara et al. [81]). *$\mathscr{B}_1$ connects all fibers with $x_{+j} > 0$, $j = 1, \ldots, J$, for the model (13.2).*

Before we give a proof of this proposition, we present a lemma.

Lemma 13.1. *Let $\mathbf{z} = \{z_{ij}\}$ be any move for (13.2). Then there exist $j_1 < j_2$ and $j_3 < j_4$ satisfying the following conditions.*

(a) $z_{1j_1} > 0$, $z_{1j_2} < 0$, $z_{1j_3} < 0$, $z_{1j_4} > 0$.
(b) $z_{1j_1} = 1$ *implies* $j_1 \neq j_4$.
(c) $z_{1j_2} = -1$ *implies* $j_2 \neq j_3$.
(d) $z_{1j} = 0$ *for* $j_1 < j < j_2$ *and* $j_3 < j < j_4$.

Proof. (a), (b), and (c) are obvious from the constraint (13.3) and $z_{1+} = 0$. We can assume without loss of generality that there exist $j_1 < j_2$ such that $z_{1j_1} > 0, z_{1j_2} < 0$, $z_{1j} \geq 0$ for $1 \leq j < j_1$ and $z_{1j} = 0$ for $j_1 < j < j_2$. Because there exist $j_2 \leq j_3 < j_4$ satisfying (a), (b), and (c), we can choose j_3 and j_4 to satisfy (d). □

The following theorem shows that a subset of $\mathscr{B}_1$ still connects all fibers with $x_{+j} > 0, \forall j$.

Theorem 13.1 (Chen et al. [33]; Hara et al. [81]). *The set of moves*

$$\mathscr{B}_1^* = \{\mathbf{z} \in \mathscr{B}_1 \mid j_2 = j_1 + 1, j_3 = j_4 - 1\} \tag{13.5}$$

connects every fiber satisfying $x_{+j} > 0$, $j = 1, \ldots, J$, for the univariate logistic regression model (13.2).

This theorem was first introduced by Chen et al. [33] without an explicit proof and Chen et al. [35] discussed this problem from an algebraic viewpoint. An explicit proof is given by Hara et al. [81]. However, the proof is complicated and omitted here. Refer to [81] for details of the proof.

13.2 Discrete Logistic Regression Model with More than One Covariate

In this section we extend the argument in the previous section to the model with more than one covariate. Let $\mathscr{I}_0$ denote the set of success and failure and $\mathscr{I}_1 = \{1, \ldots, I_1\}, \ldots, \mathscr{I}_K = \{1, \ldots, I_K\}$ be the sets of levels of K covariates. For $i_k \in \mathscr{I}_k$, $k = 0, \ldots, K$, denote $\boldsymbol{i}_{1:K} := (i_1, \ldots, i_K)$ and $\boldsymbol{i} := (i_0, \boldsymbol{i}_{1:K})'$. Let

$$\boldsymbol{i} \mid i_0 = 1 \ := \ (1, \boldsymbol{i}_{1:K})', \qquad \boldsymbol{i} \mid i_0 = 2 \ := \ (2, \boldsymbol{i}_{1:K})'.$$

Then $x(\boldsymbol{i} \mid i_0 = 1)$ and $x(\boldsymbol{i} \mid i_0 = 2)$ are the frequencies of successes and failures, respectively, for a level $\boldsymbol{i}_{1:K}$. Let $p(\boldsymbol{i} \mid i_0 = 1)$ be the probability of success for a level $(i_1, \ldots, i_K)$ and $p(\boldsymbol{i} \mid i_0 = 2) = 1 - p(\boldsymbol{i} \mid i_0 = 1)$. Let $x(\boldsymbol{i} \mid i_0 = 1)$ be distributed as a binomial distribution

$$x(\boldsymbol{i} \mid i_0 = 1) \sim \mathrm{Bin}(x^{1:K}(\boldsymbol{i}_{1:K}), p(\boldsymbol{i} \mid i_0 = 1)),$$

where $x^{1:K}(\boldsymbol{i}_{1:K}) := x(\boldsymbol{i} \mid i_0 = 1) + x(\boldsymbol{i} \mid i_0 = 2)$. Denote $\boldsymbol{\beta} := (\beta_0, \ldots, \beta_K)'$. The model is described as

$$\log \frac{p(\boldsymbol{i} \mid i_0 = 1)}{1 - p(\boldsymbol{i} \mid i_0 = 1)} = (1, \boldsymbol{i}_{1:K})\boldsymbol{\beta}. \tag{13.6}$$

A sufficient statistic for this model is

$$x^0(1) := \sum_{k=1}^{K} \sum_{i_k=1}^{I_k} x(1, \boldsymbol{i}_{1:K}), \quad \sum_{i_k=1}^{I_k} i_k x^{0k}(1, i_k), \quad k = 1, \ldots, K,$$

where $x^{0k}(1i_k) = \sum_{l \neq k} \sum_{i_l=1}^{I_l} x(1, \boldsymbol{i}_{1:K})$. Note that $x^{1:K}(\boldsymbol{i}_{1:K})$ are also fixed by a sampling scheme for every $(i_1, \ldots, i_K)$. In the same way as the model with one covariate, we need a set of moves connecting contingency tables sharing

$$\boldsymbol{t}_K = \left\{ x^0(1),\ \sum_{i_k=1}^{I_k} i_k x^{0k}(1i_k),\ k = 1, \ldots, K,\ x^{1:K}(\boldsymbol{i}_{1:K}),\ \boldsymbol{i}_{1:K} \in \mathscr{I}_1 \times \cdots \times \mathscr{I}_K \right\}.$$

Such a set of moves is equivalent to a Markov basis of the Poisson logistic regression model $x(\boldsymbol{i}) \sim Po(\lambda(\boldsymbol{i}))$ where

$$\lambda(\boldsymbol{i} \mid i_0 = 1) = \lambda p(\boldsymbol{i} \mid i_0 = 1), \quad \lambda(\boldsymbol{i} \mid i_0 = 2) = \lambda(1 - p(\boldsymbol{i} \mid i_0 = 1)), \tag{13.7}$$

with $p(\boldsymbol{i})$ satisfying (13.6).

Let

$$A_k = \begin{pmatrix} 1 & 1 & \cdots & 1 \\ 1 & 2 & \cdots & I_k \end{pmatrix} = (\boldsymbol{a}_{k,1}, \ldots, \boldsymbol{a}_{k,I_k}),$$

where $\boldsymbol{a}_{k,i_k} = (1, i_k)'$ are column vectors. The configuration of the model (13.7) is also described as the Lawrence lifting of the Segre product $A_1 \otimes \cdots \otimes A_K$:

$$\Lambda(A_1 \otimes \cdots \otimes A_K) = \begin{pmatrix} A_1 \otimes \cdots \otimes A_K & 0 \\ E_{I_1 \cdots I_K} & E_{I_1 \cdots I_K} \end{pmatrix},$$

where

$$A_k \otimes A_l = \left(\boldsymbol{a}_{k,i_k} \oplus \boldsymbol{a}_{l,i_l},\ i_k = 1, \ldots, I_k, i_l = 1, \ldots, I_l \right), \quad \boldsymbol{a}_{k,i_k} \oplus \boldsymbol{a}_{l,i_l} = \begin{pmatrix} \boldsymbol{a}_{k,i_k} \\ \boldsymbol{a}_{l,i_l} \end{pmatrix}$$

and $E_{I_1 \cdots I_K}$ is the $(I_1 \cdots I_K) \times (I_1 \cdots I_K)$ identity matrix. Then any move in this model $\boldsymbol{z} = \{z(\boldsymbol{i})\}$ satisfies

$$z^0(1) = \sum_{k=2}^{K} \sum_{i_k=1}^{I_k} z(\boldsymbol{i} \mid i_0 = 1) = 0, \quad \sum_{i_k=1}^{I_k} i_k z^{0k}(1i_k) = 0, \quad k = 1, \ldots, K,$$

$$z^{1:K}(\boldsymbol{i}_{1:K}) = z(\boldsymbol{i} \mid i_0 = 1) + z(\boldsymbol{i} \mid i_0 = 2) = 0,$$

where $z^{0k}(1i_k) = \sum_{l \neq 0, l \neq k} \sum_{i_l=1}^{I_l} z(\boldsymbol{i})$. As an extension of moves in Definition 13.2, we introduce the following class of degree 4 moves.

Definition 13.2. Let $\boldsymbol{j} \in \mathscr{I}_1 \times \cdots \times \mathscr{I}_K$. Let $\boldsymbol{e}(\boldsymbol{j})$ be an integer array with 1 at the cell $(1,\boldsymbol{j})$, -1 at the cell $(2,\boldsymbol{j})$, and 0 everywhere else. Define $\mathscr{B}_K$ by the set of moves $\boldsymbol{z} = \{z(\boldsymbol{i})\}$ satisfying

1. $\boldsymbol{z} = \boldsymbol{e}(\boldsymbol{j}_1) - \boldsymbol{e}(\boldsymbol{j}_2) - \boldsymbol{e}(\boldsymbol{j}_3) + \boldsymbol{e}(\boldsymbol{j}_4)$.
2. $\boldsymbol{j}_1 - \boldsymbol{j}_2 = \boldsymbol{j}_3 - \boldsymbol{j}_4 \neq 0$.

Example 13.1. We give some examples of moves for $K = 2$. Let $\boldsymbol{j}_l := (j_{l1}, j_{l2})$, $l = 1,\ldots,4$. Then the following integer arrays are $(i_0 = 1)$-slices of moves in $\mathscr{B}_2$.

(1) $j_{12} = \cdots = j_{42}$

	j_{11}	j_{21}	j_{31}	j_{41}
j_{12}	1	−1	−1	1

(2) $j_{12} = \cdots = j_{42}$, $j_{21} = j_{31}$

	j_{11}	j_{21}	j_{41}
j_{12}	1	−2	1

(3) $j_{12} = j_{22}$, $j_{32} = j_{42}$

	j_{11}	j_{21}	j_{31}	j_{41}
j_{12}	1	−1	0	0
j_{32}	0	0	−1	1

(4) $j_{12} = j_{22}$, $j_{21} = j_{31}$

	j_{11}	j_{21}	j_{41}
j_{12}	1	−1	0
j_{32}	0	−1	1

(5) $(j_{21}, j_{22}) = (j_{31}, j_{32})$

	j_{11}	j_{21}	j_{41}
j_{12}	1	0	0
j_{22}	0	−2	0
j_{42}	0	0	1

(6) $j_{12} = j_{42}$, $j_{21} = j_{31}$

	j_{11}	j_{21}	j_{41}
j_{22}	0	−1	0
j_{12}	1	0	1
j_{32}	0	−1	0

Theorem 13.2 (Hara et al. [81]). *$\mathscr{B}_2$ connects every fiber satisfying $x^{1:2}(\boldsymbol{i}_{1:2}) > 0$, $\forall \boldsymbol{i}_{1:2}$.*

Hara et al. [81] gave a proof of this theorem. This theorem can also be proved by the distance-reducing argument. However, the proof is complicated and omitted here. Refer to [81] for details.

It is also possible to extend the argument to the model with three dummy variables; that is, $K = 3$ and $I_1 = I_2 = I_3 = 2$. In this case $\boldsymbol{t}_3$ is written as

$$\boldsymbol{t}_3 := \left\{x^0(1),\ x^{0k}(11) + 2x^{0k}(12),\ x^{1:3}(\boldsymbol{i}_{1:3}),\ i_k = 1,2,\ k = 1,2,3\right\}.$$

Because $x^0(1) = x^{0k}(11) + x^{0k}(12)$, a table sharing $\boldsymbol{t}$ is equivalent to a table sharing

$$\left\{x^0(1),\ x^{0k}(11),\ x^{0k}(12),\ x^{1:3}(\boldsymbol{i}_{1:3}),\ i_k = 1,2,\ k = 1,2,3\right\}.$$

Therefore a move $\boldsymbol{z} = (z(\boldsymbol{i}))$ satisfies

$$z^1(1) = 0,\ z^{0k}(11) = 0,\ z^{0k}(12) = 0,\ z^{1:3}(\boldsymbol{i}_{1:3}) = 0,\ i_k = 1,2. \tag{13.8}$$

Theorem 13.3. *Assume that* $I_1 = I_2 = I_3 = 2$. *Then* $\mathscr{B}_3$ *connects every fiber satisfying* $x^{1:3}(\boldsymbol{i}_{1:3}) > 0,\ \forall i_1, i_2, i_3$.

Chen et al. [35] gave an algebraic proof for this theorem. Here we give another proof of the theorem by the distance-reducing argument.

Proof. Let $\boldsymbol{x}, \boldsymbol{y}$ $(\boldsymbol{y} \neq \boldsymbol{x})$ be two tables in the same fiber $\mathscr{F}_{t_3}$ and let $\boldsymbol{z} = \boldsymbol{y} - \boldsymbol{x}$.
From (13.8), $\boldsymbol{z}^{012}(\boldsymbol{i}_{0:2} \mid i_0 = 1) = \{z^{012}(1i_1i_2), (i_1, i_2) \in \{1,2\} \times \{1,2\}\}$ satisfies

$$\boldsymbol{z}^{012}(1, i_1, i_2) = \begin{array}{|cc|} z^{012}(111) & z^{012}(112) \\ z^{012}(121) & z^{012}(122) \\ \hline \end{array} = \begin{array}{|cc|} 0 & 0 \\ 0 & 0 \\ \hline \end{array} \quad \text{or} \quad \begin{array}{|cc|} a & -a \\ -a & a \\ \hline \end{array}$$

for $a \neq 0$.

Case 1. Suppose that

$$\boldsymbol{z}^{012}(1i_1i_2) = \begin{array}{|cc|} 0 & 0 \\ 0 & 0 \\ \hline \end{array}.$$

Without loss of generality, we can assume that $z(1111) > 0$. This implies that

$$z(1112) < 0, \quad z(2112) > 0, \quad z(2111) < 0.$$

Because $z^{03}(11) = 0$, there exist i_1 and i_2 such that $(i_1, i_2) \neq (1, 1)$,

$$z(1i_1i_21) < 0, \quad z(2i_1i_21) > 0.$$

Then

$$z(1i_1i_22) > 0, \quad z(2i_1i_22) < 0$$

from the assumption. Let

$$\boldsymbol{z}_0 = (1111)(1i_1i_22)(2i_1i_21)(2112) - (1112)(1i_1i_21)(2i_1i_22)(2111) \in \mathscr{B}_3.$$

Then $\boldsymbol{x} - \boldsymbol{z}_0 \in \mathscr{F}_{\boldsymbol{t}}$ or $\boldsymbol{y} + \boldsymbol{z}_0 \in \mathscr{F}_{\boldsymbol{t}}$ and the distance is reduced by eight.

Case 2. The case that

$$\boldsymbol{z}^{012}(1i_1i_2) = \begin{array}{|cc|} a & -a \\ -a & a \\ \hline \end{array}$$

for $a > 0$.
Without loss of generality we can assume that $z(1111) > 0$ and $z(2111) < 0$.

Case 2-1. The case that $z(1121) < 0$, $z(1211) < 0$, and $z(1221) > 0$.
This assumption implies that $z(2121) > 0$, $z(2211) > 0$, and $z(2221) < 0$,

$$\begin{array}{|cc|} z(1111) & z(1121) \\ z(1211) & z(1221) \\ \hline z(2111) & z(2121) \\ z(2211) & z(2221) \\ \hline \end{array} = \begin{array}{|cc|} + & - \\ - & + \\ \hline - & + \\ + & - \\ \hline \end{array}.$$

Define a move $\boldsymbol{z}_2$ by

$$\boldsymbol{z}_2 = (1121)(1211)(2111)(2221) - (1111)(1221)(2121)(2211) \in \mathscr{B}_3.$$

Then $\boldsymbol{x} - \boldsymbol{z}_0 \in \mathscr{F}_{\boldsymbol{t}}$ or $\boldsymbol{y} + \boldsymbol{z}_0 \in \mathscr{F}_{\boldsymbol{t}}$ and the distance is reduced by eight.

Case 2-2. The case that $z(1121) < 0$, $z(1211) < 0$, and $z(1221) = 0$.
This assumption implies that $z(2121) > 0$, $z(2211) > 0$, and $z(2221) = 0$,

$$\begin{array}{|ll|}\hline z(1111) & z(1121) \\ z(1211) & z(1221) \\ \hline z(2111) & z(2121) \\ z(2211) & z(2221) \\ \hline \end{array} = \begin{array}{|cc|}\hline + & - \\ - & 0 \\ \hline - & + \\ + & 0 \\ \hline \end{array}.$$

Then the sign patterns of $i_3 = 1$ slices of $\boldsymbol{x}$ and $\boldsymbol{y}$ satisfy either

$$\begin{array}{|ll|}\hline x(1111) & x(1121) \\ x(1211) & x(1221) \\ \hline x(2111) & x(2121) \\ x(2211) & x(2221) \\ \hline \end{array} - \begin{array}{|ll|}\hline y(1111) & y(1121) \\ y(1211) & y(1221) \\ \hline y(2111) & y(2121) \\ y(2211) & y(2221) \\ \hline \end{array} = \begin{array}{|cc|}\hline + & 0+ \\ 0+ & + \\ \hline 0+ & + \\ + & 0+ \\ \hline \end{array} - \begin{array}{|cc|}\hline 0+ & + \\ + & + \\ \hline + & 0+ \\ 0+ & 0+ \\ \hline \end{array} \qquad (13.9)$$

or

$$\begin{array}{|ll|}\hline x(1111) & x(1121) \\ x(1211) & x(1221) \\ \hline x(2111) & x(2121) \\ x(2211) & x(2221) \\ \hline \end{array} - \begin{array}{|ll|}\hline y(1111) & y(1121) \\ y(1211) & y(1221) \\ \hline y(2111) & y(2121) \\ y(2211) & y(2221) \\ \hline \end{array} = \begin{array}{|cc|}\hline + & 0+ \\ 0+ & 0+ \\ \hline 0+ & + \\ + & + \\ \hline \end{array} - \begin{array}{|cc|}\hline 0+ & + \\ + & 0+ \\ \hline + & 0+ \\ 0+ & + \\ \hline \end{array}, \qquad (13.10)$$

where 0+ denotes that the cell count is nonnegative. In the case of (13.9), we can apply $\boldsymbol{z}_2$ to $\boldsymbol{x}$ and the distance is reduced by four. In the case of (13.10), we can apply $-\boldsymbol{z}_2$ to $\boldsymbol{y}$ and the distance is reduced by four.

More generally, if $\boldsymbol{z}$ has either of the following patterns of signs,

$$\begin{array}{|ll|}\hline z(i_0 1 i_2 i_3) & z(i_0 1 i_2' i_3') \\ z(i_0 2 i_2 i_3) & z(i_0 2 i_2' i_3') \\ \hline z(i_0' 1 i_2 i_3) & z(i_0' 1 i_2' i_3') \\ z(i_0' 2 i_2 i_3) & z(i_0' 2 i_2' i_3') \\ \hline \end{array} = \begin{array}{|cc|}\hline + & - \\ - & 0 \\ \hline - & + \\ + & 0 \\ \hline \end{array}, \quad \begin{array}{|cc|}\hline + & 0 \\ - & + \\ \hline - & 0 \\ + & - \\ \hline \end{array} \quad \text{or} \quad \begin{array}{|cc|}\hline + & - \\ 0 & + \\ \hline - & + \\ 0 & - \\ \hline \end{array}$$

for $i_0 \neq i_0'$ and $(i_2, i_3) \neq (i_2', i_3')$, we can show that the distance is reduced by a move in $\mathscr{B}_3$ in the same way.

Case 2-3. In the case where

$$\begin{array}{|ll|}\hline z(i_0 i_1 1 i_3) & z(i_0 i_1' 1 i_3') \\ z(i_0 i_1 2 i_3) & z(i_0 i_1' 2 i_3') \\ \hline z(i_0' i_1 1 i_3) & z(i_0' i_1' 1 i_3') \\ z(i_0' i_1 2 i_3) & z(i_0' i_1' 2 i_3') \\ \hline \end{array} = \begin{array}{|cc|}\hline + & - \\ - & 0 \\ \hline - & + \\ + & 0 \\ \hline \end{array}, \quad \begin{array}{|cc|}\hline + & 0 \\ - & + \\ \hline - & 0 \\ + & - \\ \hline \end{array} \quad \text{or} \quad \begin{array}{|cc|}\hline + & - \\ 0 & + \\ \hline - & + \\ 0 & - \\ \hline \end{array}$$

for $i_0 \neq i'_0$ and $(i_1, i_3) \neq (i'_1, i'_3)$, the distance is reduced by a move in $\mathscr{B}_3$ in the same way as Case 2-2.

Case 2-4. In the case where $z(1121) < 0$, $z(1211) \geq 0$, and $z(1221) \geq 0$,

$$\begin{array}{|ll|}\hline z(1111) & z(1121) \\ z(1211) & z(1221) \\ \hline z(2111) & z(2121) \\ z(2211) & z(2221) \\ \hline \end{array} = \begin{array}{|cc|}\hline + & - \\ 0+ & 0+ \\ \hline - & + \\ 0- & 0- \\ \hline \end{array}.$$

We have $z(1212) < 0$, because $z^{012}(121) < 0$. If $z(1112) > 0$, we have

$$\begin{array}{|ll|}\hline z(1112) & z(1121) \\ z(1212) & z(1221) \\ \hline z(2112) & z(2121) \\ z(2212) & z(2221) \\ \hline \end{array} = \begin{array}{|cc|}\hline + & - \\ - & 0+ \\ \hline - & + \\ + & 0- \\ \hline \end{array}.$$

Then in a similar way as in Cases 2-1 and 2-2, we can reduce the distance by a move

$$\mathbf{z}_3 = (1112)(1221)(2121)(2212) - (1121)(1212)(2112)(2221) \in \mathscr{B}_3.$$

If $z(1222) > 0$, we have

$$\begin{array}{|ll|}\hline z(1111) & z(1121) \\ z(1212) & z(1222) \\ \hline z(2111) & z(2121) \\ z(2212) & z(2222) \\ \hline \end{array} = \begin{array}{|cc|}\hline + & - \\ - & 0+ \\ \hline - & + \\ + & 0- \\ \hline \end{array},$$

and we can reduce the distance by a move

$$\mathbf{z}_3 = (1111)(1222)(2121)(2212) - (1121)(1212)(2111)(2222) \in \mathscr{B}_3$$

in a similar way.

We assume that $z(1112) \leq 0$ and $z(1222) \leq 0$. Because $z^{03}(12) = 0$, we have $z(1122) > 0$ and

$$\begin{array}{|ll|}\hline z(1122) & z(1121) \\ z(1212) & z(1211) \\ \hline z(2122) & z(2121) \\ z(2212) & z(2211) \\ \hline \end{array} = \begin{array}{|cc|}\hline + & - \\ - & 0+ \\ \hline - & + \\ + & 0- \\ \hline \end{array}.$$

Then we can reduce the distance by at least four by a move

$$\mathbf{z}_4 = (1122)(1211)(2121)(2212) - (1121)(1212)(2122)(2211).$$

In the case where $z(1121) \geq 0$, $z(1211) < 0$ and $z(1221) \geq 0$, the proof is similar.

Case 2-5. In the case where $z(1121) < 0$, $z(1211) \geq 0$, and $z(1221) < 0$: because $z^{012}(121) < 0$ and $z^{012}(122) > 0$, we have $z(1212) < 0$, and $z(1222) > 0$ and

$$\begin{array}{|cc|}\hline z(1111) & z(1121) \\ z(1212) & z(1222) \\ \hline z(2111) & z(2121) \\ z(2212) & z(2222) \\ \hline \end{array} = \begin{array}{|cc|}\hline + & - \\ - & + \\ \hline - & + \\ + & - \\ \hline \end{array}.$$

Hence we can reduce the distance by eight by a move

$$\mathbf{z}_5 = (1111)(1222)(2121)(2212) - (1121)(1212)(2111)(2222).$$

In the case where $z(1121) \geq 0$, $z(1211) < 0$, and $z(1221) < 0$, the proof is similar.

Case 2-6. In the case where $z(1121) < 0$, $z(1211) < 0$, and $z(1221) < 0$: because $z^{012}(122) > 0$, we have $z(1222) > 0$. If $z(1122) \leq 0$ or $z(1212) \leq 0$,

$$\begin{array}{|cc|}\hline z(1111) & z(1122) \\ z(1211) & z(1222) \\ \hline z(2111) & z(2122) \\ z(2211) & z(2222) \\ \hline \end{array} = \begin{array}{|cc|}\hline + & - \\ - & + \\ \hline - & + \\ + & - \\ \hline \end{array} \quad \text{or} \quad \begin{array}{|cc|}\hline z(1111) & z(1121) \\ z(1212) & z(1222) \\ \hline z(2111) & z(2121) \\ z(2212) & z(2222) \\ \hline \end{array} = \begin{array}{|cc|}\hline + & - \\ - & + \\ \hline - & + \\ + & - \\ \hline \end{array}.$$

we can reduce the distance by four by

$$\mathbf{z}_{6a} = (1111)(1222)(2122)(2211) - (1121)(1212)(2111)(2222) \in \mathscr{B}_3$$

or

$$\mathbf{z}_{6b} = (1111)(1222)(2121)(2212) - (1121)(1212)(2111)(2222) \in \mathscr{B}_3.$$

Assume that $z(1122) > 0$ and $z(1212) > 0$. Because $z^{03}(12) = 0$, we have $z(1112) < 0$ and

$$\begin{array}{|cc|}\hline z(1111) & z(1121) \\ z(1112) & z(1122) \\ \hline z(2111) & z(2121) \\ z(2112) & z(2122) \\ \hline \end{array} = \begin{array}{|cc|}\hline + & - \\ - & + \\ \hline - & + \\ + & - \\ \hline \end{array}.$$

Then we can reduce the distance by eight by a move

$$\mathbf{z}_{6c} = (1111)(1122)(2121)(2112) - (1121)(1112)(2111)(2122) \in \mathscr{B}_3. \qquad \square$$

We conjecture that for any K the set of moves $\mathscr{B}_K$ connects every fiber with positive response marginals for the logistic regression with K covariates. However, the theoretical proof seems to be difficult at this point. Recently Kashimura et al. [94] showed that it is impossible to extend the proof of Theorem 13.2 given in Hara et al. [81] to the model with $K > 2$ covariates.

13.3 Numerical Examples

13.3.1 *Exact Tests of Logistic Regression Model*

Table 13.2 refers to coronary heart disease incidence in Framingham, Massachusetts [2, 41]. A sample of male residents, aged 40 through 50, were classified on blood pressure and serum cholesterol concentration. In the (1,1) cell 2/53 means that there are 53 cases, 2 of whom exhibited heart disease. We examine the goodness-of-fit of the model (13.6) with $K = 2$,

$$\log \frac{p(\boldsymbol{i} \mid i_0 = 1)}{1 - p(\boldsymbol{i} \mid i_0 = 1)} = \beta_0 + \beta_1 i_1 + \beta_2 i_2, \tag{13.11}$$

where $I_1 = 7$ and $I_2 = 8$. We first test the null hypotheses $H_{\beta_1} : \beta_1 = 0$ and $H_{\beta_2} : \beta_2 = 0$ versus (13.11) using the (twice log) likelihood ratio statistics L_{β_1} and L_{β_2}. Then we have $L_{\beta_1} = 18.09$ and $L_{\beta_2} = 22.56$ and the asymptotic p-values are 2.1×10^{-5} and 2.0×10^{-6}, respectively, from the asymptotic distribution χ^2_1. We estimated the exact distribution of L_{β_1} and L_{β_2} via MCMC with the sets of moves $\mathscr{B}_1$ and $\mathscr{B}_1^*$ defined in Sect. 13.1.

Figures 13.1 and 13.2 represent histograms of sampling distributions of L_{β_1} and L_{β_2}. The solid lines in the figures represent the density function of the asymptotic chi-square distribution with degree of freedom one. The estimated p-values and their standard errors are essentially 0 for all cases. Therefore both H_{β_1} and H_{β_2} are rejected. We can see from the figures that there is almost no difference between two histograms computed with $\mathscr{B}_1$ and $\mathscr{B}_1^*$.

Table 13.2 Data on coronary heart disease incidence

		Serum cholesterol (mg/100ml)						
	Blood	1	2	3	4	5	6	7
	pressure	< 200	200–209	210–219	220–244	245–259	260–284	> 284
1	< 117	2/53	0/21	0/15	0/20	0/14	1/22	0/11
2	117–126	0/66	2/27	1/25	8/69	0/24	5/22	1/19
3	127–136	2/59	0/34	2/21	2/83	0/33	2/26	4/28
4	137–146	1/65	0/19	0/26	6/81	3/23	2/34	4/23
5	147–156	2/37	0/16	0/6	3/29	2/19	4/16	1/16
6	157–166	1/13	0/10	0/11	1/15	0/11	2/13	4/12
7	167–186	3/21	0/5	0/11	2/27	2/5	6/16	3/14
8	> 186	1/5	0/1	3/6	1/10	1/7	1/7	1/7

Source : Cornfield [41]

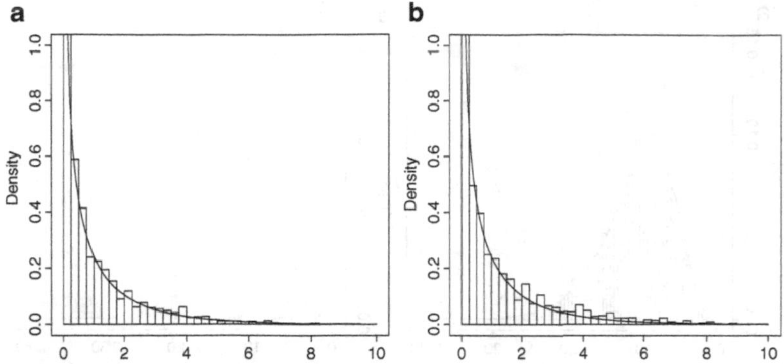

Fig. 13.1 Histograms of L_{β_1} via MCMC with $\mathscr{B}_1$ and $\mathscr{B}_1^*$ (**a**) A histogram with $\mathscr{B}_1$ (**b**) A histogram with $\mathscr{B}_1^*$

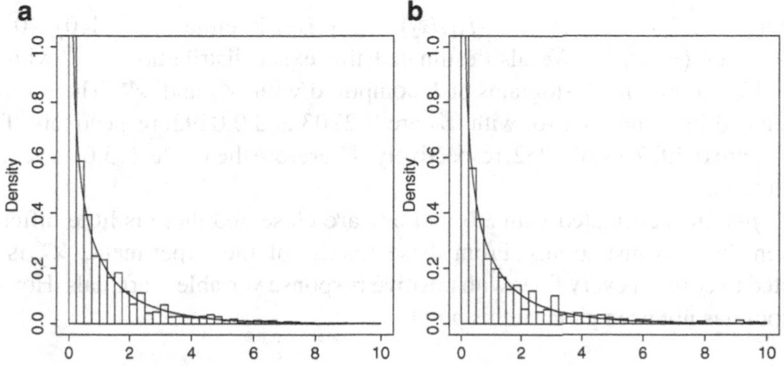

Fig. 13.2 Histograms of L_{β_2} via MCMC with $\mathscr{B}_1$ and $\mathscr{B}_1^*$ (**a**) A histogram with $\mathscr{B}_1$ (**b**) A histogram with $\mathscr{B}_1^*$

Next we set (13.11) as the null hypothesis and test it against the following ANOVA type logit model,

$$\log \frac{p(\boldsymbol{i})}{1 - p(\boldsymbol{i})} = \mu + \alpha_{1,i_1} + \alpha_{2,i_2}, \tag{13.12}$$

where $\sum_{i_1=1}^{I_1} \alpha_{1,i_1} = 0$ and $\sum_{i_2=1}^{I_2} \alpha_{2,i_2} = 0$ by likelihood ratio statistic L. The value of L is 13.08 and the asymptotic p-value is 0.2884 from the asymptotic distribution χ^2_{11}. We computed the exact distribution of L via MCMC with $\mathscr{B}_2$. As an extension of $\mathscr{B}_1^*$ in Theorem 13.1 to the bivariate model (13.6), we define $\mathscr{B}_2^*$ by the set of moves

$$\boldsymbol{z} = \boldsymbol{e}_{i_{11}i_{11}} - \boldsymbol{e}_{i_{21}i_{22}} - \boldsymbol{e}_{i_{31}i_{32}} + \boldsymbol{e}_{i_{41}i_{42}}$$

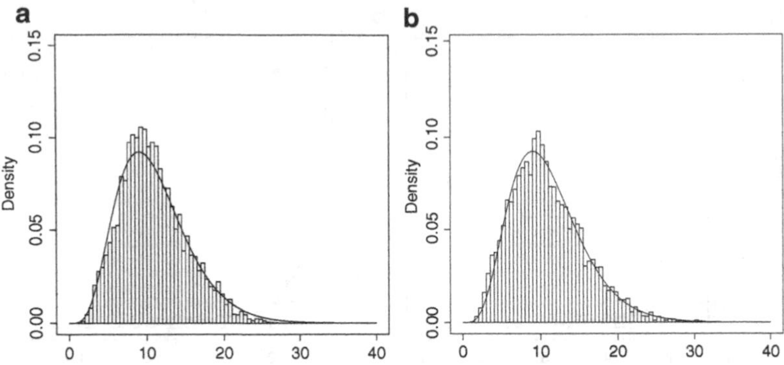

Fig. 13.3 Histograms of L via MCMC with $\mathscr{B}_2$ and $\mathscr{B}_2^*$ (**a**) A histogram with $\mathscr{B}_2$ (**b**) A histogram with $\mathscr{B}_2^*$

satisfying $(i_{11}, i_{12}) - (i_{21}, i_{22}) = (i_{31}, i_{32}) - (i_{41}, i_{42})$ is either of $(\pm 1, 0)$, $(0, \pm 1)$, $(\pm 1, \pm 1)$, or $(\pm 1, \mp 1)$. We also estimated the exact distribution of L with $\mathscr{B}_2^*$. Figure 13.3 represents histograms of L computed with $\mathscr{B}_2$ and $\mathscr{B}_2^*$. The estimated p-value and its standard error with $\mathscr{B}_2$ are 0.2703 and 0.0292, respectively. Those with $\mathscr{B}_2^*$ are 0.2977 and 0.0252, respectively. Therefore the model (13.6) is accepted in both tests.

The p-values estimated with $\mathscr{B}_2$ and $\mathscr{B}_2^*$ are close and there is little difference between the two histograms. From these results of the experiment, $\mathscr{B}_2^*$ is also expected to connect every fiber with positive response variable marginals. However, the proof has not been given at this point.

13.4 Connecting Zero-One Tables with Graver Basis

In some practical problems, the cell counts have upper bounds (e.g., Rapallo and Yoshida [126]). In this section we consider the case where cell counts are restricted to be either zero or one. The most common example is the Rasch model [127] used in the item response theory. The Rasch model can be interpreted as a logistic regression, where the number of trials is just one for each combination of covariates.

The following theorem is a basic fact on the connectivity of fibers with a zero-one restriction for the model with the configuration A.

Proposition 13.2 (Hara and Takemura [77]). *Let $\mathscr{B}_0$ denote the set of square-free moves of the Graver basis $\mathscr{B}_{\mathrm{GR}}$ of I_A. Then $\mathscr{B}_0$ is strongly distance reducing for tables of the model corresponding to I_A with the zero-one restriction.*

Proof. Let $\boldsymbol{x}$, $\boldsymbol{y}$ be two zero-one tables of the same fiber. They are connected by a conformal sum of primitive moves

$$\boldsymbol{y} = \boldsymbol{x} + \boldsymbol{z}_1 + \cdots + \boldsymbol{z}_K. \tag{13.13}$$

Because there is no cancellation of signs on the right-hand side, once an entry greater than or equal to 2 appears in an intermediate sum of the right-hand side, it cannot be canceled. Therefore it follows that $\boldsymbol{x} + \boldsymbol{z}_1 + \cdots + \boldsymbol{z}_k$ has zero-one entries for $k = 1, \ldots, K$ and $\boldsymbol{z}_1, \ldots, \boldsymbol{z}_K \in \mathscr{B}_0$. There are no sign cancellations in (13.13), thus $\boldsymbol{z}_1, \ldots, \boldsymbol{z}_K$ can be added to $\boldsymbol{x}$ in any order and $-\boldsymbol{z}_1, \ldots, -\boldsymbol{z}_K$ can be added to $\boldsymbol{y}$ in any order. Therefore $\mathscr{B}_0$ is strongly distance reducing. □

13.5 Rasch Model

The Rasch model has been extensively studied and practically used for evaluating educational and psychological tests. Suppose that I_1 persons take a test with I_2 dichotomous questions. Let $x_{ij} \in \{0, 1\}$ be a response to the jth question of the ith person. Hence the $I_1 \times I_2$ table $\boldsymbol{x} = \{x_{ij}\}$ is considered as a two-way contingency table with zero-one entries. Assume that each x_{ij} is independent. The Rasch model is expressed as

$$P(x_{ij} = 1) = \frac{\exp(\alpha_i - \beta_j)}{1 + \exp(\alpha_i - \beta_j)}, \tag{13.14}$$

where α_i is an individual's latent ability parameter and β_j is an item's difficulty parameter. Then the set of row sums $x_{i+} = \sum_{j=1}^{J} x_{ij}$ and column sums $x_{+j} = \sum_{i=1}^{I} x_{ij}$ is a sufficient statistic for α_i and β_j.

Many inference procedures have been developed (e.g., Glas and Verhelst [65]) and most of them rely on asymptotic theory. However, as Rasch [127] pointed out, a sufficiently large sample size is not necessarily expected in practice for this problem and Rasch [127] proposed using an exact testing procedure.

The conditional distribution of zero-one tables given person scores and item totals is easily shown to be uniform. In order to implement an exact test for the Rasch model via the Markov basis technique, we need a set of moves that connects every fiber of two-way zero-one tables with fixed row and column sums. It is easy to show that the set of basic moves of the two-way complete independence model

$$\begin{array}{ccc} & i & i' \\ j & 1 & -1 \\ j' & -1 & 1 \end{array}$$

connects every fiber of zero-one tables with fixed row and column sums (e.g., Ryser [130]). Many Monte Carlo procedures via the Markov basis technique to compute distribution of test statistics to test the goodness-of-fit of the Rasch model have been proposed (e.g., Besag and Clifford [24], Ponocny [122], Cobb and Chen [38]). Chen and Small [36] provided a computationally more efficient Monte Carlo procedure for implementing exact tests by using sequential importance sampling.

13.6 Many-Facet Rasch Model

The many-facet Rasch model is an extension of the Rasch model to multiple items and polytomous responses (e.g., Linacre [100]). Suppose that I_1 articles are rated by I_2 reviewers from I_3 aspects on the grade of I_4 scales from 0 to $I_4 - 1$. $x(i_1 i_2 i_3 i_4) = 1$ if the reviewer i_2 rates the article i_1 as the i_4th grade from the aspect i_3 and otherwise $x(i_1 i_2 i_3 i_4) = 0$. Then $\boldsymbol{x} = \{x(i_1 i_2 i_3 i_4)\}$ is considered as an $I_1 \times I_2 \times I_3 \times I_4$ zero-one table. We note that $\boldsymbol{x}$ satisfies $x^{123}(i_1 i_2 i_3) = \sum_{i_4=0}^{I_4-1} x(i_1 i_2 i_3 i_4) = 1$ for all i_1, i_2, and i_3. Then the three-facet Rasch model for $\boldsymbol{x}$ is expressed as

$$P(x_{i_1 i_2 i_3 i_4} = 1) = \frac{\exp\left[i_4(\beta_{i_1} - \beta_{i_2} - \beta_{i_3}) - \beta_{i_4}\right]}{\sum_{i_4=0}^{I_4-1} \exp\left[i_4(\beta_{i_1} - \beta_{i_2} - \beta_{i_3}) - \beta_{i_4}\right]}. \tag{13.15}$$

In general, the V-facet Rasch model is defined as follows. Let $\boldsymbol{x} = \{x(\boldsymbol{i})\}$, $\boldsymbol{i} = (i_1, \ldots, i_{V+1})$ be an $I_1 \times \cdots \times I_{V+1}$ zero-one table. Assume that $\mathscr{I}_{V+1} = \{0, \ldots, I_{V+1} - 1\}$ and that $\boldsymbol{x}$ satisfies

$$x(\boldsymbol{i}_{\{1,\ldots,V\}}) = \sum_{i_{V+1}=0}^{I_{V+1}-1} x(\boldsymbol{i}) = 1.$$

Then the V-facet Rasch model is expressed as

$$P(x(\boldsymbol{i}) = 1) = \frac{\exp\left[i_{V+1}(\beta_{i_1} - \beta_{i_2} - \ldots - \beta_{i_V}) - \beta_{i_{V+1}}\right]}{\sum_{i_{V+1}=0}^{I_{V+1}-1} \exp\left[i_{V+1}(\beta_{i_1} - \beta_{i_2} - \ldots - \beta_{i_V}) - \beta_{i_{V+1}}\right]}. \tag{13.16}$$

When $V = 2$, $I_3 = 2$, and $\beta_{i_3} = \text{const}$ for $i_3 \in \{0, 1\}$, the model coincides with the Rasch model (13.14). Define $\boldsymbol{t}^0$ by

$$\boldsymbol{t}^0 = \left\{ \sum_{i_{V+1}=0}^{I_{V+1}-1} i_{V+1} \cdot x(\boldsymbol{i}_{\{v,V+1\}}) \;\middle|\; \boldsymbol{i}_{\{v,V+1\}} \in \mathscr{I}_{\{v,V+1\}},\ v = 1, \ldots, V \right\}.$$

Then a sufficient statistic $\boldsymbol{t}$ is given by

$$\boldsymbol{t} = \boldsymbol{t}^0 \cup \{x(i_{V+1}) \mid i_{V+1} \in \mathscr{I}_{V+1}\}.$$

When $\beta_{i_{V+1}}$ is constant for $i_{V+1} \in \mathscr{I}_{V+1}$, $\boldsymbol{t}$ is given by

$$\boldsymbol{t} = \boldsymbol{t}^0 \cup \{x^+\},$$

Table 13.3 The number of square-free moves of the Graver basis for the three-way complete independence model

	Degree of moves				
$I_1 \times I_2 \times I_3$	2	3	4	5	6
$2\times2\times2$	12	0	0	0	0
$2\times2\times3$	33	48	0	0	0
$2\times2\times4$	64	192	96	0	0
$2\times2\times5$	105	480	480	0	0
$2\times3\times3$	90	480	396	0	0
$2\times3\times4$	174	1,632	5,436	1,152	0
$2\times3\times5$	285	3,840	23,220	33,120	720
$3\times3\times3$	243	3,438	19,008	12,312	0

where $x^+ = \sum_{i\in\mathscr{I}} x(\boldsymbol{i})$. In the case of the three-facet Rasch model (13.15), $\boldsymbol{t}$ is expressed as follows,

$$\boldsymbol{t} = \left\{ \sum_{i_4=0}^{I_4-1} i_4 x_{i_1++i_4},\ i_1 \in \mathscr{I}_1,\quad \sum_{i_4=0}^{I_4-1} i_4 x_{+i_2+i_4},\ i_2 \in \mathscr{I}_2, \right.$$
$$\left. \sum_{i_4=0}^{I_4-1} i_4 x_{++i_3 i_4},\ i_3 \in \mathscr{I}_3,\quad x_{+++i_4},\ i_4 \in \mathscr{I}_4 \right\}.$$

In order to implement exact tests for the many-facet Rasch model, we need a set of moves that connects any fiber $\mathscr{F}_{\boldsymbol{t}}$ of zero-one tables. In general, however, it is not easy to derive such a set of moves. As seen in the previous section, in the case of the Rasch model (13.14), the set of basic moves for the two-way complete independence model connects any fiber. For the many-facet Rasch model (13.16), however, the basic moves do not necessarily connect all fibers. Consider the case where $V = 3$ and $I_4 = 2$. In this case, $\boldsymbol{t}^0$ is written as

$$\boldsymbol{t}^0 = \{x_{i_1++1}, x_{+i_2+1}, x_{++i_3 1} \mid i_v \in \mathscr{I}_v, v = 1,2,3\}.$$

Because $x(i_4) = \sum_{i_v\in\mathscr{I}_v} x(i_{\{v,4\}})$ for $v = 1,2,3$, $\boldsymbol{t}^0$ is a sufficient statistic. $\boldsymbol{t}^0$ is equivalent to a sufficient statistic of three-way complete independence model for the $(i_4 = 1)$-slice of $\boldsymbol{x}$. From Proposition 13.2, the set of square-free moves of the Graver basis for the three-way complete independence model connects any fiber $\mathscr{F}_{\boldsymbol{t}}$. Table 13.3 shows the number of square-free moves of the Graver basis for the $I_1 \times I_2 \times I_3$ three-way complete independence model computed via 4ti2 [1]. We see that when the number of levels is greater than two, the sets include moves with degree greater than two. This fact does not necessarily imply that higher degree moves are required to connect every fiber for the three-way complete independence model. However, we can give an example which shows that the degree 2 moves do not connect all fibers of the three-way complete independence model.

Example 13.2 (A fiber for $3\times 3\times 3$ *three-way complete independence model).* Consider the following two zero-one tables $\boldsymbol{x}$ and $\boldsymbol{y}$ in the same fiber of the three-way complete independence model.

$$
\boldsymbol{x} := \begin{array}{c} \\ \\ j \end{array}
\begin{array}{c|ccc|} \multicolumn{4}{c}{k} \\ \multicolumn{1}{c}{} & 1 & 2 & \multicolumn{1}{c}{3} \\ \cline{2-4} 1 & 0 & 0 & 0 \\ 2 & 0 & 0 & 1 \\ 3 & 0 & 0 & 1 \\ \cline{2-4} \multicolumn{4}{c}{i=1} \end{array}
\quad
\begin{array}{|ccc|} \hline 0 & 1 & 1 \\ 0 & 1 & 1 \\ 1 & 1 & 1 \\ \hline \multicolumn{3}{c}{i=2} \end{array}
\quad
\begin{array}{|ccc|} \hline 0 & 0 & 0 \\ 0 & 0 & 1 \\ 1 & 1 & 1 \\ \hline \multicolumn{3}{c}{i=3} \end{array}
$$

$$
\boldsymbol{y} := \begin{array}{c} \\ \\ j \end{array}
\begin{array}{c|ccc|} \multicolumn{4}{c}{k} \\ \multicolumn{1}{c}{} & 1 & 2 & \multicolumn{1}{c}{3} \\ \cline{2-4} 1 & 0 & 0 & 0 \\ 2 & 0 & 0 & 0 \\ 3 & 0 & 1 & 1 \\ \cline{2-4} \multicolumn{4}{c}{i=1} \end{array}
\quad
\begin{array}{|ccc|} \hline 0 & 0 & 1 \\ 1 & 1 & 1 \\ 1 & 1 & 1 \\ \hline \multicolumn{3}{c}{i=2} \end{array}
\quad
\begin{array}{|ccc|} \hline 0 & 0 & 1 \\ 0 & 0 & 1 \\ 0 & 1 & 1 \\ \hline \multicolumn{3}{c}{i=3} \end{array}.
$$

The difference of the two tables is

$$
\boldsymbol{z} = \begin{array}{|ccc|} \hline 0 & 0 & 0 \\ 0 & 0 & 1 \\ 0 & -1 & 0 \\ \hline \end{array}
\quad
\begin{array}{|ccc|} \hline 0 & 1 & 0 \\ -1 & 0 & 0 \\ 0 & 0 & 0 \\ \hline \end{array}
\quad
\begin{array}{|ccc|} \hline 0 & 0 & -1 \\ 0 & 0 & 0 \\ 1 & 0 & 0 \\ \hline \end{array}
$$

and we can easily check that $\boldsymbol{z}$ is a move for the three-way complete independence model.

Let $\bar{\Delta}$ be the set of degenerate variables defined in Chap. 8. Then degree 2 moves for the three-way complete independence model are classified into the following four patterns.

1. $\bar{\Delta} = \{1,2,3\}:\ \begin{array}{c|cc|} & i_3 & \multicolumn{1}{c}{i_3'} \\ \cline{2-3} i_2 & 1 & 0 \\ i_2' & 0 & -1 \\ \cline{2-3} \multicolumn{3}{c}{i_1} \end{array},\quad \begin{array}{c|cc|} & i_3 & \multicolumn{1}{c}{i_3'} \\ \cline{2-3} i_2 & -1 & 0 \\ i_2' & 0 & 1 \\ \cline{2-3} \multicolumn{3}{c}{i_1'} \end{array}.$

2. $\bar{\Delta} = \{1,2\}:\ \begin{array}{c|cc|} & i_3 & \multicolumn{1}{c}{i_3'} \\ \cline{2-3} i_2 & 1 & 0 \\ i_2' & -1 & 0 \\ \cline{2-3} \multicolumn{3}{c}{i_1} \end{array},\quad \begin{array}{c|cc|} & i_3 & \multicolumn{1}{c}{i_3'} \\ \cline{2-3} i_2 & 0 & -1 \\ i_2' & 0 & 1 \\ \cline{2-3} \multicolumn{3}{c}{i_1'} \end{array}.$

3. $\bar{\Delta} = \{1,3\}:\ \begin{array}{c|cc|} & i_3 & \multicolumn{1}{c}{i_3'} \\ \cline{2-3} i_2 & 1 & -1 \\ i_2' & 0 & 0 \\ \cline{2-3} \multicolumn{3}{c}{i_1} \end{array},\quad \begin{array}{c|cc|} & i_3 & \multicolumn{1}{c}{i_3'} \\ \cline{2-3} i_2 & -1 & 1 \\ i_2' & 0 & 0 \\ \cline{2-3} \multicolumn{3}{c}{i_1'} \end{array}.$

4. $\bar{\Delta} = \{2,3\} : \begin{array}{c|cc|} & i_3 & i_3' \\ \hline i_2 & 1 & -1 \\ i_2' & -1 & 1 \\ \hline \multicolumn{3}{c}{i_1} \end{array}$.

However, it is easy to check that if we apply any move in this class to $\boldsymbol{x}$ or $\boldsymbol{y}$, -1 or 2 has to appear. Therefore we cannot apply any degree 2 moves to either $\boldsymbol{x}$ or $\boldsymbol{y}$. Hence a degree 3 move is required to connect this fiber.

This example shows that the set of basic moves does not necessarily connect every zero-one fiber for the many-facet Rasch model. As seen in Table 13.3, the number of square-free moves in the Graver basis is too large even for three-way tables. When the number of cells is greater than 100, it seems to be difficult to compute the Graver basis via 4ti2 in a practical length of time. Hence implementations of exact tests by using the Graver basis are limited to very small models at this point.

13.7 Latin Squares and Zero-One Tables for No-Three-Factor Interaction Models

Zero-one tables also appear quite often in the form of incidence matrices for combinatorial problems. Here as an example we consider Latin squares. A Latin square is an $n \times n$ table filled with n different symbols in such a way that each symbol occurs exactly once in each row and column. A 3×3 Latin square is written as

$$\boxed{\begin{array}{ccc} 1 & 2 & 3 \\ 2 & 3 & 1 \\ 3 & 1 & 2 \end{array}} \,. \tag{13.17}$$

When the symbols of an $n \times n$ Latin square are considered as co-ordinates of the third axis (sometimes called the orthogonal array representation of a Latin square), it is a particular element of a fiber for the $n \times n \times n$ no-three-factor interaction model with all two-dimensional marginals (line sums) equal to 1. For example, the 3×3 Latin square (13.17) is considered as a $3 \times 3 \times 3$ zero-one table $\boldsymbol{x} = \{x_{i_1 i_2 i_3}\}$,

$$\boldsymbol{x} = \underset{i_3=1}{i_1 \overset{i_2}{\boxed{\begin{array}{ccc} 1 & 0 & 0 \\ 0 & 0 & 1 \\ 0 & 1 & 0 \end{array}}}}\,, \quad \underset{i_3=2}{i_1 \overset{i_2}{\boxed{\begin{array}{ccc} 0 & 1 & 0 \\ 1 & 0 & 0 \\ 0 & 0 & 1 \end{array}}}}\,, \quad \underset{i_3=3}{i_1 \overset{i_2}{\boxed{\begin{array}{ccc} 0 & 0 & 1 \\ 0 & 1 & 0 \\ 1 & 0 & 0 \end{array}}}} \tag{13.18}$$

with $x_{i_1 i_2 +} = 1$, $x_{+ i_2 i_3} = 1$, $x_{i_1 + i_3} = 1$ for all i_1, i_2, and i_3. One of the reasons to consider a Markov basis for Latin squares is to generate a Latin square randomly. [60] advocated choosing a Latin square randomly from the set of Latin squares, and [92] gave a Markov basis for the set of $n \times n$ Latin squares.

Because the set of Latin squares is just a particular fiber, it may be the case that a minimal set of moves connecting all Latin squares is smaller than the set of moves connecting all zero-one tables. This is indeed the case as we show for the simple case of $n = 3$. We first present a connectivity result for $3 \times 3 \times 3$ zero-one tables with all line sums fixed.

Let $\boldsymbol{z} = \{z_{ijk}\}_{i,j,k=1,2,3}$ be a move for $3 \times 3 \times 3$ no-three-factor-interaction model. From Chap. 9 the minimal Markov basis consists of basic moves such as

$$\boldsymbol{z} = \boxed{\begin{matrix} 1 & -1 & 0 \\ -1 & 1 & 0 \\ 0 & 0 & 0 \end{matrix}}\ \boxed{\begin{matrix} -1 & 1 & 0 \\ 1 & -1 & 0 \\ 0 & 0 & 0 \end{matrix}}\ \boxed{\begin{matrix} 0 & 0 & 0 \\ 0 & 0 & 0 \\ 0 & 0 & 0 \end{matrix}} \tag{13.19}$$

and degree 6 moves such as

$$\boldsymbol{z} = \boxed{\begin{matrix} 1 & -1 & 0 \\ 0 & 1 & -1 \\ -1 & 0 & 1 \end{matrix}}\ \boxed{\begin{matrix} -1 & 1 & 0 \\ 0 & -1 & 1 \\ 1 & 0 & -1 \end{matrix}}\ \boxed{\begin{matrix} 0 & 0 & 0 \\ 0 & 0 & 0 \\ 0 & 0 & 0 \end{matrix}}\,. \tag{13.20}$$

However, these moves do not connect zero-one tables of the $3 \times 3 \times 3$ no-three-factor interaction model. We need the following type of degree 9 move, which corresponds to the difference of two Latin squares:

$$\boldsymbol{z} = \boxed{\begin{matrix} 1 & -1 & 0 \\ 0 & 1 & -1 \\ -1 & 0 & 1 \end{matrix}}\ \boxed{\begin{matrix} 0 & 1 & -1 \\ -1 & 0 & 1 \\ 1 & -1 & 0 \end{matrix}}\ \boxed{\begin{matrix} -1 & 0 & 1 \\ 1 & -1 & 0 \\ 0 & 1 & -1 \end{matrix}}\,. \tag{13.21}$$

Proposition 13.3. *The set of basic moves (13.19), degree* 6 *moves (13.20), and degree* 9 *moves (13.21) forms a Markov basis for* $3 \times 3 \times 3$ *zero-one tables for the no-three-factor interaction model.*

Proof. Consider any line sum, such as $0 = z_{+11} = z_{111} + z_{211} + z_{311}$ of a move $\boldsymbol{z}$. If $(z_{111}, z_{211}, z_{311}) \neq (0,0,0)$, then we easily see that $\{z_{111}, z_{211}, z_{311}\} = \{-1,0,1\}$. By a similar consideration as in Sect. 9.2, each i- or j- or k-slice is either a loop of degree 2 or loop of degree 3, such as

$$\boxed{\begin{matrix} 1 & -1 & 0 \\ -1 & 1 & 0 \\ 0 & 0 & 0 \end{matrix}} \quad \text{or} \quad \boxed{\begin{matrix} 1 & -1 & 0 \\ 0 & 1 & -1 \\ -1 & 0 & 1 \end{matrix}}\,. \tag{13.22}$$

Now we consider two cases: (1) there exists a slice with a loop of degree 2, and (2) all slices are loops of degree 3.

Case 1. Without loss of generality, we can assume that the $(i = 1)$-slice of $\boldsymbol{z}$ is the loop of degree 2 in (13.22). Then we can further assume that $z_{211} = -1$ and $z_{311} = 0$. Now suppose that $z_{222} = -1$. If $z_{212} = 1$ or $z_{221} = 1$, we can reduce $|\boldsymbol{z}|$ by a basic move. This implies $z_{212} = z_{221} = 0$. But then $z_{213} = z_{223} = 1$ and this contradicts the pattern of $\{z_{213}, z_{223}, z_{233}\} = \{-1,0,1\}$.

By the above consideration we have $z_{222} = 0$ and $z_{322} = -1$. By a similar consideration for the cells z_{i12} and z_{i21}, $i = 1,2,3$, we easily see that $\boldsymbol{z}$ is of the form

$$\begin{array}{|ccc|}\hline 1 & -1 & 0 \\ -1 & 1 & 0 \\ 0 & 0 & 0 \\ \hline\end{array}\quad \begin{array}{|ccc|}\hline -1 & 1 & 0 \\ 0 & 0 & 0 \\ 1 & -1 & 0 \\ \hline\end{array}\quad \begin{array}{|ccc|}\hline 0 & 0 & 0 \\ 1 & -1 & 0 \\ -1 & 1 & 0 \\ \hline\end{array},$$

which is a degree 6 move.

Case 2. It is easily seen that the only case where degree 6 moves cannot be applied is of the form of the move of degree 9 in (13.21). This proves that connectivity is guaranteed if we add degree 9 moves. □

We also want to show that degree 9 moves are needed for connectivity. Consider

$$\boldsymbol{x} = \begin{array}{|ccc|}\hline 1 & 0 & 1 \\ 0 & 1 & 0 \\ 0 & 0 & 1 \\ \hline\end{array}\quad \begin{array}{|ccc|}\hline 0 & 1 & 0 \\ 0 & 1 & 1 \\ 1 & 0 & 0 \\ \hline\end{array}\quad \begin{array}{|ccc|}\hline 0 & 0 & 1 \\ 1 & 0 & 0 \\ 1 & 1 & 0 \\ \hline\end{array}.$$

By a simple program it is easily checked that if we apply any basic move or any move of degree 6 to $\boldsymbol{x}$, -1 or 2 has to appear. Hence degree 9 moves are required to connect zero-one tables.

Now consider 3×3 Latin squares (13.18). It is well known that there is only one isotopy class of 3×3 Latin squares (Chap. III of [39]); that is, all 3×3 Latin squares are connected by the action of the direct product $S_3 \times S_3 \times S_3$ of the symmetric group S_3 which is generated by transpositions, and a transposition corresponds to a move of degree 6 in (13.20). Therefore, *the set of* 3×3 *Latin squares in the orthogonal array representation is connected by the set of moves of degree* 6 *in* (13.20). In view of Proposition 13.3, we see that we do not need basic moves or degree 9 moves for connecting 3×3 Latin squares.

There are two isotopy classes for 4×4 Latin squares (1.18 of III.1.3 of [39]) and representative elements of these two classes are connected by a basic move. Transposition of two levels for a factor corresponds to a degree 8 move of the following form.

$$\boldsymbol{z} = \begin{array}{|cccc|}\hline 1 & -1 & 0 & 0 \\ 0 & 1 & -1 & 0 \\ 0 & 0 & 1 & -1 \\ -1 & 0 & 0 & 1 \\ \hline\end{array}\quad \begin{array}{|cccc|}\hline -1 & 1 & 0 & 0 \\ 0 & -1 & 1 & 0 \\ 0 & 0 & -1 & 1 \\ 1 & 0 & 0 & -1 \\ \hline\end{array}\quad \begin{array}{|cccc|}\hline 0 & 0 & 0 & 0 \\ 0 & 0 & 0 & 0 \\ 0 & 0 & 0 & 0 \\ 0 & 0 & 0 & 0 \\ \hline\end{array}\quad \begin{array}{|cccc|}\hline 0 & 0 & 0 & 0 \\ 0 & 0 & 0 & 0 \\ 0 & 0 & 0 & 0 \\ 0 & 0 & 0 & 0 \\ \hline\end{array}.$$

Therefore the set of 4×4 Latin squares is connected by the set of basic moves and moves of degree 8 of the above form. We can apply a similar consideration to the celebrated result of 22 isotopy classes of 6×6 Latin squares derived by [60].

Part IV
Some Other Topics of Algebraic Statistics

In the last part of this book, we discuss some other topics of algebraic statistics.

In Chap. 14 we explain a relation between the Markov basis methodology and disclosure limitation problem. In particular we show that the technique of swapping records for the disclosure limitation problem is closely related to the Markov basis and results on the Markov basis for hierarchical models can be applied to the disclosure limitation problem.

In Chap. 15 we give a brief survey on the use of the Gröbner basis for the design of experiments, mainly based on recent results of the authors. As we discussed in the preface to this book, application of the Gröbner basis to the design of experiments was one of the two sources for the field of algebraic statistics.

Finally in Chap. 16 we explain how we can run a Markov chain with a lattice basis, when a Markov basis is not available, namely when it is not possible to compute a Markov basis by an algebraic algorithm within a practical amount of time. As we saw in Part III of this book, Markov bases tend to be complicated except for some nice models, such as decomposable models of contingency tables. Even if Markov bases are not available, we can run a Markov chain with a lattice basis, which is much easier to compute. With many numerical examples, we confirm that a Markov chain with a lattice basis works well.

Chapter 14
Disclosure Limitation Problem and Markov Basis

14.1 Swapping with Some Marginals Fixed

Consider a microdata set where all variables of the data set have already been categorized. Suppose that a statistical agency is considering granting access to such a microdata set to some researchers and that the data set contains some rare and risky records. Swapping of observations is one of the useful techniques of protecting these records (Dalenius and Reiss [44], Schlorer [133], Takemura [141]). If some marginals from the data set have already been published, it is desirable to perform the swapping in such a way that the swapping does not disturb the published marginal frequencies. Therefore it is important to determine whether it is possible to perform swapping of risky records under the restriction that some marginals are fixed.

Here we give an illustration with a simple hypothetical example. Suppose that a microdata set contains the following two records.

Sex	Age	Occupation	Residence
Male	55	Nurse	Tokyo
Female	50	Police officer	Tokyo

If we swap "occupation" between these two records, we obtain

Sex	Age	Occupation	Residence
Male	55	Police officer	Tokyo
Female	50	Nurse	Tokyo

.

By this swapping the one-dimensional marginals are preserved, but the two-dimensional marginal of {age, occupation} is disturbed. If we swap both age and occupation we obtain

Sex	Age	Occupation	Residence
Male	50	Police officer	Tokyo
Female	55	Nurse	Tokyo

S. Aoki et al., *Markov Bases in Algebraic Statistics*, Springer Series in Statistics 199, DOI 10.1007/978-1-4614-3719-2_14,

and the {age, occupation}-marginal is also preserved. This simple example shows that the observations can be freely swapped if we fix only the one-dimensional marginals, but observations have to be swapped in two variables together to keep two-dimensional marginals fixed.

A categorized microdata set is considered as a contingency table. For example, two records in the above example correspond to two frequencies in the following three-way subtable with residence = Tokyo:

	Police officer		Nurse	
	Male	Female	Male	Female
50	1	0	0	0
55	0	0	0	1

. (14.1)

Swapping "occupation" is equivalent to adding an integer array

	Police officer		Nurse	
	Male	Female	Male	Female
50	−1	0	1	0
55	0	1	0	−1

. (14.2)

We note that this is a move for the three-way complete independence model. Therefore all one-dimensional marginals are preserved. In this way swapping under the condition of fixed marginals is equivalent to adding a move for the model where the marginals are in the sufficient statistic.

A swapping preserving all two-dimensional marginals in the above example corresponds to adding a move of a no-three-factor interaction model as discussed in Chap. 9. As shown in Chap. 9, there is no degree 2 move for the no-three-factor interaction model and the minimum degree of moves of the model is four. This shows that there is no swapping between these two records and at least four records are required to preserve all two-dimensional marginals.

Therefore swappability of the given two records such that a given set of marginals is fixed depends on the existence of degree 2 moves in the corresponding model. In the following sections we give some necessary and sufficient conditions for swappability of two given records.

14.2 *E*-Swapping

Consider an $n \times m$ microdata set $\boldsymbol{X}$ consisting of observations on m variables for n individuals (records). As mentioned above we assume that the variables have already been categorized. Therefore we can identify the microdata set with an m-way contingency table, if we ignore the labels of the individuals. If $x(\boldsymbol{i}) = 1$, we say that the record falling into cell $\boldsymbol{i}$ is a *sample unique record*.

Let E be a nonempty proper subset of the set of variables $\Delta = \{1,\dots,m\}$. For two records of $\boldsymbol{x}$ falling into cells $\boldsymbol{i} = (\boldsymbol{i}_E, \boldsymbol{i}_{E^C})$ and $\boldsymbol{j} = (\boldsymbol{j}_E, \boldsymbol{j}_{E^C})$, $\boldsymbol{i} \neq \boldsymbol{j}$, swapping of $\boldsymbol{i}$ and $\boldsymbol{j}$ with respect to $E \subset \Delta$, or more simply *E-swapping*, means that these records are changed as

$$\{(\boldsymbol{i}_E, \boldsymbol{i}_{E^C}), (\boldsymbol{j}_E, \boldsymbol{j}_{E^C})\} \to \{(\boldsymbol{i}_E, \boldsymbol{j}_{E^C}), (\boldsymbol{j}_E, \boldsymbol{i}_{E^C})\} =: (\boldsymbol{i}', \boldsymbol{j}'). \tag{14.3}$$

Note that E-swapping is equivalent to E^C-swapping. Also note that if $\boldsymbol{i}_E = \boldsymbol{i}'_E$ or $\boldsymbol{i}_{E^C} = \boldsymbol{i}'_{E^C}$, then swapping in (14.3) results in the same set of records. In the example of the previous section, swapping "residence" does not make any difference. Therefore (14.3) results in a different set of records if and only if

$$\boldsymbol{i}_E \neq \boldsymbol{i}'_E \text{ and } \boldsymbol{i}_{E^C} \neq \boldsymbol{i}'_{E^C}. \tag{14.4}$$

From now on we say that E-swapping is *effective* if it results in a different set of records.

Proposition 14.1. *For a subset $D \subset \Delta$, D-marginals $\{x_D(\boldsymbol{i}_D) \mid \boldsymbol{i}_D \in \mathscr{I}_D\}$ are fixed by E-swapping if and only if one of the following four conditions holds.*

$$\text{(i) } D \subset E, \ \text{(ii) } D \subset E^C, \ \text{(iii) } \boldsymbol{i}_{E\cap D} = \boldsymbol{i}'_{E\cap D}, \ \text{(iv) } \boldsymbol{i}_{E^C\cap D} = \boldsymbol{i}'_{E^C\cap D}. \tag{14.5}$$

Proof. It is obvious that if one of the conditions holds, then D-marginals are not altered. On the other hand assume that none of the four conditions holds. Let $D_1 = D \cap E$ and $D_2 = D \cap E^C$. These are nonempty because (i) and (ii) do not hold. Furthermore $i_{D_1} \neq j_{D_1}$ and $i_{D_2} \neq j_{D_2}$ because (iii) and (iv) do not hold. Let $i_D = (i_{D_1}, i_{D_2})$. Then $x_D(i_D) = x_D(\boldsymbol{i}_{D_1}, \boldsymbol{i}_{D_2})$ is decreased by 1 by this swapping and this particular D-marginal changes. □

14.3 Equivalence of Degree-Two Square-Free Move of Markov Bases and Swapping of Two Records

As mentioned in Sect. 14.1, swapping records in a microdata set is equivalent to applying a move to a contingency table $\boldsymbol{x} = \{x(\boldsymbol{i})\}$. It is intuitively clear that a square-free move of degree 2 and swapping of observations of two records are equivalent. However, there is at least a conceptual difference between them, because a move is defined for a given set of marginals $\mathscr{D}$ whereas E-swapping is defined only in terms of two records and a subset E.

Now we give a proof of this equivalence. An effective E-swapping in (14.3) changes the cell frequencies of $\boldsymbol{i}$, $\boldsymbol{j}$, $\boldsymbol{i}'$, and $\boldsymbol{j}'$ into

$$x(\boldsymbol{i}) \to x(\boldsymbol{i}) - 1, \quad x(\boldsymbol{j}) \to x(\boldsymbol{j}) - 1, \quad x(\boldsymbol{i}') \to x(\boldsymbol{i}') + 1, \quad x(\boldsymbol{j}') \to x(\boldsymbol{j}') + 1. \tag{14.6}$$

Hence if this E-swapping fixes all $\mathscr{D}$-marginals, then the difference between the post-swapped table and the pre-swapped table is a square-free move of degree 2 for $\mathscr{D}$.

Next we show that any square-free move of degree 2 (14.6) for $\mathscr{D}$ can be expressed by E-swapping (14.3) for some $E \subset \Delta$. Write

$$\boldsymbol{i} = (i_1,\ldots,i_m),\ \boldsymbol{j} = (j_1,\ldots,j_m),\quad \boldsymbol{i}' = (i'_1,\ldots,i'_m),\ \boldsymbol{j} = (j'_1,\ldots,j'_m).$$

We first show that $\{i_\delta, j_\delta\} = \{i'_\delta, j'_\delta\}$ for $1 \le \delta \le m$. Because $\bigcup_t D_t = \Delta$, there exists t for any δ such that δ belongs to D_t. In the case where $\boldsymbol{i}_{D_t} = \boldsymbol{j}_{D_t}$, two records of $x_{D_t}(\boldsymbol{i}_{D_t})$ have to be preserved in $\boldsymbol{i}'_{D_t}$ and $\boldsymbol{j}'_{D_t}$. Hence $\boldsymbol{i}'_{D_t} = \boldsymbol{j}'_{D_t} = \boldsymbol{i}_{D_t} = \boldsymbol{j}_{D_t}$. On the other hand if $\boldsymbol{i}_{D_t} \neq \boldsymbol{j}_{D_t}$, each one record of both $x_{D_t}(\boldsymbol{i}_{D_t})$ and $x_{D_t}(\boldsymbol{j}_{D_t})$ has to be preserved in $\{\boldsymbol{i}'_{D_t}, \boldsymbol{j}'_{D_t}\}$, which implies $\{\boldsymbol{i}_{D_t}, \boldsymbol{j}_{D_t}\} = \{\boldsymbol{i}'_{D_t}, \boldsymbol{j}'_{D_t}\}$. Therefore we have $\{i_\delta, j_\delta\} = \{i'_\delta, j'_\delta\}$ for $1 \le \delta \le m$.

If we set

$$E = \{\delta \mid i'_\delta = j_\delta\} = \{\delta \mid i_\delta = j'_\delta\},$$

E satisfies (14.3). This completes the proof of the equivalence of E-swapping and a square-free move of degree 2 for $\mathscr{D}$.

14.4 Swappability Between Two Records

Consider two records in a categorized microdata set. In the following we recognize the two records as a contingency table $\boldsymbol{x} = \{x(\boldsymbol{i})\}$ in a degree 2 fiber that was discussed in Sect. 8.3. Consider a swapping between these two records preserving marginals in $\mathscr{D} = \{D_1,\ldots,D_r\}$. Note that if some variable has the same value in two records, swapping or no swapping of the variable does not make any difference. Therefore we should only look at variables taking different values in two records. Let

$$\bar{\Delta} = \{\delta \mid i_\delta \neq i'_\delta\} \tag{14.7}$$

denote the set of nondegenerate variables defined in Sect. 8.3. Note that (14.4) holds if and only if

$$E \cap \bar{\Delta} \neq \emptyset \ \text{ and } \ E^C \cap \bar{\Delta} \neq \emptyset. \tag{14.8}$$

Therefore E-swapping is effective if and only if $E \cap \bar{\Delta} \neq \emptyset$ and $E^C \cap \bar{\Delta} \neq \emptyset$. In particular $\bar{\Delta}$ has to contain at least two elements, because if $\bar{\Delta}$ has less than two elements swapping between $\boldsymbol{i}$ and $\boldsymbol{j}$ cannot result in a different set of records.

As an example, consider the hypothetical example of Sect. 14.1.

Sex	Age	Occupation	Residence
Male	55	Nurse	Tokyo
Female	50	Police officer	Tokyo

(14.9)

Then $\bar{\Delta} = \{1,2,3\}$. $E = \{2,3\} = \{\text{age}, \text{occupation}\}$ results in an effective swapping.

The following lemma says that the variables in $\bar{\Delta} \cap D$ have to be swapped simultaneously or otherwise stay together in order not to disturb D-marginals.

Lemma 14.1. *An effective E-swapping fixes D-marginals if and only if $\bar{\Delta} \cap D \subset E$ or $\bar{\Delta} \cap D \subset E^C$ under (14.8).*

Proof. We have to check that at least one of the four conditions in (14.5) holds if and only if $\bar{\Delta} \cap D \subset E$ or $\bar{\Delta} \cap D \subset E^C$.

Assume that one of the four conditions in (14.5) holds. If $D \subset E$, then $\bar{\Delta} \cap D \subset E$. Similarly if $D \subset E^C$, then $\bar{\Delta} \cap D \subset E^C$. Now suppose $i_{E \cap D} = j_{E \cap D}$. Then

$$\emptyset = \bar{\Delta} \cap (E \cap D) = (\bar{\Delta} \cap D) \cap E \quad \Rightarrow \quad \bar{\Delta} \cap D \subset E^C.$$

Similarly if $i_{E^C \cap D} = j_{E^C \cap D}$ then $\bar{\Delta} \cap D \subset E$.

Conversely assume that $\bar{\Delta} \cap D \subset E$ or $\bar{\Delta} \cap D \subset E^C$. In the former case $\bar{\Delta} \cap D \cap E^C = \emptyset$ and this implies (iv) $i_{E^C \cap D} = j_{E^C \cap D}$. Similarly in the latter case (iii) $i_{E \cap D} = j_{E \cap D}$ holds. □

In the above lemma, E is given. Now suppose that two records $\boldsymbol{i}, \boldsymbol{j}$ and a marginal D are given and we are asked to find a nonempty proper subset $E \subset \Delta$ such that E-swapping is effective and fixes D-marginals. As a simple consequence of Lemma 14.1 we have the following lemma.

Lemma 14.2. *Given two records $\boldsymbol{i}, \boldsymbol{j}$ and $D \subset \Delta$, we can find $E \subset \Delta$ such that E-swapping is effective and fixes D-marginals if and only if $\bar{\Delta} \cap D^C \neq \emptyset$ and $|\bar{\Delta}| \geq 2$.*

Proof. If $\bar{\Delta} \cap D^C \neq \emptyset$ and $|\bar{\Delta}| \geq 2$, then choose $s \in \bar{\Delta} \cap D^C$ and let $E = \{s\}$ be a one-element set. Then E satisfies the requirement.

If $|\bar{\Delta}| \leq 1$, there is no E-swapping resulting in a different set of records as mentioned above. On the other hand if $\bar{\Delta} \cap D^C = \emptyset$ or $\bar{\Delta} \subset D$, then by Lemma 14.1 $\bar{\Delta} \subset E$. But this contradicts $E^C \cap \bar{\Delta} \neq \emptyset$ in (14.8) and there exists no E satisfying the requirement. □

Based on the above preparations we now consider the following problem. Let two records $\boldsymbol{i}, \boldsymbol{j}$ and a set of marginals $\mathscr{D} = \{D_1, \ldots, D_r\}$ be given. We are asked to find E such that E-swapping fixes all marginals of $\mathscr{D}$ and results in a different set of records.

Theorem 14.1 (Takemura and Hara [144]). *Let $\mathscr{G}^{\mathscr{D}}$ be the independence graph with respect to $\mathscr{D}$ and let $\mathscr{G}(\bar{\Delta})$ denote the subgraph of $\mathscr{G}^{\mathscr{D}}$ induced by $\bar{\Delta} \subset \Delta$. Given two records $\boldsymbol{i}, \boldsymbol{j}$ and a generating class $\mathscr{D}$, we can find $E \subset \Delta$ such that E-swapping is effective and fixes all D-marginals, $\forall D \in \mathscr{D}$, if and only if $\mathscr{G}(\bar{\Delta})$ is not connected.*

Proof. As discussed in Sect. 14.1 the variables δ and δ' belonging to some $D \in \mathscr{D}$ either have to be swapped out simultaneously or stay together. It follows that any variable in a connected component of $\mathscr{G}(\bar{\Delta})$ has to be swapped out simultaneously or stay together simultaneously. Therefore there exists no $E \subset \Delta$ such that E-swapping is effective and fixes all D-marginals when $\mathscr{G}(\bar{\Delta})$ is connected.

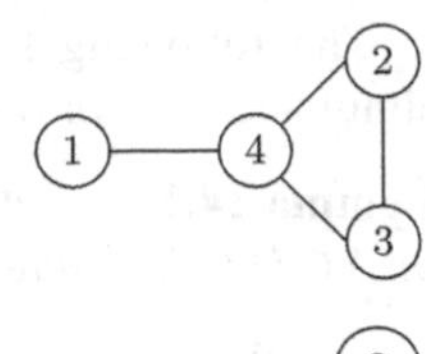

Fig. 14.1 $G^{\mathscr{D}}$

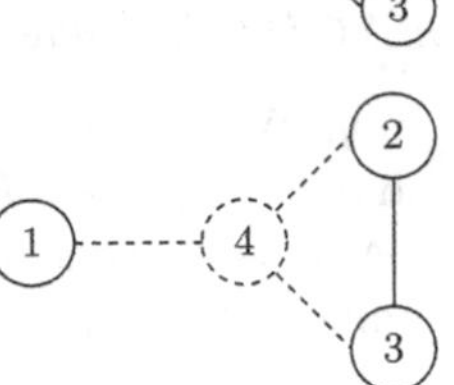

Fig. 14.2 $G(\bar{\Delta})$

Conversely assume that $\mathscr{G}(\bar{\Delta})$ is not connected. Let $\gamma_{\bar{\Delta}}$ be a connected component of $\mathscr{G}_{\bar{\Delta}}$. Then for any two vertices $\{\delta,\delta'\}$ such that $\delta \in \gamma_{\bar{\Delta}}$ and $\delta' \in \bar{\Delta} \setminus \gamma_{\bar{\Delta}}$ there exists no $D \in \mathscr{D}$ satisfying $\{\delta,\delta'\} \subset D$. Therefore if we set $E = \gamma_{\bar{\Delta}}$, E-swapping is effective and fixes all D-marginals. □

This theorem is essentially equivalent to Theorem 8.2. Theorem 8.2 says that a necessary and sufficient condition on degree 2 fibers to have more than one element is that $\mathscr{G}(\bar{\Delta})$ be disconnected. Having more than one element in a degree 2 fiber means that it is possible to swap any two of the elements.

As an example consider $\mathscr{D}$ consisting of all two-element sets of Δ; that is, all two-dimensional marginals are fixed. Then $\mathscr{G}(\mathscr{D})$ is complete and hence $\mathscr{G}(\bar{\Delta})$ is also complete. In particular $\mathscr{G}(\bar{\Delta})$ is connected and Theorem 14.1 says that we cannot find an effective swapping fixing all two-dimensional marginals.

As an illustrative example again consider the table in (14.9). Suppose that we want to fix the following set of two-dimensional marginals

$$\mathscr{D} = \{\{1,4\},\{2,3\},\{2,4\},\{3,4\}\},$$

where $1 = $ sex, $2 = $ age, $3 = $ occupation, and $4 = $ residence. In this case $\mathscr{D}$ is also the set of edges of $\mathscr{G}^{\mathscr{D}}$ (Fig. 14.1). Then $\mathscr{G}(\bar{\Delta})$ has the vertices $1,2,3$, and only one edge $\{2,3\}$ (Fig. 14.2). Therefore $\mathscr{G}(\bar{\Delta})$ is not connected. Again let $E = \{2,3\}$. We see that E-swapping is effective and fixes all marginals in $\mathscr{D}$. On the other hand if

$$\mathscr{D} = \{\{1,2\},\{2,3\},\{3,4\}\},$$

then it is easy to see that $\mathscr{G}^{\mathscr{D}}$ is connected and no effective swapping is possible if we fix all marginals for this $\mathscr{D}$.

Let $\mathscr{S}^{\mathscr{D}}$ be the set of the minimal vertex separators of $\mathscr{G}^{\mathscr{D}}$. Let $\mathrm{adj}(\delta)$, $\delta \in \Delta$ denote the set of vertices that are adjacent to δ. Define $\mathrm{adj}(A)$ for $A \subset \Delta$ by $\mathrm{adj}(A) = \bigcup_{\delta \in A} \mathrm{adj}(\delta) \setminus A$. Then we obtain the following lemma.

Lemma 14.3. *$\mathscr{G}(\bar{\Delta})$ is not connected if and only if there exist $S \in \mathscr{S}^{\mathscr{D}}$ and two connected components γ_α and γ_β of $\mathscr{G}^{\mathscr{D}}(\Delta \setminus S)$ satisfying*

$$S \cap \bar{\Delta} = \emptyset, \quad \gamma_\alpha \cap \bar{\Delta} \neq \emptyset, \quad \gamma_\beta \cap \bar{\Delta} \neq \emptyset. \tag{14.10}$$

Proof. Assume that $\mathscr{G}(\bar{\Delta})$ is not connected. Let $\gamma_{\bar{\Delta},1}$ and $\gamma_{\bar{\Delta},2}$ be any two connected components of $\mathscr{G}(\bar{\Delta})$. For any pair of vertices (α,β) such that $\alpha \in \gamma_{\bar{\Delta},1}$ and $\beta \in \gamma_{\bar{\Delta},2}$, $\mathrm{adj}(\gamma_{\bar{\Delta},1})$ is a (α,β)-separator (not necessarily minimal) in $\mathscr{G}^{\mathscr{D}}$. Note that $\mathrm{adj}(\gamma_{\bar{\Delta},1}) \cap \bar{\Delta} = \emptyset$. Hence there exists $S_{\alpha,\beta} \in \mathscr{S}^{\mathscr{D}}$ such that $S_{\alpha,\beta} \subset \mathrm{adj}(\gamma_{\bar{\Delta},1})$ and $S_{\alpha,\beta} \cap \bar{\Delta} = \emptyset$. Therefore there exists a minimal (α,β)-separator such that $S_{\alpha,\beta} \cap \bar{\Delta} = \emptyset$.

Because each of $\gamma_{\bar{\Delta},1}$ and $\gamma_{\bar{\Delta},2}$ is a connected component, $S_{\alpha,\beta}$ satisfying $S_{\alpha,\beta} \cap \bar{\Delta} = \emptyset$ also separates any pair of vertices in $\gamma_{\bar{\Delta},1}$ and $\gamma_{\bar{\Delta},2}$ other than (α,β). Hence $S_{\alpha,\beta}$ separates $\gamma_{\bar{\Delta},1}$ and $\gamma_{\bar{\Delta},2}$ in $\mathscr{G}^{\mathscr{D}}$. This implies that $\gamma_{\bar{\Delta},1}$ and $\gamma_{\bar{\Delta},2}$ belong to different connected components of $\mathscr{G}^{\mathscr{D}}(\Delta \setminus S_{\alpha,\beta})$. Therefore (14.10) is satisfied.

On the other hand if there exist S, γ_α, and γ_β satisfying (14.10), it is obvious that $\mathscr{G}(\bar{\Delta})$ is not connected. □

By the above lemma, we have the following corollary.

Corollary 14.1. *Given two records* $\boldsymbol{i},\boldsymbol{i}'$ *and a generating class* $\mathscr{D}$, *we can find* $E \subset \Delta$ *such that* E*-swapping is effective and fixes all* D*-marginals,* $\forall D \in \mathscr{D}$, *if and only if there exist* $S \in \mathscr{S}^{\mathscr{D}}$ *and two connected components* γ_α *and* γ_β *of* $\mathscr{G}^{\mathscr{D}}(\Delta \setminus S)$ *satisfying* (14.10); that is,

$$\boldsymbol{i}_S = \boldsymbol{i}'_S, \quad \boldsymbol{i}_{\gamma_\alpha} \neq \boldsymbol{i}'_{\gamma_\alpha}, \quad \boldsymbol{i}_{\gamma_\beta} \neq \boldsymbol{i}'_{\gamma_\beta}. \tag{14.11}$$

Proof. Note that $S \cap \bar{\Delta} = \emptyset$ if and only if $\boldsymbol{i}_S = \boldsymbol{i}'_S$. Similarly $\gamma_\alpha \cap \bar{\Delta} \neq \emptyset$ if and only if $\boldsymbol{i}_{\gamma_\alpha} \neq \boldsymbol{i}'_{\gamma_\alpha}$ and $\gamma_\beta \cap \bar{\Delta} \neq \emptyset$ if and only if $\boldsymbol{i}_{\gamma_\beta} \neq \boldsymbol{i}'_{\gamma_\beta}$. Therefore the corollary follows from Lemma 14.3. □

14.5 Searching for Another Record for Swapping

So far we have considered some necessary and sufficient conditions for E-swapping between two records $\boldsymbol{i},\boldsymbol{i}'$ to be effective and fix D-marginals for general hierarchical models. In this section we provide a simple algorithm to find another record that is swappable for a particular sample unique record $\boldsymbol{i}$ by using the results in the previous section.

Given a particular record $\boldsymbol{i}$, by Corollary 14.1, we could scan through the microdata set for another record $\boldsymbol{j}$ satisfying the conditions of Corollary 14.1. Instead of checking the conditions of Corollary 14.1 for each $\boldsymbol{j}$, we could first construct the list $\mathscr{S}^{\mathscr{D}}$ of minimal vertex separators S and the connected components γ_α, γ_β of $\mathscr{G}^{\mathscr{D}}(\Delta \setminus S)$. Then for a particular triple $(S,\gamma_\alpha,\gamma_\beta)$ we could check whether there exists another record $\boldsymbol{j}$ satisfying (14.11) of Corollary 14.1. Actually it is straightforward to check the existence of $\boldsymbol{j}$ satisfying (14.11). Because we require $\boldsymbol{i}_S = \boldsymbol{j}_S$, we only need to look at the slice of the contingency table given the value of $\boldsymbol{i}_S$. Then in this slice we look at the $\{i_{\gamma_\alpha}, i_{\gamma_\beta}\}$-marginal table. By the requirement

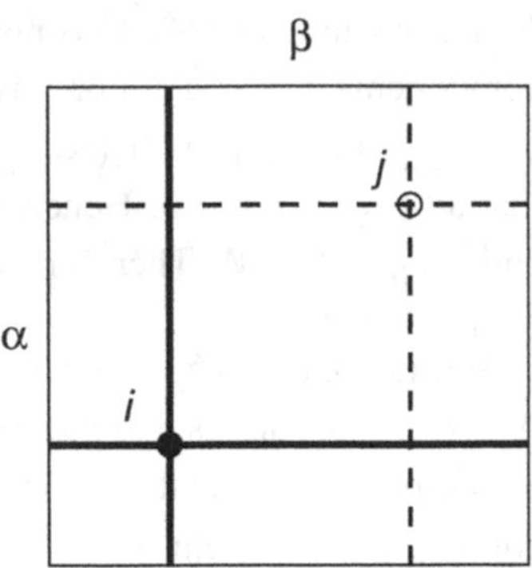

Fig. 14.3 j swappable with i in a diagonal position

$\boldsymbol{i}_{\gamma_\alpha} \neq \boldsymbol{j}_{\gamma_\alpha}, \boldsymbol{i}_{\gamma_\beta} \neq \boldsymbol{j}_{\gamma_\beta}$, we omit the "row" $\boldsymbol{i}_{\gamma_\alpha}$ and the "column" $\boldsymbol{i}_{\gamma_\beta}$ from the marginal table. If the resulting table is nonempty, then we can find another record $\boldsymbol{j}$ in a diagonal position to $\boldsymbol{i}$ and we can swap observations in $\boldsymbol{j}$ and $\boldsymbol{i}$. See Fig. 14.3.

More precisely, for γ_α, γ_β, S, write $\gamma_{\alpha,\beta} = \gamma_\alpha \cup \gamma_\beta \cup S$. Define a subtable $\bar{\boldsymbol{x}}_{\gamma_{\alpha,\beta}}(\boldsymbol{i}'_{\gamma_{\alpha,\beta}} \mid \boldsymbol{i}_{\gamma_{\alpha,\beta}})$ by

$$\begin{aligned}\bar{\boldsymbol{x}}_{\gamma_{\alpha,\beta}}(\boldsymbol{i}'_{\gamma_{\alpha,\beta}} \mid \boldsymbol{i}_{\gamma_{\alpha,\beta}}) &:= \left\{\bar{x}_{\gamma_{\alpha,\beta}}(\boldsymbol{i}'_{\gamma_{\alpha,\beta}} \mid \boldsymbol{i}_{\gamma_{\alpha,\beta}})\right\} \\ &= \left\{x_{\gamma_{\alpha,\beta}}(\boldsymbol{i}'_{\gamma_{\alpha,\beta}}) \mid \boldsymbol{i}'_{\gamma_\alpha} \neq \boldsymbol{i}_{\gamma_\alpha}, \boldsymbol{i}'_{\gamma_\beta} \neq \boldsymbol{i}_{\gamma_\beta}, \boldsymbol{i}'_S = \boldsymbol{i}_S\right\}.\end{aligned}$$

$\bar{\boldsymbol{x}}_{\gamma_{\alpha,\beta}}(\boldsymbol{i}'_{\gamma_{\alpha,\beta}} \mid \boldsymbol{i}_{\gamma_{\alpha,\beta}}) \neq \boldsymbol{0}$ denotes that there exists at least one positive count in $\bar{\boldsymbol{x}}_{\gamma_{\alpha,\beta}}(\boldsymbol{i}'_{\gamma_{\alpha,\beta}} \mid \boldsymbol{i}_{\gamma_{\alpha,\beta}})$.

Lemma 14.4. *There exists a record $\boldsymbol{i}'$ with $\boldsymbol{i}_S = \boldsymbol{i}'_S$, $\boldsymbol{i}_{\gamma_\alpha} \neq \boldsymbol{i}'_{\gamma_\alpha}$, and $\boldsymbol{i}_{\gamma_\beta} \neq \boldsymbol{i}'_{\gamma_\beta}$ if and only if $\bar{\boldsymbol{x}}_{\gamma_{\alpha,\beta}}(\boldsymbol{i}'_{\gamma_{\alpha,\beta}} \mid \boldsymbol{i}_{\gamma_{\alpha,\beta}}) \neq \boldsymbol{0}$.*

The proof of this lemma is obvious and omitted. Therefore it remains to compute the set of minimal vertex separators $\mathscr{S}^{\mathscr{D}}$ and the connected components of $\mathscr{G}^{\mathscr{D}}(\Delta \setminus S)$. Shiloach and Vishkin [138] proposed an algorithm for computing connected components of a graph. On listing minimal vertex separators there exist algorithms by Berry et al. [21] and Kloks and Kratsch [95]. The input of their algorithms is $\mathscr{G}^{\mathscr{D}}$. However, in our case generating class $\mathscr{D}$ is given in advance. It may be possible to obtain more efficient algorithms if we also use the information of $\mathscr{D}$ as the input.

The following algorithm searches for another record $\boldsymbol{j}$ that is swappable for a sample unique record $\boldsymbol{i}$ and swaps them if one exists.

Algorithm 14.1 (Finding $\boldsymbol{j}$ swappable for a given $\boldsymbol{i}$)
Input : $\boldsymbol{x}$, $\mathscr{D}$, $\mathscr{S}^{\mathscr{D}}$, $\boldsymbol{i}$
Output : a post-swapped table $\boldsymbol{x}' = \{x'(\boldsymbol{i})\}$

```
begin
    x' ← x ;
    for every S ∈ 𝒮^𝒟 do
    begin
        compute connected components of 𝒢^𝒟(Δ \ S) ;
        for every pair of connected components (γ_α, γ_β) do
        begin
            if x̄_{γα,β}(i'_{γα,β} | i_{γα,β}) ≠ 0 then
            begin
                select a marginal cell i'_{γα,β} such that x̄_{γα,β}(i'_{γα,β} | i_{γα,β}) ≠ 0 ;
                select a cell j ∈ ℐ such that j_{γα,β} = i'_{γα,β} ;
                E ← γ_α;
                E-swapping between i and j;
                x'(i) ← x(i) − 1;
                x'(j) ← x(j) − 1;
                x'(j_E, i_{E^C}) ← x(j_E, i_{E^C}) + 1;
                x'(i_E, j_{E^C}) ← x(i_E, j_{E^C}) + 1;
                exit ;
            end if
        end for
    end for
    if x' = x then i is not swappable ;
end
```

Chapter 15
Gröbner Basis Techniques for Design of Experiments

15.1 Design Ideals

Consider fractional factorial designs of m controllable factors. We assume that the levels of each factor are coded as elements of a field k, which is a finite extension of the field $\mathbb{Q}$ of rational numbers. In the original paper [120], only the case of $k = \mathbb{Q}$ is considered. However, coding the levels of factors with more than two levels by complex numbers is considered in [119]. We see it briefly in Sect. 15.4. This chapter is mainly based on [18].

A fractional factorial design without replication is defined as a finite subset of k^m. In the algebraic arguments, this subset is considered as the set of solutions of polynomial equations, called an algebraic variety, and the set of polynomials vanishing on all the solutions is called an ideal. This ideal, *design ideal*, is a key item in this chapter.

Now we define design ideals. Let $\mathscr{D}$ be the full factorial design of m factors. We call a (proper) subset $\mathscr{F} \subsetneq \mathscr{D}$ a fractional factorial design. Let $k[x_1,\ldots,x_m]$ be the polynomial ring of indeterminates $x_1,\ldots,x_m$ with the coefficients in k. Then the set of polynomials vanishing on the points of $\mathscr{F}$

$$I(\mathscr{F}) = \{f \in k[x_1,\ldots,x_m] \mid f(x_1,\ldots,x_m) = 0 \text{ for all } (x_1,\ldots,x_m) \in \mathscr{F}\}$$

is the design ideal of $\mathscr{F}$.

In this chapter, we suppose there are n runs (i.e., points) in a fractional factorial design $\mathscr{F} \subset \mathscr{D}$. A general method to derive a basis (i.e., a set of generators) of $I(\mathscr{F})$ is to make use of the algorithm for calculating the intersection of the ideals. By definition, the design ideal of the design consisting of a single point, $(a_1,\ldots,a_m) \in k^m$, is written as

$$\langle x_1 - a_1,\ldots,x_m - a_m\rangle \subset k[x_1,\ldots,x_m].$$

Therefore the design ideal of the n-run design, $\mathscr{F} = \{(a_{i1},\ldots,a_{im}),\ i = 1,\ldots,n\}$, is given as

$$I(\mathscr{F}) = \bigcap_{i=1}^{n} \langle x_1 - a_{i1},\ldots,x_m - a_{im}\rangle. \qquad (15.1)$$

S. Aoki et al., *Markov Bases in Algebraic Statistics*, Springer Series in Statistics 199, DOI 10.1007/978-1-4614-3719-2_15,

To calculate the intersection of ideals, we can use the theory of Gröbner bases. In fact, by introducing the indeterminates $t_1,\dots,t_n$ and the polynomial ring $k[x_1,\dots,x_m,t_1,\dots,t_n]$, (15.1) is written as

$$I(\mathscr{F}) = I^* \cap k[x_1,\dots,x_m],$$

where

$$I^* = \langle t_i(x_1 - a_{i1}),\dots,t_i(x_m - a_{im}),\ i = 1,\dots,n,\ t_1 + \cdots + t_n - 1\rangle$$

is an ideal of $k[x_1,\dots,x_m,t_1,\dots,t_n]$. Therefore we can obtain a basis of $I(\mathscr{F})$ as the reduced Gröbner basis of I^* with respect to a term order satisfying $\{t_1,\dots,t_n\} \succ \{x_1,\dots,x_m\}$. This is an elimination theory, one of the important applications of Gröbner bases we have seen in Sect. 3.4.

15.2 Identifiability of Polynomial Models and the Quotient with Respect to the Design Ideal

As one of the merits of considering a design ideal, we consider the identifiability or the confounding of polynomial models. To define these concepts, we revisit a design matrix discussed in Chap. 11, where we defined a design matrix for the two-level case in Definition 11.1 and the three-level case in Definition 11.2. We extend these and give a general definition. We write a monomial of $\{x_1,\dots,x_m\}$ as $\boldsymbol{x}^{\boldsymbol{\alpha}} = x_1^{\alpha_1}\cdots x_m^{\alpha_m}$. The polynomial model is written as

$$f(\boldsymbol{x}) = \sum_{\boldsymbol{\alpha}\in\mathscr{L}} \theta_{\boldsymbol{\alpha}} \boldsymbol{x}^{\boldsymbol{\alpha}}, \tag{15.2}$$

where $\mathscr{L}$ is a set of exponents and $\boldsymbol{\theta} = (\theta_{\boldsymbol{\alpha}})_{\boldsymbol{\alpha}\in\mathscr{L}}$ is a parameter.

Definition 15.1. Let $\mathscr{F} \subset \mathscr{D}$ be a fractional factorial design of m factors with n runs. Let $\mathscr{F} = \{\boldsymbol{a}_i = (a_{i1},\dots,a_{im}) \in k^m,\ i = 1,\dots,n\}$. Consider the polynomial model (15.2). Then the matrix

$$A = [\boldsymbol{a}_i^{\boldsymbol{\alpha}}]_{i=1,\dots,n;\ \boldsymbol{\alpha}\in\mathscr{L}}$$

is called a design matrix for $\mathscr{L}$.

Note that the definition of the design matrix above differs slightly from the definition in Chap. 11; that is, we defined A as the transpose of A in Chap. 11. Although it is somewhat confusing, we use the conventional definition of experimental design only in this chapter, whereas it is better to transpose A for clarifying the relation between the design matrix and the configuration matrix of the toric ideals. Note also that there are $|\mathscr{L}|$ columns in A. Write as $\boldsymbol{y} = (y_1,\dots,y_n)'$ an observation vector for the design $\mathscr{F}$. Then if A is of full rank, the usual least square estimator of the parameter in (15.2) is written as

$$\hat{\boldsymbol{\theta}} = (A'A)^{-1}A'\mathbf{y}.$$

In this sense, we define the identifiability of the polynomial model as follows.

Definition 15.2. The polynomial model (15.2) is called identifiable by $\mathscr{F}$ if the design matrix for $\mathscr{L}$ is of full rank.

An important identifiable model is a saturated model. In the saturated model in which $|\mathscr{L}| = n$ holds, A is a square full rank matrix and the estimator of the parameter is $\hat{\boldsymbol{\theta}} = A^{-1}\mathbf{y}$. In this case, (15.2) is an interpolatory polynomial.

Example 15.1 (Regular 2_{III}^{5-2} design). Let $\mathscr{F}$ be a regular 2_{III}^{5-2} design of $m = 5$ factors with two levels defined as $x_1x_2x_4 = x_1x_3x_5 = 1$. $\mathscr{F}$ contains $n = 8$ runs and is displayed as follows.

Run\factor	x_1	x_2	x_3	x_4	x_5
1	1	1	1	1	1
2	1	1	−1	1	−1
3	1	−1	1	−1	1
4	1	−1	−1	−1	−1
5	−1	1	1	−1	−1
6	−1	1	−1	−1	1
7	−1	−1	1	1	−1
8	−1	−1	−1	1	1

The design matrix for the main effect model, that is, a polynomial model written as

$$f(\boldsymbol{x}) = \theta_{00000} + \theta_{10000}x_1 + \theta_{01000}x_2 + \theta_{00100}x_3 + \theta_{00010}x_4 + \theta_{00001}x_5, \qquad (15.3)$$

is given as

$$A = \begin{pmatrix} 1 & 1 & 1 & 1 & 1 & 1 \\ 1 & 1 & 1 & -1 & 1 & -1 \\ 1 & 1 & -1 & 1 & -1 & 1 \\ 1 & 1 & -1 & -1 & -1 & -1 \\ 1 & -1 & 1 & 1 & -1 & -1 \\ 1 & -1 & 1 & -1 & -1 & 1 \\ 1 & -1 & -1 & 1 & 1 & -1 \\ 1 & -1 & -1 & -1 & 1 & 1 \end{pmatrix}.$$

Inasmuch as this A is of full rank, the polynomial model (15.3) is identifiable by $\mathscr{F}$. Another polynomial model

$$f(\boldsymbol{x}) = \theta_{00000} + \theta_{10000}x_1 + \theta_{01000}x_2 + \theta_{00100}x_3 + \theta_{00010}x_4 + \theta_{00001}x_5$$
$$+\theta_{01100}x_2x_3 + \theta_{00110}x_3x_4 \qquad (15.4)$$

is also identifiable by $\mathscr{F}$ because the design matrix for this model,

$$A = \begin{pmatrix} 1 & 1 & 1 & 1 & 1 & 1 & 1 & 1 \\ 1 & 1 & 1 & -1 & 1 & -1 & -1 & -1 \\ 1 & 1 & -1 & 1 & -1 & 1 & -1 & -1 \\ 1 & 1 & -1 & -1 & -1 & -1 & 1 & 1 \\ 1 & -1 & 1 & 1 & -1 & -1 & 1 & -1 \\ 1 & -1 & 1 & -1 & -1 & 1 & -1 & 1 \\ 1 & -1 & -1 & 1 & 1 & -1 & -1 & 1 \\ 1 & -1 & -1 & -1 & 1 & 1 & 1 & -1 \end{pmatrix},$$

is of full rank. The polynomial model (15.4) is one of the saturated models. It is also one of the interpolatory polynomials and the parameter $\boldsymbol{\theta}$ is estimated as $\hat{\boldsymbol{\theta}} = \frac{1}{8}A'\mathbf{y}$ because $A^{-1} = \frac{1}{8}A'$. Note that this A is an Hadamard matrix. On the other hand, the polynomial model

$$f(\boldsymbol{x}) = \theta_{00000} + \theta_{10000}x_1 + \theta_{01000}x_2 + \theta_{00100}x_3 + \theta_{00010}x_4 + \theta_{00001}x_5 + \theta_{11000}x_1x_2$$

is not identifiable by $\mathscr{F}$ because the design matrix for this model is not of full rank.

The design we considered in Example 15.1 is a regular two-level design, thus we have already considered an identifiability of models in Chap. 11. Of course, the arguments in this chapter are valid for general designs. We see another example.

Example 15.2. Consider a fractional factorial design of three factors $x_1, x_2 \in \{-1,1\}$, $x_3 \in \{-1,0,1\}$ given as follows.

Run\factor	x_1	x_2	x_3
1	1	1	1
2	1	−1	0
3	1	−1	−1
4	−1	1	0
5	−1	1	−1
6	−1	−1	1

For this 6-run design, we can consider several identifiable polynomial saturated models with six parameters such as

$$f_1(\boldsymbol{x}) = \theta_{000} + \theta_{100}x_1 + \theta_{010}x_2 + \theta_{001}x_3 + \theta_{002}x_3^2 + \theta_{011}x_2x_3 \tag{15.5}$$

or

$$f_2(\boldsymbol{x}) = \theta_{000} + \theta_{010}x_2 + \theta_{001}x_3 + \theta_{002}x_3^2 + \theta_{011}x_2x_3 + \theta_{012}x_2x_3^2. \tag{15.6}$$

The design matrices for these models are given as

$$A_1 = \begin{pmatrix} 1 & 1 & 1 & 1 & 1 & 1 \\ 1 & 1 & -1 & 0 & 0 & 0 \\ 1 & 1 & -1 & -1 & 1 & 1 \\ 1 & -1 & 1 & 0 & 0 & 0 \\ 1 & -1 & 1 & -1 & 1 & -1 \\ 1 & -1 & -1 & 1 & 1 & -1 \end{pmatrix} \quad \text{and} \quad A_2 = \begin{pmatrix} 1 & 1 & 1 & 1 & 1 & 1 \\ 1 & -1 & 0 & 0 & 0 & 0 \\ 1 & -1 & -1 & 1 & 1 & -1 \\ 1 & 1 & 0 & 0 & 0 & 0 \\ 1 & 1 & -1 & 1 & -1 & 1 \\ 1 & -1 & 1 & 1 & -1 & -1 \end{pmatrix}.$$

Any submodel of these saturated models is also an identifiable polynomial model.

Now we consider the relation between the identifiability and the design ideal. The key item is the set of standard monomials defined in Sect. 3.2. For a design $\mathscr{F}$, we fix a term order τ and consider a Gröbner basis G_τ of $I(\mathscr{F})$. Then the set of standard monomials is defined as

$$\begin{aligned} \mathrm{Est}_\tau(\mathscr{F}) &= \{\boldsymbol{x}^{\boldsymbol{\alpha}} \mid \boldsymbol{x}^{\boldsymbol{\alpha}} \text{ is not divisible by any of the leading terms} \\ &\qquad \text{of the elements of the Gröbner basis of } I(\mathscr{F})\} \\ &= \{\boldsymbol{x}^{\boldsymbol{\alpha}} \mid \boldsymbol{x}^{\boldsymbol{\alpha}} \notin \langle \mathrm{LT}(g),\ g \in I(\mathscr{F})\rangle\}. \end{aligned}$$

Then from Theorem 3.1 in Sect. 3.2 (i.e., the fact that $\mathrm{Est}_\tau(\mathscr{F})$ is a basis of the vector space $k[x_1,\ldots,x_m]/I(\mathscr{F})$), we have the following results.

Theorem 15.1. *Let $\mathscr{F}$ be a design with n runs and τ be a term order. Write*

$$\mathrm{Est}_\tau(\mathscr{F}) = \{\boldsymbol{x}^{\boldsymbol{\alpha}} \mid \boldsymbol{\alpha} \in \mathscr{L}\}, \tag{15.7}$$

where $\mathscr{L}$ is the set of exponents in the elements in $\mathrm{Est}_\tau(\mathscr{F})$. *Then*

1. *$|\mathscr{L}| = n$ holds.*
2. *The polynomial model* (15.2) *is identifiable by $\mathscr{F}$ if $\mathscr{L}$ is defined as* (15.7).

This result is known as the first application of Gröbner basis theory by Buchberger and Hironaka in the 1960s to statistics. Pistone and Wynn [120] revisit this result in the statistical framework, saying "This important point does not seem to be stated explicitly in the statistical literature," and show many examples with actual computations using MAPLE software.

Theorem 15.1 shows how to construct an identifiable polynomial model from Gröbner basis theory. In fact, although the dimension is independent of the order, the elements of $\mathrm{Est}_\tau(\mathscr{F})$ depend on the chosen term order. Let us check this fact by the previous example.

Example 15.3 (Continuation of Example 15.2). Consider the 6-run design of Example 15.2. Under the lexicographic term order with $x_1 \succ x_2 \succ x_3$, the Gröbner basis of $I(\mathscr{F})$ is calculated as

$$\{\underline{x_2^2} - 1,\ \underline{x_3^3} - x_3,\ \underline{x_1} - x_2x_3^2 - x_2x_3 + x_2\},$$

where the leading terms are underlined. Therefore the set of standard monomials is

$$\mathrm{Est}_{\mathrm{lex}}(\mathscr{F}) = \{1,\ x_2,\ x_3,\ x_3^2,\ x_2x_3,\ x_2x_3^2\},$$

which gives the saturated polynomial model (15.6). On the other hand, under the graded reverse lexicographic term order of $x_1 \succ x_2 \succ x_3$, the Gröbner basis of $I(\mathscr{F})$ is calculated as

$$\Big\{\underline{x_1^2} - 1,\ \underline{x_2^2} - 1,\ \underline{x_3^3} - x_3,\ \underline{x_1x_2} - x_3^2 - x_3 + 1,\ \underline{x_1x_3} - x_2x_3 - x_1 + x_2,$$
$$\underline{x_2x_3^2} + x_2x_3 - x_1 - x_2\Big\},$$

which yields the set of standard monomials

$$\mathrm{Est}_{\mathrm{grevlex}}(\mathscr{F}) = \{1,\ x_1,\ x_2,\ x_3,\ x_3^2,\ x_2x_3\}.$$

This basis gives the saturated polynomial model (15.5).

As we saw in Example 15.3, we have several identifiable polynomial models in general by varying term order. In application, it seems efficient to select a term order by considering the model structures that we want to use. For example, when main effects are more important than interaction effects, a term order that reflects the total order of terms such as the graded reverse lexicographic term order may be used. (See the model (15.5), for example.) On the other hand, when one effect dominates all the others, a simple lexicographic term order may be used (see the model (15.6), for example). The next interesting question is whether all the identifiable polynomial models can be obtained by the algebraic approach. Unfortunately, the answer is no. One of the counterexamples from [118] is given below.

Example 15.4. Consider the design $\mathscr{F} = \{(0,0),(0,-1),(1,0),(1,1),(-1,1)\}$. Varying term order τ, we have two sets as $\mathrm{Est}_\tau(\mathscr{F})$, $\{1,x_1,x_1^2,x_2,x_1x_2\}$ and $\{1,x_2,x_2^2,x_1,x_1x_2\}$. However, there does not exist τ that gives another identifiable model $\{1,x_1,x_2,x_1^2,x_2^2\}$ as $\mathrm{Est}_\tau(\mathscr{F})$.

In [22], an interesting subset of the hierarchical polynomial models, namely, *corner cut models*, is considered and called a design *generic* if all the corner cut models of size n are identifiable.

Another look of the set $\mathrm{Est}_\tau(\mathscr{F})$ is the representative of an equivalence class congruent modulo $I(\mathscr{F})$. In fact, the vector space $k[x_1,\ldots,x_m]/I(\mathscr{F})$ is the set of classes of remainders of the polynomials in $k[x_1,\ldots,x_m]$ with respect to the division by G_τ. Thus, for $f \in k[x_1,\ldots,x_m]$, the equivalence class of f in $k[x_1,\ldots,x_m]/I(\mathscr{F})$ is

$$\{g \in k[x_1,\ldots,x_m] \mid f - g \in I(\mathscr{F})\}.$$

Each of the equivalence classes in $k[x_1,\ldots,x_m]/I(\mathscr{F})$ can be seen as an aliasing class in the sense that only one term from each class can be included in the same identifiable model. We summarize this point.

Definition 15.3. Two models, f and g, are confounded (or aliased) under the design $\mathscr{F}$ if $f-g \in I(\mathscr{F})$.

The methods of constructing the design matrix in Chap. 11 are based on the idea of choosing each column of the design matrix so that each corresponding term of the polynomial model is in a different equivalence class. In Chap. 11, we identified a monomial $\boldsymbol{x}^{\boldsymbol{a}} = x_1^{a_1} \cdots x_m^{a_m}$ with a main or an interaction effect between the m factors $x_1, \ldots, x_m$. For example, of two level factors, $\boldsymbol{x}^{\boldsymbol{a}}$ is identified with a main effect if $\sum_{i=1}^m a_i = 1$, and a two-factor interaction effect if $\sum_{i=1}^m a_i = 2$, and so on. Then two main or interaction effects are confounded in the design $\mathscr{F}$ if $\boldsymbol{x}^{\boldsymbol{a}_1}\boldsymbol{x}^{\boldsymbol{a}_2}$ is identically equal to $+1$ or -1 for all the points in $\boldsymbol{x} \in \mathscr{F}$. This confounding relation is expressed in terms of the design ideal as follows.

Proposition 15.1. *Let $c \in \{-1, +1\}$. Then the following two conditions are equivalent.*

$$\text{(i) } \boldsymbol{x}^{\boldsymbol{a}_1}\boldsymbol{x}^{\boldsymbol{a}_2} = c \ \text{ for all } \boldsymbol{x} \in \mathscr{F} \qquad \text{(ii) } \boldsymbol{x}^{\boldsymbol{a}_1} - c\boldsymbol{x}^{\boldsymbol{a}_2} \in I(\mathscr{F})$$

In general, we have to calculate a Gröbner basis to judge whether a given polynomial belongs to a given ideal, that is, to solve the ideal membership problem.

15.3 Regular Two-Level Designs

We saw in Sect. 15.2 that the identifiability of the terms in polynomial models can be treated algebraically by considering the Gröbner basis of the design ideal. As we have seen in Example 15.2, these theories can be used for arbitrary design $\mathscr{F}$, regardless of being regular or nonregular.

In the statistical literature, however, the theory of regular fractional factorial designs is well developed. For example, an elegant theory based on linear algebra over the finite field GF(2) is well established for regular two-level fractional factorial designs. See [123]. On the other hand, it is very difficult to derive theoretical results for general nonregular fractional factorial designs. One of the merits of the algebraic approach is that we need not distinguish whether the design is regular because the design is characterized simply as a set of points. In fact, many important concepts such as resolution and aberration, which are originally defined for regular designs, can be generalized naturally to nonregular designs by an algebraic approach. See [153] or [152], for example. As another approach to deal with nonregular designs, some classes or criteria of nonregular designs are considered from the viewpoint of algebra. See [14] and [6], for example.

Nevertheless, it is instructive to consider the simple setting of regular designs to understand the practicality of Gröbner basis theory in designs of experiments. In this section, we focus on regular fractional factorial designs with two-level factors.

Similarly to Example 15.1, we code two levels of factors as $\{-1,+1\}$. Therefore for a monomial $\boldsymbol{x}^{\boldsymbol{a}} = x_1^{a_1}\cdots x_m^{a_m}$ of the indeterminates $x_1,\ldots,x_m$ representing m factors, it is sufficient to consider $\boldsymbol{a} = (a_1,\ldots,a_m) \in \{0,1\}^m$. The full factorial design of m factors with two levels is expressed as

$$\mathscr{D} = \{(x_1,\ldots,x_m) \mid x_1^2 = \cdots = x_m^2 = 1\} = \{-1,+1\}^m,$$

and the design ideal of $\mathscr{D}$ is written as

$$I(\mathscr{D}) = \langle x_1^2-1,\ldots,x_m^2-1\rangle.$$

Without loss of generality, we consider a regular 2^{m-s} fractional factorial design $\mathscr{F} \subset \mathscr{D}$ generated by s defining relations

$$\{\boldsymbol{x}^{\boldsymbol{a}_\ell} = 1,\ \ell = 1,\ldots,s\}, \tag{15.8}$$

such that $\boldsymbol{x}^{\boldsymbol{a}_\ell} = 1$ for all $\boldsymbol{x} \in \mathscr{F}$. One of the expressions of the design ideal for the regular design is obtained directly by this defining relation.

Proposition 15.2. *The design ideal for the regular two-level fractional factorial design* $\mathscr{F}$ *defined by* (15.8) *is written as*

$$I(\mathscr{F}) = \langle x_1^2-1,\ldots,x_m^2-1,\boldsymbol{x}^{\boldsymbol{a}_1}-1,\ldots,\boldsymbol{x}^{\boldsymbol{a}_s}-1\rangle. \tag{15.9}$$

Note that the basis in the expression (15.9) is not a Gröbner basis in general. In the arguments of Sect. 15.1, we used elimination theory as a general method to obtain a basis of the design ideal and obtain a reduced Gröbner basis as a result. However, one of the obvious bases can be obtained directly from the defining relation for the regular fractional factorial designs without calculating a Gröbner basis.

Example 15.5 (2_{III}^{7-4} design). Consider the design known as the orthogonal array $L_8(2^7)$ of resolution III with the defining relation

$$x_1x_2x_3 = x_1x_4x_5 = x_2x_4x_6 = x_1x_2x_4x_7 = 1$$

given as follows.

Run\factor	x_1	x_2	x_3	x_4	x_5	x_6	x_7
1	1	1	1	1	1	1	1
2	1	1	1	−1	−1	−1	−1
3	1	−1	−1	1	1	−1	−1
4	1	−1	−1	−1	−1	1	1
5	−1	1	−1	1	−1	1	−1
6	−1	1	−1	−1	1	−1	1
7	−1	−1	1	1	−1	−1	1
8	−1	−1	1	−1	1	1	−1

From (15.9), the design ideal for this design is expressed as

$$I(\mathscr{F}) = \langle x_1^2 - 1, \ldots, x_7^2 - 1, x_1x_2x_3 - 1, x_1x_4x_5 - 1, x_2x_4x_6 - 1, x_1x_2x_4x_7 - 1\rangle.$$

On the other hand, the reduced Gröbner basis of $I(\mathscr{F})$ is

$$\{x_7^2 - 1,\ x_6^2 - 1,\ x_5^2 - 1,\ x_3 - x_5x_6,\ x_2 - x_5x_7,\ x_1 - x_6x_7,\ x_4 - x_5x_6x_7\} \tag{15.10}$$

under the lexicographic term order with $x_1 \succ \cdots \succ x_7$, and

$$\begin{array}{l}\{x_7^2 - 1,\ x_6^2 - 1,\ x_5^2 - 1,\ x_4^2 - 1,\ x_3^2 - 1,\ x_2^2 - 1,\ x_1^2 - 1,\\ x_1x_2 - x_3,\ x_1x_3 - x_2,\ x_2x_3 - x_1,\ x_1x_4 - x_5,\ x_1x_5 - x_4,\ x_4x_5 - x_1,\\ x_2x_4 - x_6,\ x_2x_6 - x_4,\ x_4x_6 - x_2,\ x_3x_4 - x_7,\ x_3x_7 - x_4,\ x_4x_7 - x_3,\\ x_2x_5 - x_7,\ x_2x_7 - x_5,\ x_5x_7 - x_2, x_3x_5 - x_6,\ x_3x_6 - x_5,\ x_5x_6 - x_3,\\ x_1x_6 - x_7,\ x_1x_7 - x_6,\ x_6x_7 - x_1\}\end{array} \tag{15.11}$$

under the graded reverse lexicographic term order with $x_1 \succ \cdots \succ x_7$. As we have seen, the set of standard monomials for these Gröbner bases,

$$\mathrm{Est}_{\mathrm{lex}} = \{1,\ x_5,\ x_6,\ x_7,\ x_5x_6,\ x_5x_7,\ x_6x_7,\ x_5x_6x_7\}$$

and

$$\mathrm{Est}_{\mathrm{grevlex}} = \{1,\ x_1,\ x_2,\ x_3,\ x_4,\ x_5,\ x_6,\ x_7\}$$

present the interpolatory polynomials. Moreover, the expression (15.10) or (15.11) enables us to identify the confounding relations between the factors. For example, we see that two-factor interactions x_2x_3 and x_4x_5 are confounded because $x_2x_3 - x_4x_5 \in I(\mathscr{F})$. This is verified from (15.10) as

$$x_2x_3 - x_4x_5 = x_3(x_2 - x_5x_7) + x_5x_7(x_3 - x_5x_6) - x_5(x_4 - x_5x_6x_7)$$

and from (15.11) as

$$x_2x_3 - x_4x_5 = (x_2x_3 - x_1) - (x_4x_5 - x_1).$$

15.4 Indicator Functions

In this section, we briefly introduce an indicator function, which was first defined by [61].

Definition 15.4. The indicator function of a design $\mathscr{F} \subset \mathscr{D}$ is a polynomial $f \in k[x_1, \ldots, x_m]$ satisfying

$$f(\boldsymbol{x}) = \begin{cases} 1, \text{ if } \boldsymbol{x} \in \mathscr{F}, \\ 0, \text{ if } \boldsymbol{x} \in \mathscr{D} \setminus \mathscr{F}. \end{cases}$$

There is a one-to-one correspondence between the indicator function and the design $\mathscr{F}$ under the appropriate constraints. For example, if each factor has two levels coded as $\{-1,+1\}$, the indicator function has a unique square-free representation under the constraint $x_i^2 = 1$, $i = 1,\dots,m$.

General cases are also considered in [119]. For the case that x_i is an n_i-level factor for $i = 1,\dots,m$, [119] introduces a complex coding, $x_i \in \Omega_{n_i} = \{\omega_0,\dots,\omega_{n_i-1}\}$ where Ω_{n_i} is the set of the n_ith roots of the unity, and considers the polynomials in the complex field $\mathbb{C}$. Note that all fractional factorial designs can be obtained by adding further polynomial equations, called *generating equations*, to $\{x_i^{n_i} - 1 = 0, i = 1,\dots,m\}$, in order to restrict the number of solutions. The generating equation is a generalized concept of defining relations of regular fractional factorial designs. The indicator function forms a generating equation by itself. For example, in the two-level case, the design ideal of the fractional factorial design $\mathscr{F}$ is written as

$$I(\mathscr{F}) = \langle x_1^2 - 1,\dots,x_m^2 - 1, f(\boldsymbol{x}) - 1\rangle,$$

where $f(\boldsymbol{x})$ is the indicator function of $\mathscr{F}$.

The indicator function is a polynomial, therefore it can be incorporated into the theory of computational algebraic statistics naturally. In fact, many important results in the field of computational algebraic statistics are related to the indicator functions. It is also shown that some concepts of designed experiments such as confounding, resolution, orthogonality, and estimability are related to the structure of the indicator function of a design. In addition, because the indicator function is defined for any design, some classical notions for regular designs, such as confounding and resolution, can be generalized to nonregular designs naturally by the notion of the indicator function. See [62] or [154].

We show some basic results on the indicator functions of the two-level regular fractional factorial designs. Consider 2^{m-s} fractional factorial design $\mathscr{F}$ generated by s defining relations (15.8). The s defining relations generate an additive group $\{\boldsymbol{x}^{\boldsymbol{a}} = 1 \mid \boldsymbol{a} \in A_{\mathscr{F}}\}$, where

$$A_{\mathscr{F}} = \left\{ \boldsymbol{a} \;\middle|\; \boldsymbol{a} = \sum_{\ell=1}^{s} u_\ell \boldsymbol{a}_\ell,\ u_\ell \in \{0,1\} \text{ for } \ell = 1,\dots,s \right\}. \tag{15.12}$$

The summation above is as in GF(2). Then the indicator function of $\mathscr{F}$ is written as

$$f(\boldsymbol{x}) = \frac{1}{2^s}(1 + \boldsymbol{x}^{\boldsymbol{a}_1})\cdots(1 + \boldsymbol{x}^{\boldsymbol{a}_s}) = \frac{1}{2^s}\sum_{\boldsymbol{a}\in A_{\mathscr{F}}} \boldsymbol{x}^{\boldsymbol{a}}. \tag{15.13}$$

Note that the coefficient $b_{\boldsymbol{a}}$ of the monomial $\boldsymbol{x}^{\boldsymbol{a}}$ equals 2^{-s} for all $\boldsymbol{a} \in A_{\mathscr{F}}$ and 0 otherwise.

Example 15.6 (Continuation of Example 15.5). The indicator function of the 2^{7-4}_{III} design in Example 15.5 is written as

$$f(\boldsymbol{x}) = \frac{1}{16}(1+x_1x_2x_3)(1+x_1x_4x_5)(1+x_2x_4x_5)(1+x_1x_2x_4x_7).$$

The set of the exponents in the expansion of the right-hand side forms a group (15.12).

Next we consider the design ideal for adding factors, as a simple application of the indicator function. The additional factors may be real controllable factors, whose levels are determined by some defining relations. For the purpose of Markov bases in Chap. 11, the additional factors are formal and correspond to interaction effects included in a null hypothesis.

Let $\mathscr{F}_1$ be a fractional factorial design of the factors $x_1,\dots,x_m$. Consider adding factors $y_1,\dots,y_k$ to $\mathscr{F}_1$. We suppose the levels of the additional factors are determined by the defining relations among $x_1,\dots,x_m$ as

$$y_1 = e_1\boldsymbol{x}^{\boldsymbol{b}_1},\dots,y_k = e_k\boldsymbol{x}^{\boldsymbol{b}_k},$$

where $e_1,\dots,e_k \in \{-1,1\}$. Write this new design of $x_1,\dots,x_m,y_1,\dots,y_k$ as $\mathscr{F}_2$. The run sizes of $\mathscr{F}_1$ and $\mathscr{F}_2$ are the same.

Let f_1 and f_2 be the indicator functions of $\mathscr{F}_1$ and $\mathscr{F}_2$, respectively. Then we have

$$f_2(x_1,\dots,x_m,y_1,\dots,y_k) = \frac{1}{2^k}(1+e_1y_1\boldsymbol{x}^{\boldsymbol{b}_1})\cdots(1+e_ky_k\boldsymbol{x}^{\boldsymbol{b}_k})f_1(x_1,\dots,x_m). \quad (15.14)$$

In fact, for $(x_1,\dots,x_m,y_1,\dots,y_k) \in \mathscr{F}_2$, $(x_1,\dots,x_m) \in \mathscr{F}_1$ and

$$e_1y_1\boldsymbol{x}^{\boldsymbol{b}_1} = \cdots = e_ky_k\boldsymbol{x}^{\boldsymbol{b}_k} = 1$$

hold, which yields $f_2 = 1$. Conversely, if $(x_1,\dots,x_m,y_1,\dots,y_k) \notin \mathscr{F}_2$, then $(x_1,\dots, x_m) \notin \mathscr{F}_1$ or some of $e_1y_1\boldsymbol{x}^{\boldsymbol{b}_1},\dots,e_ky_k\boldsymbol{x}^{\boldsymbol{b}_k}$ has to be -1, which yields $f_2 = 0$. Note that (15.14) generalizes the indicator function of regular fractional factorial designs (15.13), by taking $f_1 \equiv 1$, that is, by assuming the full factorial design for $x_1,\dots,x_m$.

From the above result, we have an expression of $I(\mathscr{F}_2)$ as

$$I(\mathscr{F}_2) = \langle x_1^2-1,\dots,x_m^2-1,y_1^2-1,\dots,y_k^2-1,f_1-1,f_2-1\rangle.$$

If we fix the term order τ on $x_1,\dots,x_m$ and σ on $x_1,\dots,x_m,y_1,\dots,y_k$, $\mathrm{Est}_\tau(\mathscr{F}_1)$ and $\mathrm{Est}_\sigma(\mathscr{F}_2)$ are defined. $\mathrm{Est}_\tau(\mathscr{F}_1)$ and $\mathrm{Est}_\sigma(\mathscr{F}_2)$ contain the same number of monomials because the run sizes of $\mathscr{F}_1$ and $\mathscr{F}_2$ are the same. In particular, if we use a lexicographic term order σ with $\{y_1,\dots,y_k\} \succ_\sigma \{x_1,\dots,x_m\}$, then $\mathrm{Est}_\tau(\mathscr{F}_1) = \mathrm{Est}_\sigma(\mathscr{F}_2)$ holds. We end this chapter with the following example.

Example 15.7 (The indicator function and the standard monomials under the lexicographic term order for adding factors). Let $\mathscr{F}_1$ be a 2^{4-1}_{IV} fractional factorial design of x_1,x_2,x_3,x_4 defined by $x_1x_2x_3x_4 = 1$ and $\mathscr{F}_2$ be a fractional factorial design of x_1,x_2,x_3,x_4,y_1,y_2 by adding $y_1 = x_1x_2, y_2 = x_1x_3$ to $\mathscr{F}_1$.

$\mathscr{F}_1$

run\factor	x_1	x_2	x_3	x_4
1	1	1	1	1
2	1	1	−1	−1
3	1	−1	1	−1
4	1	−1	−1	1
5	−1	1	1	−1
6	−1	1	−1	1
7	−1	−1	1	1
8	−1	−1	−1	−1

$\mathscr{F}_2$

run\factor	x_1	x_2	x_3	x_4	y_1	y_2
1	1	1	1	1	1	1
2	1	1	−1	−1	1	−1
3	1	−1	1	−1	−1	1
4	1	−1	−1	1	−1	−1
5	−1	1	1	−1	−1	−1
6	−1	1	−1	1	−1	1
7	−1	−1	1	1	1	−1
8	−1	−1	−1	−1	1	1

The indicator function of $\mathscr{F}_1$ is

$$f_1(x_1,x_2,x_3,x_4) = \frac{1}{2}(1 + x_1x_2x_3x_4)$$

from (15.13). Then the indicator function of $\mathscr{F}_2$ is calculated as

$$\begin{aligned} f_2(x_1,x_2,x_3,x_4,y_1,y_2) &= \frac{1}{4}(1 + y_1x_1x_2)(1 + y_2x_1x_3)f_1(x_1,x_2,x_3,x_4) \\ &= \frac{1}{8}(1 + y_1x_1x_2)(1 + y_2x_1x_3)(1 + x_1x_2x_3x_4) \end{aligned}$$

from (15.14). Inasmuch as the reduced Gröbner basis of $I(\mathscr{F}_1)$ under the lexicographic term order with $x_1 \succ \cdots \succ x_4$ is

$$\{\underline{x_1} - x_2x_3x_4,\ \underline{x_2^2} - 1,\ \underline{x_3^2} - 1,\ \underline{x_4^2} - 1\},$$

the set of the standard monomials for this Gröbner basis is given as

$$\mathrm{Est}_{\mathrm{lex}}(\mathscr{F}_1) = \{1,\ x_2,\ x_3,\ x_4,\ x_2x_3,\ x_2x_4,\ x_3x_4\}.$$

On the other hand, because the reduced Gröbner basis of $I(\mathscr{F}_2)$ under the lexicographic term order with $y_1 \succ y_2 \succ x_1 \succ \cdots \succ x_4$ is

$$\{\underline{y_1} - x_3x_4,\ \underline{y_2} - x_2x_4,\ \underline{x_1} - x_2x_3x_4,\ \underline{x_2^2} - 1,\ \underline{x_3^2} - 1,\ \underline{x_4^2} - 1\},$$

we see

$$\mathrm{Est}_{\mathrm{lex}}(\mathscr{F}_2) = \{1,\ x_2,\ x_3,\ x_4,\ x_2x_3,\ x_2x_4,\ x_3x_4\} = \mathrm{Est}_{\mathrm{lex}}(\mathscr{F}_1).$$

Of course, such a result is due to the lexicographic term order. Under the graded reverse lexicographic term order, on the other hand, the reduced Gröbner bases are calculated as

$$\{\underline{x_1^2}-1,\ \underline{x_2^2}-1,\ \underline{x_3^2}-1,\ \underline{x_4^2}-1,\ \underline{x_1x_2}-x_3x_4,\ \underline{x_1x_3}-x_2x_4,\ \underline{x_1x_4}-x_2x_3\}$$

for $I(\mathscr{F}_1)$ and

$$\begin{array}{l}\{\underline{y_1^2}-1,\ \underline{y_2^2}-1,\ \underline{x_1^2}-1,\ \underline{x_2^2}-1,\ \underline{x_3^2}-1,\ \underline{x_4^2}-1,\ \underline{y_1y_2}-x_1x_4,\\ \underline{y_1x_1}-x_2,\ \underline{y_1x_2}-x_1,\ \underline{y_1x_3}-x_4,\ \underline{y_1x_4}-x_3,\\ \underline{y_2x_1}-x_3,\ \underline{y_2x_2}-x_4,\ \underline{y_2x_3}-x_1,\ \underline{y_2x_4}-x_3,\\ \underline{x_1x_2}-y_1,\ \underline{x_1x_3}-y_2,\ \underline{x_1x_4}-x_2x_3,\ \underline{x_2x_4}-y_2,\ \underline{x_3x_4}-y_1\}\end{array}$$

for $I(\mathscr{F}_2)$, respectively, and therefore the sets of the standard monomials are given as

$$\mathrm{Est}_{\mathrm{grevlex}}(\mathscr{F}_1)=\{1,x_1,x_2,x_3,x_4,x_2x_3,x_2x_4,x_3x_4\}$$

and

$$\mathrm{Est}_{\mathrm{grevlex}}(\mathscr{F}_2)=\{1,y_1,y_2,x_1,x_2,x_3,x_4,x_2x_3\},$$

respectively.

Chapter 16
Running Markov Chain Without Markov Bases

16.1 Performing Conditional Tests When a Markov Basis Is Not Available

As discussed in the previous chapters, explicit forms of Markov bases are known only for some simple structured models. Furthermore general algorithms for Markov basis computation often fail to produce Markov bases even for moderate-sized models in a practical amount of time. Hence so far we could not perform exact tests based on Markov basis methodology for many important practical problems, such as no-three-factor interaction models with many levels and logistic regression models with many covariates.

Some methodologies alternative to the Markov basis approach have been proposed. The sequential importance sampling (SIS) developed by Yuguo Chen and others (Chen and Small [36], Chen et al. [32], Chen et al. [34]) provides an algorithm for producing contingency tables by filling the cells of a table starting from the empty table. In SIS we never subtract a frequency from an existing table. Hence the problem of negative frequency is avoided in SIS. In the examples given in the above papers, SIS is found to work efficiently.

More recently Dobra [53] proposed a dynamic Markov basis, where elements of a Markov basis are computed dynamically during a Monte Carlo simulation.

In this section we propose another method based on a lattice basis for problems, where a Markov basis is not known [72].

16.2 Sampling Contingency Tables with a Lattice Basis

For a configuration matrix A, let

$$\ker_{\mathbb{Z}} A = \ker A \cap \mathbb{Z}^{|\mathscr{I}|} = \{z \in \mathbb{Z}^{|\mathscr{I}|} \mid Az = 0\}$$

S. Aoki et al., *Markov Bases in Algebraic Statistics*, Springer Series in Statistics 199, DOI 10.1007/978-1-4614-3719-2_16,

denote the integer kernel of A. By definition a move is an element of $\ker_{\mathbb{Z}} A$. Let $d = \dim\ker A = |\mathscr{I}| - \operatorname{rank} A$ be the dimension of linear space spanned by the elements of $\ker A$ in $\mathbb{R}^{|\mathscr{I}|}$. As mentioned in Sect. 4.3, the integer lattice $\ker_{\mathbb{Z}} A$ possesses a lattice basis $\mathscr{L} = \{\boldsymbol{z}_1, \ldots, \boldsymbol{z}_d\}$ and every move $\boldsymbol{z} \in \ker_{\mathbb{Z}} A$ is written as a unique integer combination of $\boldsymbol{z}_1, \ldots, \boldsymbol{z}_d$ (e.g., Schrijver [134]). As we mentioned, the computation of a lattice basis for a given A is relatively easy. Also, for many statistical models, where a Markov basis is hard to obtain, we can more easily identify a lattice basis.

A Markov basis is defined as a set of moves connecting every fiber. Let $k[\boldsymbol{u}]$, $\boldsymbol{u} = \{u(\boldsymbol{i}), \boldsymbol{i} \in \mathscr{I}\}$ be a polynomial ring and let $I_{\mathscr{L}} = \langle \boldsymbol{u}^{\boldsymbol{z}} \mid \boldsymbol{z} \in \mathscr{L}\rangle$ be the ideal generated by a lattice basis $\mathscr{L}$. The toric ideal I_A is obtained from $I_{\mathscr{L}}$ by taking a saturation [105, 139],

$$
\begin{aligned}
I_A &= \left(I_{\mathscr{L}} : \left\langle \prod_{\boldsymbol{i}\in\mathscr{I}} u(\boldsymbol{i}) \right\rangle \right) \\
&= \left\{ v \in k[\boldsymbol{u}] \mid \left(\prod_{\boldsymbol{i}\in\mathscr{I}} u(\boldsymbol{i}) \right)^m v \in I_{\mathscr{L}} \text{ for some } m > 0 \right\}.
\end{aligned}
$$

In this way a Markov basis is computed from a lattice basis. Intuitively this fact also shows that when the frequency of each cell is sufficiently large, the fiber is connected by the lattice basis $\mathscr{L}$. However, a lattice basis itself does not guarantee the connectivity of every fiber. By definition every move is written as an integer combination of elements of a lattice basis. Hence, if we generate moves in such a way that every integer combination of elements of a lattice basis has a positive probability, then we can indeed guarantee the connectivity of every fiber.

Usually a lattice basis contains exactly d elements. Here we allow redundancy of a lattice basis: we call a finite set $\mathscr{L}$ of moves a lattice basis if every move is written by an integral combination of the elements of $\mathscr{L}$.

Let $\mathscr{L} = \{\boldsymbol{z}_1, \ldots, \boldsymbol{z}_K\}$, $K \geq d$, be a lattice basis. Then any move $\boldsymbol{z} \in \ker_{\mathbb{Z}} A$ is expressed as

$$
\boldsymbol{z} = \alpha_1 \boldsymbol{z}_1 + \cdots + \alpha_K \boldsymbol{z}_K, \quad \alpha_1, \ldots, \alpha_K \in \mathbb{Z}.
$$

Then we can generate a move $\boldsymbol{z}$ by generating the integer coefficients $\alpha_1, \ldots, \alpha_K$. Here we use the following two methods to generate $\alpha_1, \ldots, \alpha_K$. Both methods generate all integer combinations of elements of $\mathscr{L}$ with positive probabilities and hence guarantee the connectivity of all the fibers.

Algorithm 16.1

Step 1. Generate $|\alpha_1|, \ldots, |\alpha_K|$ *from a Poisson distribution with mean* λ,

$$
|\alpha_k| \overset{\text{iid}}{\sim} Po(\lambda)
$$

and exclude the case $|\alpha_1| = \cdots = |\alpha_K| = 0$.

Step 2. $\alpha_k \leftarrow |\alpha_k|$ *or* $\alpha_k \leftarrow -|\alpha_k|$ *with probability* $\frac{1}{2}$ *for* $k = 1, \ldots, K$.

Algorithm 16.2

Step 1. Generate $|\alpha| = \sum_{i=1}^{K} |\alpha_i|$ *from geometric distribution with parameter* p

$$|\alpha| \sim Geom(p)$$

and allocate $|\alpha|$ *to* $\alpha_1, \dots, \alpha_K$ *according to multinomial distribution*

$$\alpha_1, \dots, \alpha_K \sim \mathrm{Mult}(|\alpha|; 1/K, \dots, 1/K).$$

Step 2. $\alpha_k \leftarrow |\alpha_k|$ *or* $\alpha_k \leftarrow -|\alpha_k|$ *with probability* $\frac{1}{2}$ *for* $k = 1, \dots, K$.

16.3 A Lattice Basis for Higher Lawrence Configuration

Let $\Lambda^{(r)}$ be the rth Lawrence configuration in (9.10). Many practical statistical models including the no-three-factor interaction model and the discrete logistic regression model discussed in the previous chapters have Lawrence configurations. In general a Markov basis for the Lawrence configuration is very difficult to compute (e.g., Chen et al. [33], Hara et al. [81]). On the other hand, because the computation of a lattice basis is easy, Algorithms 16.1 and 16.2 are available even for such models. Furthermore we can compute a lattice basis of $\Lambda^{(r)}(A)$ from a lattice basis of A by Proposition 16.1. This proposition is closely related to Proposition 4.3 for $r = 2$ and was first discussed in Santos and Sturmfels [131].

Proposition 16.1. *Let the column vectors of* B *form a lattice basis of* A. *Then the column vectors of*

$$B^{(r)} = \begin{pmatrix} \overbrace{\begin{matrix} B & 0 & \cdots & 0 \end{matrix}}^{r-1} \\ \begin{matrix} 0 & B & \ddots & \vdots \end{matrix} \\ \begin{matrix} \vdots & \ddots & \ddots & 0 \end{matrix} \\ \begin{matrix} 0 & \cdots & 0 & B \end{matrix} \\ \begin{matrix} -B & -B & \cdots & -B \end{matrix} \end{pmatrix} \tag{16.1}$$

form a lattice basis of higher Lawrence configuration $\Lambda^{(r)}(A)$.

Proof. We can interpret the rth Lawrence lifting as r slices of the original contingency table corresponding to A. The number of the cells for $\Lambda^{(r)}(A)$ is $|\mathscr{I}| = rm$, where m is the number of cells (columns) of A. Let

$$\boldsymbol{z} = \begin{pmatrix} \boldsymbol{z}_1 \\ \vdots \\ \boldsymbol{z}_r \end{pmatrix} = \boldsymbol{y} - \boldsymbol{x} = \begin{pmatrix} \boldsymbol{y}_1 - \boldsymbol{x}_1 \\ \vdots \\ \boldsymbol{y}_r - \boldsymbol{x}_r \end{pmatrix}$$

be a move of $\Lambda^{(r)}(A)$. We can express $z_1 = B\boldsymbol{\alpha}_1$. Then using the rth slice as a "base level," we can write

$$\boldsymbol{z} = \begin{pmatrix} B \\ 0 \\ \vdots \\ 0 \\ -B \end{pmatrix} \boldsymbol{\alpha}_1 + \begin{pmatrix} 0 \\ z_2 \\ \vdots \\ z_{r-1} \\ z_r + B\boldsymbol{\alpha}_1 \end{pmatrix}.$$

Note that the first block of $\boldsymbol{z}$ is now eliminated. Performing the same operation recursively to other blocks we are left with the $(r-1)$th slice and rth slice, which is similar to Proposition 4.3. □

In this proposition we only used the last slice as the base level. A more symmetric lattice basis can be obtained by columns of all pairwise differences of slices, for example, for $r = 3$,

$$\begin{pmatrix} B & B & 0 \\ -B & 0 & B \\ 0 & -B & -B \end{pmatrix}.$$

The lattice bases in the above propositions may contain redundant elements. However, the set of moves including redundant elements is sometimes preferable for moving around the fiber. In general the computation of a lattice basis of A is easier than the computation of a lattice basis of $\Lambda^{(r)}(A)$. Sometimes we can compute a Markov basis for A even when it is difficult to compute a Markov basis of $\Lambda^{(r)}(A)$. If a Markov basis for A is known, we can use it as a lattice basis for A and apply the above propositions for obtaining a lattice basis of $\Lambda^{(r)}(A)$.

16.4 Numerical Experiments

In this section we apply the proposed method to the no-three-factor interaction model and the discrete logistic regression model and show the usefulness of the proposed method.

16.4.1 No-Three-Factor Interaction Model

As discussed in Chap. 9, the structure of the no-three-factor interaction model

$$\log p_{ijk} = \mu_{\{1,2\}}(ij) + \mu_{\{1,3\}}(ik) + \mu_{\{2,3\}}(jk)$$

is complicated and the closed-form expression of Markov bases for this model of general tables is not yet obtained at present. Even by using 4ti2, it is difficult to compute a Markov basis for contingency tables larger than $5 \times 5 \times 5$ tables within a practical amount of time.

This model has the higher Lawrence configuration in (9.10) such that A is a configuration for the two-way complete independence model. The set of basic moves of form

	i_1	i_1'
i_2	1	−1
i_2'	−1	1

is known to be a Markov basis for the two-way complete independence model. By using this fact and Proposition 16.1, we can compute a lattice basis as a set of degree 4 moves,

i_3

	i_1	i_1'
i_2	1	−1
i_2'	−1	1

i_3'

	i_1	i_1'
i_2	−1	1
i_2'	1	−1

.

In this experiment we compute an exact distribution of the (twice log) likelihood ratio (LR) statistic of the goodness-of-fit test for the no-three-factor interaction model against the three-way saturated model

$$\log p_{i_1 i_2 i_3} = \mu_{\{1,2,3\}}(i_1 i_2 i_3).$$

We computed the sampling distribution of the LR statistic for $I \times I \times I$ ($I = 3,5,10$) three-way contingency tables. Then the degree of freedom of the asymptotic χ^2 distribution of the LR statistic is $(I-1)^3$. We set the sample size as $5I^3$. For $3 \times 3 \times 3$ tables, the number of burn-in samples and iterations are $(\text{burn-in}, \text{iteration}) = (1{,}000, 10{,}000)$. In $3 \times 3 \times 3$ tables, as discussed in Sect. 9.1, a minimal Markov basis is known and we also compute a sampling distribution by a Markov basis. In other cases, we set $(\text{burn-in}, \text{iteration}) = (10{,}000, 100{,}000)$.

Figure 16.1 presents the results for $3 \times 3 \times 3$ tables. Left, center, and right figures are histograms, paths, and correlograms of the LR statistic, respectively. Solid lines in the left figures are asymptotic χ^2 distributions with 8 degrees of freedom. α_k is generated from $Po(\lambda)$, $\lambda = 1, 10, 50$. We can see from the figures that the proposed methods show comparative performance to the sampling with a Markov basis. Although the sampling distribution and the path are somewhat unstable for $\lambda = 50$, in other cases the sampling distributions are similar and the paths are stable after the burn-in period. Unless we set λ extremely high, the performance of the proposed method is robust against the distribution of α_k.

Figure 16.2 presents the results for $5 \times 5 \times 5$ and $10 \times 10 \times 10$ tables. In these cases the Markov basis cannot be computed via 4ti2 within a practical amount of time by an Intel Core 2 Duo 3.0 GHz CPU machine. So we compute sampling

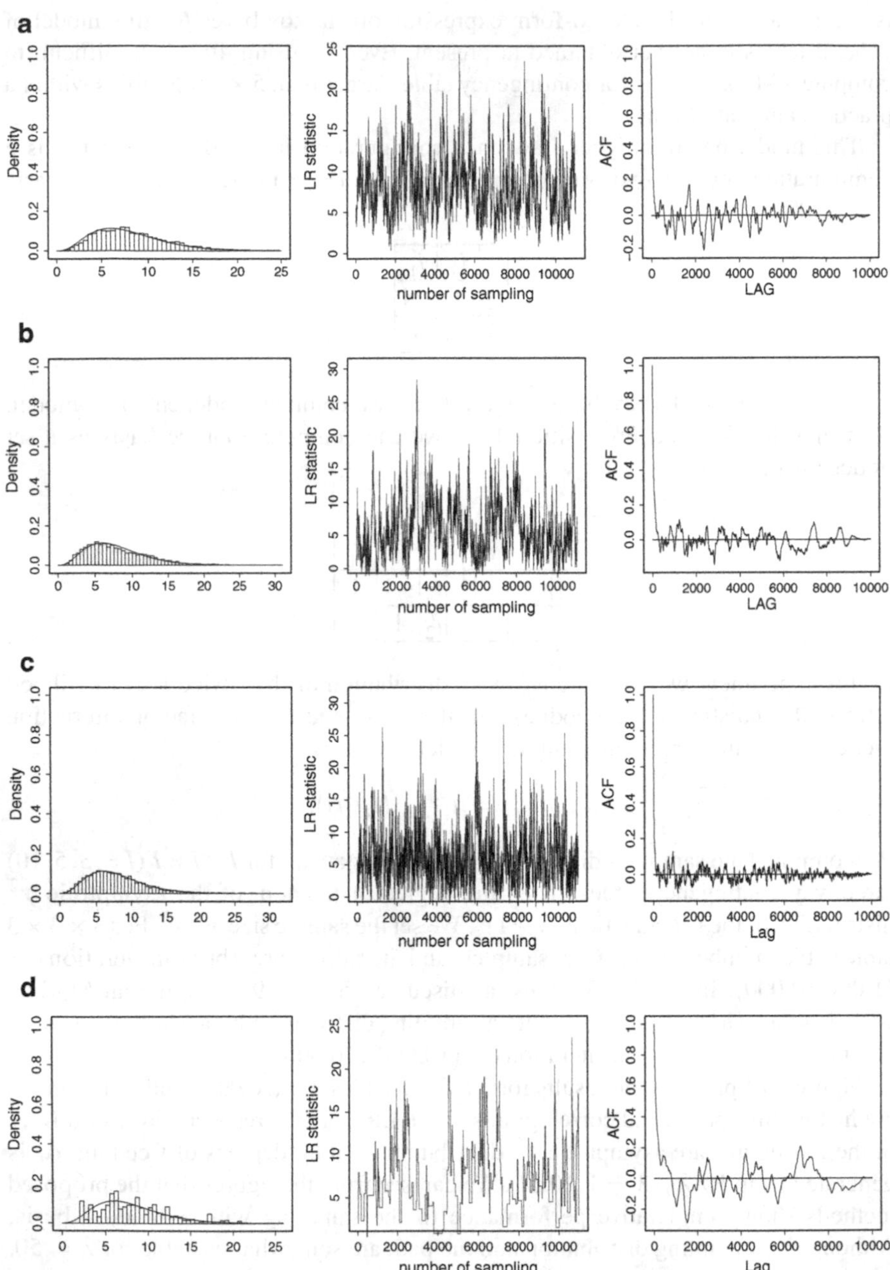

Fig. 16.1 Histograms, paths of LR statistics, and correlograms for $3 \times 3 \times 3$ no-three-factor interaction model ((burn-in, iteration) = (1,000,10,000)): (**a**) a Markov basis; (**b**) a lattice basis with $Po(1)$; (**c**) a lattice basis with $Po(10)$; (**d**) a lattice basis with $Po(50)$

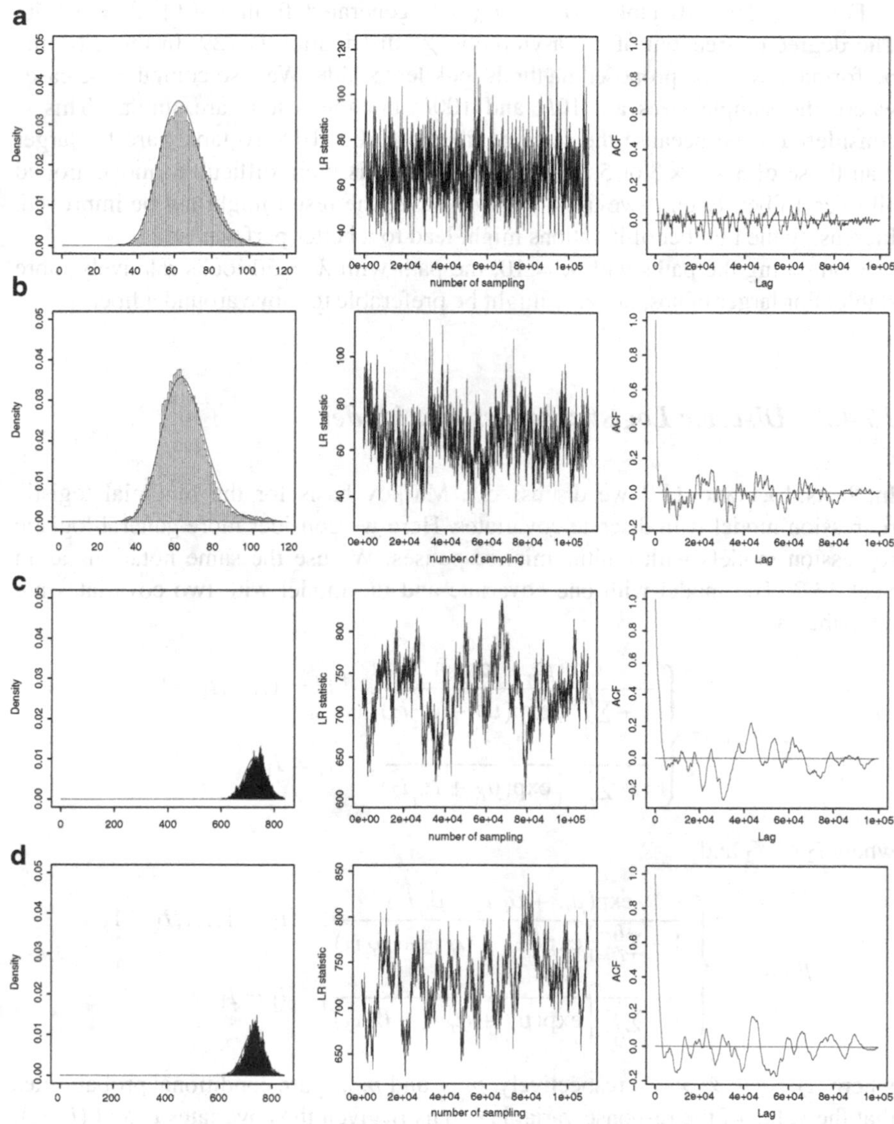

Fig. 16.2 Histograms, paths of LR statistics, and correlograms of paths for no-three-factor interaction model ((burn-in, iteration) = (10,000,100,000)): **(a)** $5 \times 5 \times 5$, a lattice basis with *Geom*(0.1); **(b)** $5 \times 5 \times 5$, a lattice basis with *Geom*(0.5); **(c)** $10 \times 10 \times 10$, a lattice basis with *Po*(10); **(d)** $10 \times 10 \times 10$, a lattice basis with *Po*(50)

distributions by using a lattice basis. For $5 \times 5 \times 5$ tables, $\alpha_1, \ldots, \alpha_K$ are generated from $Geom(p)$, $p = 0.1, 0.5$. The degree of freedom of the asymptotic χ^2 distribution is 64. Also in this case we can see that the proposed methods perform well. The approximation of the sampling distributions to the asymptotic χ^2 distribution is good and the paths are stable after the burn-in period.

For $10 \times 10 \times 10$ tables, $\alpha_1, \dots, \alpha_K$ are generated from $Po(\lambda)$, $\lambda = 10, 50$. The degree of freedom of the asymptotic χ^2 distribution is 729. In this case the performances of the proposed methods look less stable. We also compute the cases where the sample sizes are $10I^3$ and $100I^3$ but the results are similar. This is considered to be because the sizes of fibers of $10 \times 10 \times 10$ tables are far larger than those of $3 \times 3 \times 3$ or $5 \times 5 \times 5$ tables and it is more difficult to move around all over a fiber. Even if we use a Markov basis, the result might not be improved. Increasing the number of iterations might lead to a better performance.

Comparing the paths with $\lambda = 10$, the path with $\lambda = 50$ looks relatively more stable. For larger tables, larger λ might be preferable to move around a fiber.

16.4.2 Discrete Logistic Regression Model

In Sects. 13.1 and 13.2 we discussed a Markov basis for the binomial logistic regression model with discrete covariates. Here we consider more general logistic regression models with multinomial responses. We use the same notations as in Sect. 13.2. The model with one covariate and the model with two covariates are described as

$$p_{i_1|i_2} = \begin{cases} \dfrac{\exp(\mu_{i_1} + \alpha_{i_1} i_2)}{1 + \sum_{i_1'=1}^{I_1-1} \exp(\mu_{i_1'} + \alpha_{i_1'} i_2)}, & i_1 = 1, \dots, I_1 - 1, \\ \dfrac{1}{1 + \sum_{i_1'=1}^{I_1-1} \exp(\mu_{i_1'} + \alpha_{i_1'} i_2)}, & i_1 = I_1, \end{cases}$$

where $i_2 \in \mathscr{I}_2$ and

$$p_{i_1|i_2 i_3} = \begin{cases} \dfrac{\exp(\mu_{i_1} + \alpha_{i_1} i_2 + \beta_{i_1} i_3)}{1 + \sum_{i_1'=1}^{I_1-1} \exp(\mu_{i_1'} + \alpha_{i_1'} i_2 + \beta_{i_1'} i_3)}, & i_1 = 1, \dots, I_1 - 1, \\ \dfrac{1}{1 + \sum_{i_1'=1}^{I_1-1} \exp(\mu_{i_1'} + \alpha_{i_1'} i_2 + \beta_{i_1'} i_3)}, & i_1 = I_1, \end{cases}$$

where $(i_2, i_3) \in \mathscr{I}_2 \times \mathscr{I}_3$, respectively. $p_{i_1|i_2}$ and $p_{i_1|i_2 i_3}$ are conditional probabilities that the value of the response variable equals i_1 given the covariates i_2 and (i_2, i_3), respectively. $\mathscr{I}_2$ and $\mathscr{I}_2 \times \mathscr{I}_3$ are designs for covariates. As discussed in Sects. 13.1 and 13.2, the structure of Markov bases for the discrete logistic regression model is complicated even for the case where responses are binary $I_1 = 2$ and covariates are equally spaced. Table 16.1 presents the highest degrees and the numbers of moves in the minimal Markov bases of binomial logistic regression models with one covariate computed by 4ti2. Even for models with one covariate, if a covariate has more than 20 levels, it is difficult to compute Markov bases of models via 4ti2 within a practical amount of time by a computer with a 32-bit processor.

Table 16.1 The highest degrees and the number of moves in a minimal Markov basis for binomial logistic regression models with one covariate

	Number of levels of a covariate						
	10	11	12	13	14	15	16
Maximum degree	18	20	22	24	26	28	30
Number of moves	1,830	3,916	8,569	16,968	34,355	66,066	123,330

The logistic regression model with r responses has the rth Lawrence configuration (9.10) where A is a configuration for the Poisson regression model. The computation of Markov bases of Poisson regression model is relatively easy. Therefore a lattice basis can be computed by Proposition 16.1 and we can apply the proposed method to these models.

In the experiment we computed the LR statistics for the goodness-of-fit test of a binomial or trinomial logistic regression model with two covariates against a model with three covariates

$$p_{i_1|i_2i_3i_4} = \begin{cases} \dfrac{\exp(\mu_{i_1} + \alpha_{i_1} i_2 + \beta_{i_1} i_3 + \gamma_{i_1} i_4)}{1 + \sum_{i'_1=1}^{I_1-1} \exp(\mu_{i'_1} + \alpha_{i'_1} i_2 + \beta_{i'_1} i_3 + \gamma_{i'_1} i_4)}, & i_1 = 1, \ldots, I_1 - 1, \\[2ex] \dfrac{1}{1 + \sum_{i'_1=1}^{I_1-1} \exp(\mu_{i'_1} + \alpha_{i'_1} i_2 + \beta_{i'_1} i_3 + \gamma_{i'_1} i_4)}, & i_1 = I_1, \end{cases}$$

where $(i_2, i_3) \in \mathscr{I}_2 \times \mathscr{I}_3$, $i_4 \in \mathscr{I}_4$. We assume that $\mathscr{I}_2 \times \mathscr{I}_3$ are 4×4 and 10×10 checkered designs as described in the following figure for the 4×4 case, where only (i_2, i_3) in dotted patterns have positive frequencies.

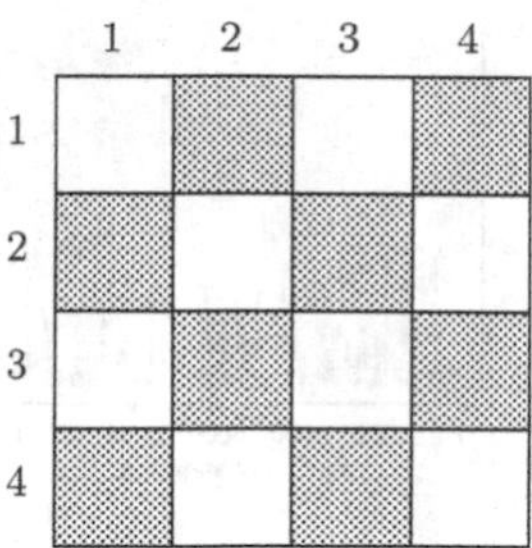

We also assume that $\mathscr{I}_4 = \{1,2,3,4,5\}$. The degrees of freedom of the asymptotic χ^2 distribution of the LR statistic is 1. We set the sample sizes for 4×4 and 10×10 designs at 200 and 625, respectively. We also set $(\text{burn-in}, \text{iteration}) = (1{,}000, 10{,}000)$.

Figures 16.3 and 16.4 present the results for a binomial and a trinomial logistic regression model with a 4×4 checkered design, respectively. Solid lines in the left

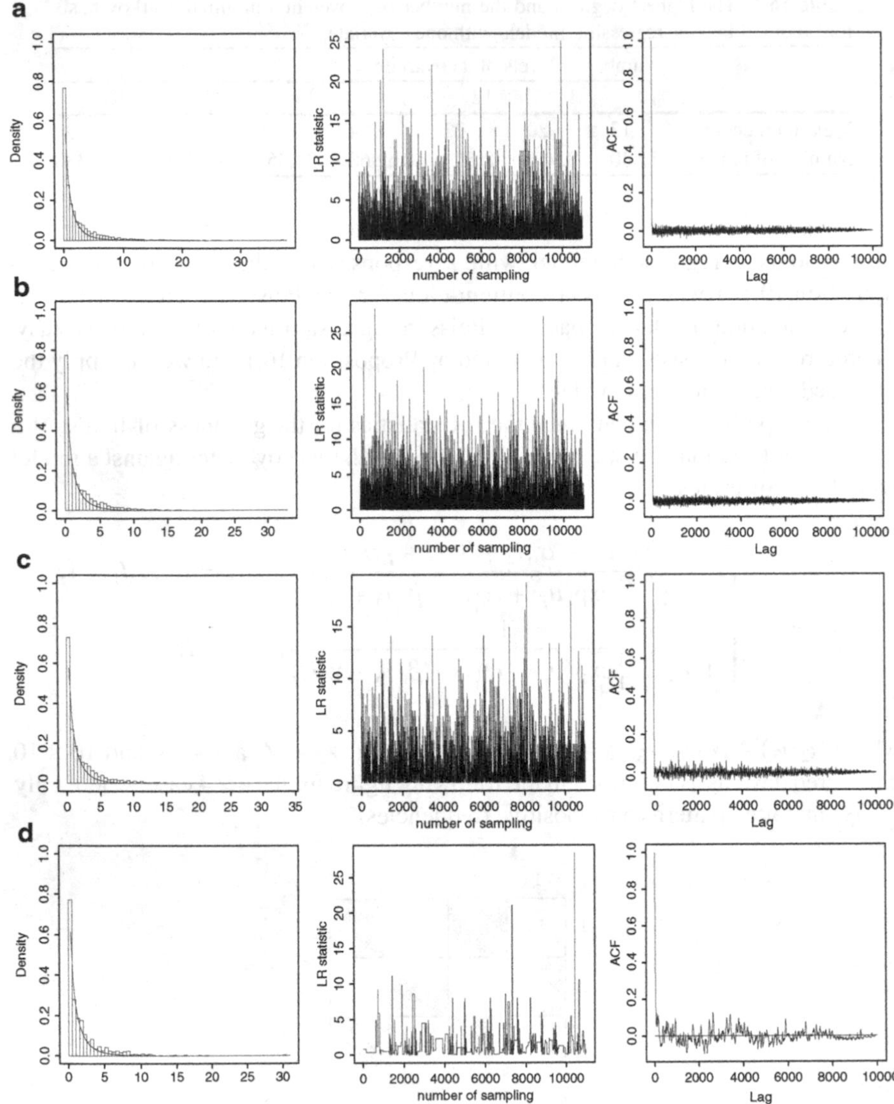

Fig. 16.3 Histograms, paths of LR statistics, and correlograms of paths for discrete logistic regression model ((burn-in, iteration) = (1,000,10,000)): (**a**) a Markov basis; (**b**) a lattice basis with $Po(1)$; (**c**) a lattice basis with $Po(10)$; (**d**) a lattice basis with $Po(50)$

figures are asymptotic χ^2 distributions. α_k is generated from $Po(\lambda)$, $\lambda = 1,10,50$. We can compute Markov bases in these models by 4ti2. So we also present the results for Markov bases. The figures show that the proposed methods show the

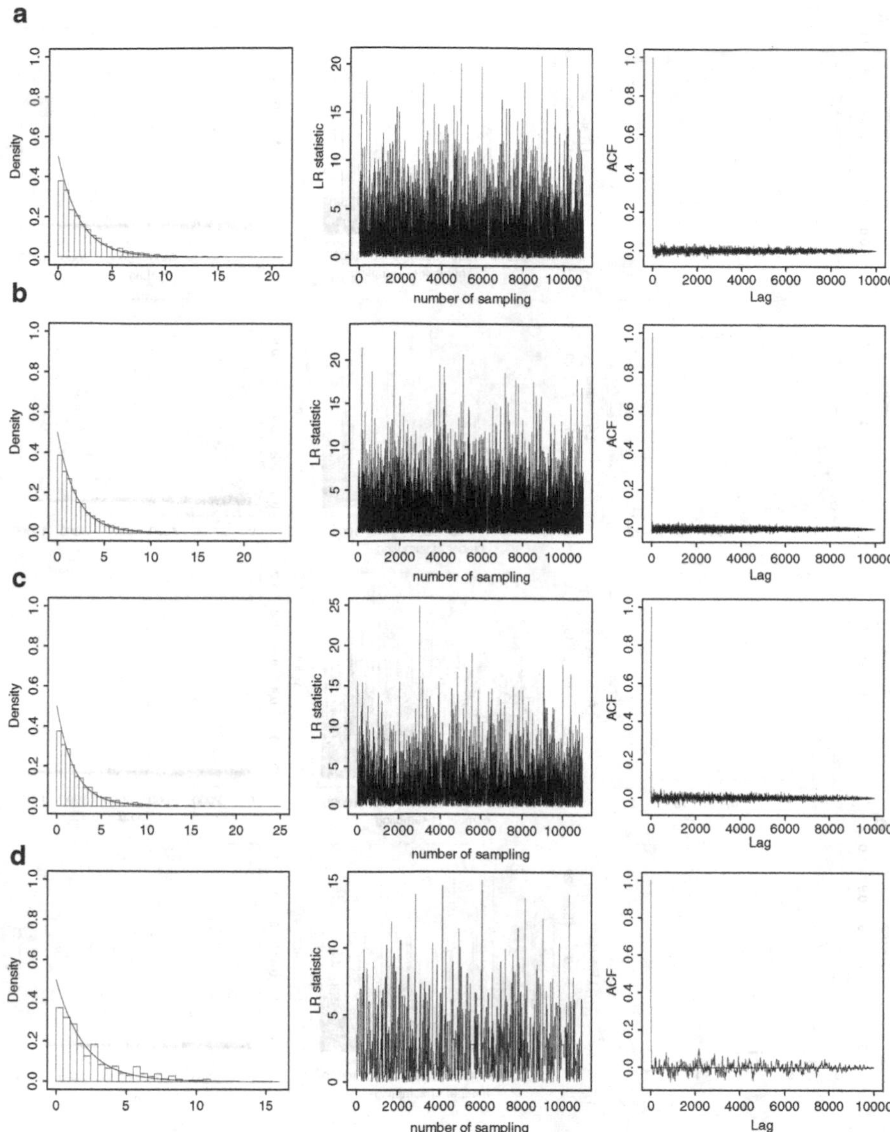

Fig. 16.4 Histograms, paths of LR statistics, and correlograms of paths for trinomial discrete logit model ((burn-in, iteration) = (1,000,10,000)): (**a**) a Markov basis; (**b**) a lattice basis with $Po(1)$; (**c**) a lattice basis with $Po(10)$; (**d**) a lattice basis with $Po(50)$

comparative performance to a Markov basis also in these models. We note that the paths are also stable even for the case where $\alpha_1, \ldots, \alpha_K$ are generated from $Po(50)$.

Figure 16.5 presents the results for a 10×10 checkered pattern. In this case Markov bases cannot be computed via 4ti2 by a computer with a 32-bit processor.

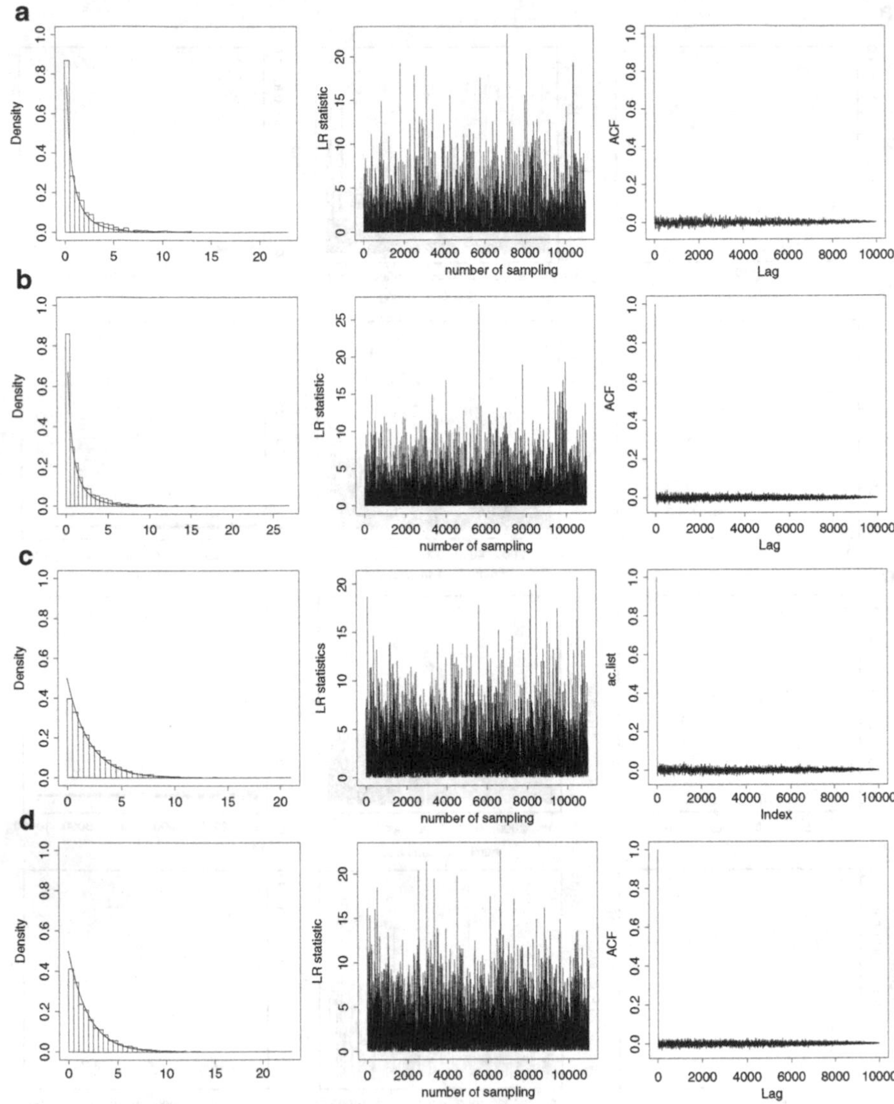

Fig. 16.5 Histograms, paths of LR statistics, and correlograms of paths for discrete logistic regression model ((burn-in, iteration) = (1,000,10,000)): (**a**) binomial, a lattice basis with $Geom(0.1)$; (**b**) binomial, a lattice basis with $Geom(0.5)$; (**c**) trinomial, a lattice basis with $Geom(0.1)$; (**d**) trinomial, a lattice basis with $Geom(0.5)$

α_k is generated from $Geom(p)$, $\lambda = 0.1, 0.5$. Also in these cases the results look stable. These results show that the proposed method is practical for the logistic regression models.

References

1. 4ti2 team: 4ti2 — a software package for algebraic, geometric and combinatorial problems on linear spaces. Available at www.4ti2.de
2. Agresti, A.: Categorical Data Analysis, 1st edn. Wiley, New York (1990)
3. Agresti, A.: A survey of exact inference for contingency tables. Stat. Sci. **7**, 131–177 (1992)
4. Agresti, A.: Categorical Data Analysis, 2nd edn. Wiley, New York (2002)
5. Agresti, A.: An Introduction to Categorical Data Analysis. In: Wiley Series in Probability and Statistics, 2nd edn. Wiley-Interscience [John Wiley & Sons], Hoboken, NJ (2007)
6. Aoki, S.: Some optimal criteria of model-robustness for two-level non-regular fractional factorial designs. Ann. Inst. Statist. Math. **62**(4), 699–716 (2010)
7. Aoki, S., Hibi, T., Ohsugi, H., Takemura, A.: Gröbner bases of nested configurations. J. Algebra **320**, 2583–2593 (2008)
8. Aoki, S., Hibi, T., Ohsugi, H., Takemura, A.: Markov basis and Gröbner basis of Segre-Veronese configuration for testing independence in group-wise selections. Ann. Inst. Statist. Math. **62**(2), 299–321 (2010)
9. Aoki, S., Takemura, A.: The list of indispensable moves of the unique minimal markov basis for $3 \times 4 \times k$ and $4 \times 4 \times 4$ contingency tables with fixed two-dimensional marginals. Tech. Rep. METR 2003-38, University of Tokyo (2003)
10. Aoki, S., Takemura, A.: Minimal basis for a connected Markov chain over $3 \times 3 \times K$ contingency tables with fixed two-dimensional marginals. Aust. N. Z. J. Stat. **45**(2), 229–249 (2003)
11. Aoki, S., Takemura, A.: Markov chain Monte Carlo exact tests for incomplete two-way contingency table. J. Stat. Comput. Simulat. **75**(10), 787–812 (2005)
12. Aoki, S., Takemura, A.: The largest group of invariance for Markov bases and toric ideals. J. Symbolic Comput. **43**, 342–358 (2008)
13. Aoki, S., Takemura, A.: Minimal invariant markov basis for sampling contingency tables with fixed marginals. Ann. Inst. Statist. Math. **60**, 229–256 (2008)
14. Aoki, S., Takemura, A.: Some characterizations of affinely full-dimensional factorial designs. J. Statist. Plann. Infer. **139**(10), 3525–3532 (2009)
15. Aoki, S., Takemura, A.: Statistics and Gröbner bases—the origin and development of computational algebraic statistics. In: Selected papers on probability and statistics, Amer. Math. Soc. Transl. Ser. 2, vol. 227, pp. 125–145. Amer. Math. Soc., Providence, RI (2009)
16. Aoki, S., Takemura, A.: Markov basis for design of experiments with three-level factors. In: Algebraic and Geometric Methods in Statistics, pp. 225–238. Cambridge University Press, Cambridge (2010)
17. Aoki, S., Takemura, A.: Markov chain Monte Carlo tests for designed experiments. J. Statist. Plann. Infer. **140**(3), 817–830 (2010)

S. Aoki et al., *Markov Bases in Algebraic Statistics*, Springer Series in Statistics 199, DOI 10.1007/978-1-4614-3719-2,

18. Aoki, S., Takemura, A.: Design and analysis of fractional factorial experiments from the viewpoint of computational algebraic statistics. J. Statist. Theor. Pract. **6**(1), 147–161 (2012).
19. Aoki, S., Takemura, A., Yoshida, R.: Indispensable monomials of toric ideals and markov bases. J. Symbolic Comput. **43**, 490–507 (2008)
20. Badsberg, J.H., Malvestuto, F.M.: An implementaition of the iterative proportional fitting procecure by propagation trees. Comput. Statist. Data. Anal. **37**, 297–322 (2001)
21. Berry, A., Bordat, J.P., Cogis, O.: Generating all the minimal separators of a graph. Int. J. Found. Comput. Sci. **11**(3), 397–403 (2000)
22. Berstein, Y., Maruri-Aguilar, H., Onn, S., Riccomagno, E., Wynn, H.: Minimal average degree aberration and the state polytope for experimental designs. Ann. Inst. Statist. Math. **62**(4), 673–698 (2010)
23. Berstein, Y., Onn, S.: The Graver complexity of integer programming. Ann. Comb. **13**(3), 289–296 (2009)
24. Besag, J., Clifford, P.: Generalized monte carlo significance tests. Biometrika **76**, 633–642 (1989)
25. Bishop, Y.M.M., Fienberg, S.E.: Incomplete two-dimensional contingency tables. Biometrics **25**, 119–128 (1969)
26. Bishop, Y.M.M., Fienberg, S.E., Holland, P.W.: Discrete Multivariate Analysis: Theory and Practice. The MIT Press, Cambridge, Massachusetts (1975)
27. Boffi, G., Rossi, F.: Lexicographic Gröbner bases for transportation problems of format $r \times 3 \times 3$. J. Symbolic Comput. **41**(3-4), 336–356 (2006)
28. Briales, E., Campillo, A., Marijuán, C., Pisón, P.: Minimal systems of generators for ideals of semigroups. J. Pure Appl. Algebra **124**(1-3), 7–30 (1998)
29. Bruns, W., Hemmecke, R., Ichim, B., Köppe, M., Söger, C.: Challenging computations of Hilbert bases of cones associated with algebraic statistics. Exp. Math. **20**(1), 25–33 (2011)
30. Buneman, P.: A characterization on rigid circuit graphs. Discrete Math. **9**, 205–212 (1974)
31. Charalambous, H., Katsabekis, A., Thoma, A.: Minimal systems of binomial generators and the indispensable complex of a toric ideal. Proc. Amer. Math. Soc. **135**(11), 3443–3451 (2007)
32. Chen, Y., Diaconis, P., Holmes, S.P., Liu, J.S.: Sequential Monte Carlo methods for statistical analysis of tables. J. Amer. Statist. Assoc. **100**(469), 109–120 (2005)
33. Chen, Y., Dinwoodie, I., Dobra, A., Huber, M.: Lattice points, contingency tables, and sampling. In: Integer Points in Polyhedra—Geometry, Number Theory, Algebra, Optimization, Contemp. Math., vol. 374, pp. 65–78. Amer. Math. Soc., Providence, RI (2005)
34. Chen, Y., Dinwoodie, I., Sullivant, S.: Sequential importance sampling for multiway tables. Ann. Statist. **34**(1), 523–545 (2006)
35. Chen, Y., Dinwoodie, I., Yoshida, R.: Markov chain, quotient ideals and connectivity with positive margins. In: Gibilisco, P., Riccomagno, E., Rogantin, M.P., Wynn, H.P. (eds.) Algebraic and Geometric Methods in Statistics, pp. 99–110. Cambridge University Press, Cambridge (2008)
36. Chen, Y., Small, D.: Exact tests for the Rasch model via sequential importance sampling. Psychometrika **70**, 11–30 (2005)
37. Christensen, R.: Log-linear Models and Logistic Regression, 2nd edn. Springer Texts in Statistics. Springer, New York (1997)
38. Cobb, G.W., Chen, Y.P.: An application of Markov chain Monte Carlo to community ecology. Amer. Math. Monthly **110**, 265–288 (2003)
39. Colbourn, C.J., Dinitz, J.H. (eds.): Handbook of Combinatorial Designs. In: Discrete Mathematics and its Applications, 2nd edn. (Boca Raton). Chapman & Hall/CRC, Boca Raton, FL (2007)
40. Condra, L.W.: Reliability Improvement with Design of Experiments. Marcel Dekker, New York (1993)
41. Cornfield, J.: Joint dependence of risk of coronary heart disease on serum cholesterol and systolic blood pressure: A discriminant function analysis. Fed. Proc. **21**(11), 58–61 (1962)
42. Cox, D., Little, J., O'Shea, D.: Ideals, varieties, and algorithms. In: Undergraduate Texts in Mathematics, 3rd edn. Springer, New York (2007)

43. Crow, J.E.: Eighty years ago: The beginnings of population genetics. Genetics **119**, 473–476 (1988)
44. Dalenius, T., Reiss, S.P.: Data-swapping: A technique for disclosure control. J. Statist. Plann. Infer. **6**(1), 73–85 (1982)
45. De Loera, J.A., Hemmecke, R., Onn, S., Weismantel, R.: *n*-fold integer programming. Discrete Optim. **5**(2), 231–241 (2008)
46. De Loera, J.A., Onn, S.: The complexity of three-way statistical tables. SIAM J. Comput. **33**(4), 819–836 (2004)
47. De Loera, J.A., Onn, S.: All linear and integer programs are slim 3-way transportation programs. SIAM J. Optim. **17**(3), 806–821 (2006)
48. De Negri, E., Hibi, T.: Gorenstein algebras of Veronese type. J. Algebra **193**(2), 629–639 (1997)
49. Develin, M., Sullivant, S.: Markov bases of binary graph models. Ann. Comb. **7**(4), 441–466 (2003)
50. Diaconis, P., Sturmfels, B.: Algebraic algorithms for sampling from conditional distributions. Ann. Statist. **26**(1), 363–397 (1998)
51. Dirac, G.A.: On rigid circuit graphs. Abh. Math. Sem. Univ. Hamburg **25**, 71–76 (1961)
52. Dobra, A.: Markov bases for decomposable graphical models. Bernoulli **9**(6), 1093–1108 (2003)
53. Dobra, A.: Dynamic Markov bases. J. Comput. Graph. Stat. **21**(2), 496–517 (2012)
54. Dobra, A., Sullivant, S.: A divide-and-conquer algorithm for generating Markov bases of multi-way tables. Comput. Statist. **19**(3), 347–366 (2004)
55. Drton, M., Sturmfels, B., Sullivant, S.: Lectures on Algebraic Statistics. In: Oberwolfach Seminars, vol. 39. Birkhäuser Verlag, Basel (2009)
56. Eaton, M.L.: Group Invariance Applications in Statistics. In: NSF-CBMS Regional Conference Series in Probability and Statistics, 1. Institute of Mathematical Statistics, Hayward, CA (1989)
57. Edwards, D.: Introduction to Graphical Modelling, 2nd edn. Springer, New York (2000)
58. Ferguson, T.S.: Mathematical Statistics: A Decision Theoretic Approach. In: Probability and Mathematical Statistics, Vol. 1. Academic Press, New York (1967)
59. Fienberg, S.E.: The Analysis of Cross-classified Categorical Data, 2nd edn. Springer, New York (2007)
60. Fisher, R.A., Yates, F.: The 6×6 Latin squares. Proc. Camb. Phil. Soc. **30**, 492–507 (1934)
61. Fontana, R., Pistone, G., Rogantin, M.P.: Algebraic analysis and generation of two-level designs. Statistica Applicata **9**, 15–29 (1997)
62. Fontana, R., Pistone, G., Rogantin, M.P.: Classification of two-level factorial fractions. J. Stat. Plann. Infer. **87**, 149–172 (2000)
63. García-Sánchez, P.A., Ojeda, I.: Uniquely presented finitely generated commutative monoids. Pacific J. Math. **248**(1), 91–105 (2010)
64. Gavril, F.: The intersection graphs of subtrees in trees are exactly the chordal graphs. J. Combin. Theory(B) **116**, 47–56 (1974)
65. Glas, C.A.W., Verhelst, N.D.: Testing the Rasch model. In: Rasch Models: Their Foundations, Recent Developments and Applicatio ns, pp. 69–95. Springer, New York (1995)
66. Goodman, L.A.: The analysis of cross-classified data: independence, quasi-independence and interactions in contingency tables with or without missing entries. J. Am. Stat. Assoc. **63**, 1091–1131 (1968)
67. Guo, S.W., Thompson, E.A.: Performing the exact test of Hardy-Weinberg proportion for multiple alleles. Biometrics **48**, 361–372 (1992)
68. Haberman, S.J.: A warning on the use of chi-squared statistics with frequency tables with small expected cell counts. J. Amer. Statist. Assoc. **83**(402), 555–560 (1988)
69. Häggström, O.: Finite Markov Chains and Algorithmic Applications. In: London Mathematical Society Student Texts, vol. 52. Cambridge University Press, Cambridge (2002)
70. Hamada, M., Nelder, J.A.: Generalized linear models for quality-improvement experiments. J. Qual. Tech. **29**, 292–304 (1997)

71. Hara, H., Aoki, S., Takemura, A.: Minimal and minimal invariant Markov bases of decomposable models for contingency tables. Bernoulli **16**, 208–233 (2010)
72. Hara, H., Aoki, S., Takemura, A.: Running Markov chain without Markov basis. In: Hibi, T. (ed.) Proceedings of the Second CREST-SBM International Conference, Harmony of Gröbner Bases and the Modern Industrial Society. World Scientfic, Singapore, 45–62 (2012)
73. Hara, H., Sei, T., Takemura, A.: Hierarchical subspace models for contingency tables. J. Multivariate Anal. **103**, 19–34 (2012)
74. Hara, H., Takemura, A.: Boundary cliques, clique trees and perfect sequences of maximal cliques of a chordal graph (2006) Preprint. arXiv:cs.DM/0607055
75. Hara, H., Takemura, A.: Improving on the maximum likelihood estimators of the means in Poisson decomposable graphical models. J. Multivariate Anal. **98**, 410–434 (2007)
76. Hara, H., Takemura, A.: Bayes admissible estimation of the means in poisson decomposable graphical models. J. Stat. Plann. Infer. **139**, 1297–1319 (2009)
77. Hara, H., Takemura, A.: Connecting tables with zero-one entries by a subset of a Markov basis. In: Algebraic Methods in Statistics and Probability II, Contemp. Math., vol. 516, pp. 199–213. Amer. Math. Soc., Providence, RI (2010)
78. Hara, H., Takemura, A.: A localization approach to improve iterative proportional scaling in gaussian graphical models. Comm. Stat. Theor. Meth. **39**, 1643–1654 (2010)
79. Hara, H., Takemura, A., Yoshida, R.: Markov bases for subtable sum problems. J. Pure Appl. Algebra **213**, 1507–1521 (2009)
80. Hara, H., Takemura, A., Yoshida, R.: A Markov basis for conditional test of common diagonal effect in quasi-independence model for square contingency tables. Comput. Stat. Data Anal. **53**, 1006–1014 (2009)
81. Hara, H., Takemura, A., Yoshida, R.: On connectivity of fibers with positive marginals in multiple logistic regression. J. Multivariate Anal. **101**, 909–925 (2010)
82. Hastings, W.K.: Monte carlo sampling methods using markov chains and their applications. Biometrika **57**, 97–109 (1970)
83. Hemmecke, R., Malkin, P.N.: Computing generating sets of lattice ideals and Markov bases of lattices. J. Symbolic Comput. **44**(10), 1463–1476 (2009)
84. Hemmecke, R., Nairn, K.A.: On the Gröbner complexity of matrices. J. Pure Appl. Algebra **213**(8), 1558–1563 (2009)
85. Hemmecke, R., Takemura, A., Yoshida, R.: Computing holes in semi-groups and its applications to transportation problems. Contrib. Discrete Math. **4**(1), 81–91 (2009)
86. Hibi, T.: Gröbner Bases. Asakura Shoten, Tokyo (2003). (In Japanese)
87. Hirotsu, C.: Two-way change-point model and its application. Aust. J. Stat. **39**(2), 205–218 (1997)
88. Hoşten, S., Sullivant, S.: Gröbner bases and polyhedral geometry of reducible and cyclic models. J. Combin. Theory Ser. A **100**(2), 277–301 (2002)
89. Hoşten, S., Sullivant, S.: A finiteness theorem for Markov bases of hierarchical models. J. Combin. Theory Ser. A **114**(2), 311–321 (2007)
90. Huber, M., Chen, Y., Dobra, I.D.A., Nicholas, M.: Monte carlo algorithms for Hardy-Weinberg proportions. Biometrics **62**, 49–53 (2006)
91. Irving, R.W., Jerrum, M.R.: Three-dimensional statistical data security problems. SIAM J. Comput. **23**(1), 170–184 (1994)
92. Jacobson, M.T., Matthews, P.: Generating uniformly distributed random Latin squares. J. Combin. Des. **4**, 405–437 (1996)
93. JST CREST Hibi team: Gröbner Bases, Statistics and Sofstware Systems. Kyoritsu Shuppan, Tokyo (2011). (In Japanese)
94. Kashimura, T., Numata, Y., Takemura, A.: Separation of integer points by a hyperplane under some weak notions of discrete convexity (2010). Preprint. arXiv:1002.2839
95. Kloks, D., Kratsch, D.: Listing all minimal separators of a graph. SIAM J. Comput. **27**(3), 605–613 (1998)
96. Kozlov, D.: Combinatorial Algebraic Topology. In: Algorithms and Computation in Mathematics, vol. 21. Springer, Berlin (2008)

97. Lauritzen, S.L.: Graphical Models. Oxford University Press, Oxford (1996)
98. Lehmann, E.L., Romano, J.P.: Testing Statistical Hypotheses. In: Springer Texts in Statistics, 3rd edn. Springer, New York (2005)
99. Leimer, H.G.: Optimal decomposition by clique separators. Discrete Math. **113**, 99–123 (1993)
100. Linacre, J.M.: Many-Facet Rasch Measurement. MESA Press, Chicago (1989)
101. Malvestuto, F.M., Moscarini, M.: Decomposition of a hypergraph by partial-edge separators. Theoret. Comput. Sci. **237**, 57–79 (2000)
102. Mantel, N.: Incomplete contingency tables. Biometrics **26**, 291–304 (1970)
103. Martin, B., Parker, D., Zenick, L.: Minimize slugging by optimizing controllable factors on topaz windshield molding. In: Fifth Symposium on Taguchi Methods, pp. 519–526. American Supplier Institute, Inc., Dearborn, MI (1987)
104. McCullagh, P., Nelder, J.A.: Generalized Linear Models. In: Monographs on Statistics and Applied Probability. Chapman & Hall, London (1983)
105. Miller, E., Sturmfels, B.: Combinatorial Commutative Algebra. In: Graduate Texts in Mathematics, vol. 227. Springer, New York (2005)
106. National Center for University Entrance Examinations website: http://www.dnc.ac.jp/ (in Japanese)
107. Ninomiya, Y.: Construction of consevative test for change-point problem in two-dimensional random fields. J. Multivariate Anal. **89**, 219–242 (2004)
108. Oguma, T., Palmer, L.J., Birben, E., Sonna, L.A., Asano, K., Lilly, C.M.: Role of prostanoid dp receptor variants in susceptiblity to asthma. New Engl. J. Med. **351**, 1752–1763 (2004)
109. Ohsugi, H., Hibi, T.: Compressed polytopes, initial ideals and complete multipartite graphs. Illinois J. Math. **44**(2), 391–406 (2000)
110. Ohsugi, H., Hibi, T.: Indispensable binomials of finite graphs. J. Algebra Appl. **4**(4), 421–434 (2005)
111. Ohsugi, H., Hibi, T.: Toric ideals arising from contingency tables. In: Proceedings of the Ramanujan Mathematical Society's Lecture Notes Series pp. 87–111 (2006)
112. Ohsugi, H., Hibi, T.: Two way subtable sum problems and quadratic Gröbner bases. Proc. Amer. Math. Soc. **137**(5), 1539–1542 (2009)
113. Ohsugi, H., Hibi, T.: Non-very ample configurations arising from contingency tables. Ann. Inst. Statist. Math. **62**(4), 639–644 (2010)
114. Ohsugi, H., Hibi, T.: Toric rings and ideals of nested configurations. J. Commut. Algebra **2**(2), 187–208 (2010)
115. Ojeda, I., Vigneron-Tenorio, A.: Indispensable binomials in semigroup ideals. Proc. Amer. Math. Soc. **138**(12), 4205–4216 (2010)
116. Pachter, L., Sturmfels, B.: Algebraic Statistics for Computational Biology. Cambridge University Press, Cambridge (2005)
117. Petrović, S., Stokes, E.: Betti numbers of Stanley-Reisner rings determine hierarchical Markov degrees (2012). To appear in Journal of Algebraic Combinatorics
118. Pistone, G., Riccomagno, E., Wynn, H.P.: Algebraic Statistics: Computational Commutative Algebra in Statistics. Chapman & Hall Ltd, Boca Raton (2001)
119. Pistone, G., Rogantin, M.P.: Indicator function and complex coding for mixed fractional factorial designs. J. Statist. Plann. Infer. **138**(3), 787–802 (2008)
120. Pistone, G., Wynn, H.P.: Generalised confounding with Gröbner bases. Biometrika **83**(3), 653–666 (1996)
121. Ploog, D.M.: The behavior of squirrel monkeys (*saimiri sciureus*) as revealed by sociometry, bioacoustics, and brain stimulation. In: Altmann, S. (ed.) Social Communication Among Primates, pp. 149–184. Chicago Press, Chicago (1967)
122. Ponocny, I.: Nonparametric goodness-of-fit tests for the Rasch model. Psychometrika **66**, 437–460 (2001)
123. Rahul., M., Wu, C.F.J.: A Modern Theory of Factorial Designs. In: Springer Series in Statistics. Springer, New York (2006)

124. Rapallo, F.: Algebraic Markov bases and MCMC for two-way contingency tables. Scand. J. Statist. **30**(2), 385–397 (2003)
125. Rapallo, F.: Markov bases and structural zeros. J. Symbolic Comput. **41**(2), 164–172 (2006)
126. Rapallo, F., Yoshida, R.: Markov bases and subbases for bounded contingency tables. Ann. Inst. Statist. Math. **62**(4), 785–805 (2010)
127. Rasch, G.: Probabilistic Models for Some Intelligence and Attainment Tests. University of Chicago Press, Chicago (1980)
128. Ripley, B.D.: Stochastic Simulation. In: Wiley Series in Probability and Mathematical Statistics: Applied Probability and Statistics. Wiley, New York (1987)
129. Rotman, J.J.: An Introduction to the Theory of Groups. In: Graduate Texts in Mathematics, vol. 148, 4th edn. Springer, New York (1995)
130. Ryser, H.J.: Combinatorial properties of matrices of zeros and ones. Canad. J. Math. **9**, 371–377 (1957)
131. Santos, F., Sturmfels, B.: Higher Lawrence configurations. J. Combin. Theory Ser. A **103**(1), 151–164 (2003)
132. Scheffé, H.: The Analysis of Variance. Wiley, New York (1959)
133. Schlörer, J.: Security of statistical databases: multidimensional transformation. ACM Trans. Database Syst. **6**(1), 95–112 (1981)
134. Schrijver, A.: Theory of Linear and Integer Programming. In: Wiley-Interscience Series in Discrete Mathematics. Wiley, Chichester (1986).
135. Sei, T., Aoki, S., Takemura, A.: Perturbation method for determining the group of invariance of hierarchical models. Adv. Appl. Math. **43**(4), 375–389 (2009)
136. Seress, Á.: Permutation Group Algorithms. In: Cambridge Tracts in Mathematics, vol. 152. Cambridge University Press, Cambridge (2003)
137. Shibata, Y.: On the tree representation of chordal graphs. J. Graph Theor. **12**, 421–428 (1988)
138. Shiloach, Y., Vishkin, U.: An $O(\log n)$ parallel connectivity algorithm. J. Algorithms **3**, 57–67 (1975)
139. Sturmfels, B.: Gröbner Bases and Convex Polytopes. In: University Lecture Series, vol. 8. American Mathematical Society, Providence, RI (1996)
140. Sundberg, R.: Some results about decomposable (or Markov-type) models for multidimensional contingency tables: Distribution of marginals and partitioning of tests. Scand. J. Statist. **2**(2), 71–79 (1975)
141. Takemura, A.: Local recording and record swapping by maximum weight matching for disclosure control of microdata sets. J. Offic. Stat. **18**(2), 275–289 (2002)
142. Takemura, A., Aoki, S.: Some characterizations of minimal Markov basis for sampling from discrete conditional distributions. Ann. Inst. Statist. Math. **56**(1), 1–17 (2004)
143. Takemura, A., Aoki, S.: Distance reducing Markov bases for sampling from a discrete sample space. Bernoulli **11**(5), 793–813 (2005)
144. Takemura, A., Hara, H.: Conditions for swappability of records in a microdata set when some marginals are fixed. Comput. Stat. **22**, 173–185 (2007)
145. Takemura, A., Yoshida, R.: A generalization of the integer linear infeasibility problem. Discrete Optim. **5**, 36–52 (2008)
146. Takemura, A., Yoshida, R.: Saturation points on faces of a rational polyhedral cone. In: Integer Points in Polyhedra—Geometry, Number Theory, Representation Theory, Algebra, Optimization, Statistics, Contemp. Math., vol. 452, pp. 147–161. Amer. Math. Soc., Providence, RI (2008)
147. Vidmar, N.: Effects of decision alternatives on the verdicts and social perceptions of simulated jurors. J. Pers. Soc. Psych. **22**, 211–218 (1972)
148. Vlach, M.: Conditions for the existence of solutions of the three-dimensional planar transportation problem. Discrete Appl. Math. **13**(1), 61–78 (1986)
149. Wilson, R. J.: Introduction to Graph Theory. 3rd ed., Longman, New York (1985)
150. White, H.C.: An Anatomy of Kinship. Prentice-Hall, London (1963)
151. Wu, C.F.J., Hamada, M.: Experiments: Planning, Analysis, and Parameter Design Optimization. In: Wiley Series in Probability and Statistics: Texts and References Section. Wiley, New York (2000)

152. Xu, H.: Minimum moment aberration for nonregular designs and supersaturated designs. Statist. Sinica **13**, 691–708 (2003)
153. Xu, H., Wu, C.F.J.: Generalized minimum aberration for asymmetrical fractional factorial designs. Ann. Statist. **29**, 1066–1077 (2001)
154. Ye, K.Q.: Indicator function and its application in two-level factorial designs. Ann. Statist. **31**, 984–994 (2003)

Index

S. Aoki et al., *Markov Bases in Algebraic Statistics*, Springer Series in Statistics 199, DOI 10.1007/978-1-4614-3719-2,